AP* Achiever Advanced Placement Chemistry Exam Preparation Guide

to accompany

Chemistry
Raymond Chang

Ninth Edition

Prepared by
Reen Gibb
Brookline High School
Carol Murphree
Acton-Boxborough Regional High School
Lisa McGaw
University of Central Oklahoma
Steve Bertenshaw
Cheshire High School

* Pre-AP, AP and Advanced Placement program are registered trademarks of the College Entrance Examination Board, which was not involved in the production of and does not endorse these products.

 **Higher Education**

Boston   Burr Ridge, IL   Dubuque, IA   New York   San Francisco   St. Louis
Bangkok   Bogotá   Caracas   Kuala Lumpur   Lisbon   London   Madrid   Mexico City
Milan   Montreal   New Delhi   Santiago   Seoul   Singapore   Sydney   Taipei   Toronto

The McGraw·Hill Companies

AP* Achiever Advanced Placement Chemistry Exam Preparation Guide to accompany
CHEMISTRY, NINTH EDITION
RAYMOND CHANG

Published by McGraw-Hill Higher Education, an imprint of The McGraw-Hill Companies, Inc., 1221 Avenue of the Americas,
New York, NY 10020. Copyright © 2007 by The McGraw-Hill Companies, Inc. All rights reserved.

1 2 3 4 5 6 7 8 9 0 QPD/QPD 0 9 8 7

ISBN: 978-0-07-328659-4
MHID: 0-07-328659-1

www.mhhe.com

About the Authors

Reen Gibb, Brookline High School, has been teaching Advanced Placement Chemistry for over 25 years. She has been a workshop leader for the College Board since 2001 and a reader since 2003. She has been a Curriculum Consultant for WGBH NOVA science film series, a Curriculum Advisor for Science NOW, PBS, and a Curriculum Advisor for ZOOM series, PBS. She is a graduate of Tufts and Worcester Polytechnic Institute. Reen was responsible for writing the chapter reviews and the end of chapter multiple choice questions.

Carol Murphree, Acton-Boxborough Regional High School, has been teaching Advanced Placement Chemistry for over 15 years. She has been an AP Faculty Consultant Educational Testing Service/College Board since 1999, an AP workshop leader since 2000, and served as an AP Consultant reviewing resource material for AP Central. Carol is a graduate of Smith College, Georgetown University, and Northeastern University. Carol was responsible for writing the chapter reviews and the end of chapter multiple choice questions.

Lisa McGaw, University of Central Oklahoma, has served as a reader for the AP Chemistry exam since 1994, table leader for the exam since 1998, and a question leader since 2004. She received the Southwest Regional Award for AP Chemistry in 1997. Lisa is a graduate of Stephen F. Austin State University and West Texas A&M University. Lisa was responsible for writing the multiple choice questions in each of the practice exams.

Steve Bertenshaw graduated from the University of Illinois with a B.S. in chemistry and biology, an M.S. in chemistry, and a Ph.D. in organic chemistry. He worked as a research scientist at Monsanto Corporation and at Bristol-Meyers Squibb. He has been teaching chemistry, AP chemistry, biotechnology, and biology at Cheshire High School in Connecticut since 2000. Steve was responsible for writing the comprehensive questions at the end of each chapter and the Free-Response Questions in the practice exams.

Table of Contents

Introduction

Chapter 1   Chemistry: The Study of Change *1*

Chapter 2   Atoms, Molecules, and Ions *11*

Chapter 3   Mass Relationships in Chemical Reactions *24*

Chapter 4   Reactions in Aqueous Solutions *43*

Chapter 5   Gases *69*

Chapter 6   Thermochemistry *87*

Chapter 7   Quantum Theory and the Electronic Structure of Atoms *108*

Chapter 8   The Periodic Classification of the Elements *127*

Chapter 9   Chemical Bonding I: Basic Concepts *143*

Chapter 10   Chemical Bonding II: Molecular Geometry and Hybridization of Atomic Orbitals *159*

Chapter 11   Intermolecular Forces, Liquids, and Solids *175*

Chapter 12   Physical Properties of Solutions *190*

Chapter 13   Chemical Kinetics *210*

Chapter 14   Chemical Equilibria *235*

Chapter 15   Acids and Bases *257*

Chapter 16   Acid-Base Equilibria and Solubility Equilibria *282*

Chapter 17   The Chemistry in the Atmosphere 326

Chapter 18   Entropy, Free Energy, and Equilibrium *328*

Chapter 19   Electrochemistry *348*

Chapter 20   Metallurgy and the Chemistry of Metals 373

Chapter 21   Nonmetallic Elements and Their Compounds 375

Chapter 22   *Transition Metal Chemistry and Coordination Compounds*   *376*

Chapter 23   *Nuclear Chemistry*   *380*

Chaper 24   *Organic Chemistry*   *389*

Chapter 25   *Synthetic and Natural Organic Polymers*   *393*

Advanced Placement Chemistry Exam I   *395*

Answers Explanations to Practice Exam I   *423*

Advanced Placement Chemistry Exam II   *447*

Answers Explanations to Practice Exam II   *475*

ABOUT THIS BOOK

This book is intended to help you *review* for the Advanced Placement Chemistry exam. It assumes you have had almost two full years of chemistry instruction and are now preparing to take the AP Chemistry exam. The organization of this review book is based on *Chemistry*, by Raymond Chang (McGraw-Hill). The charts, graphs, and chapter references are taken from the textbook. However, you do not need any particular textbook in order to use this review book to prepare for the AP exam.

We have incorporated a number of features to facilitate your review process. These include:

> *Take Note* boxes include concrete tips on how the subject matter is treated on the AP Chemistry exam.
> *Sample problems* illustrate important concepts frequently tested on the AP Chemistry exam. The step-by-step solutions to the Example problems will allow you to follow the reasoning that leads to the solution. Then you will be able to solve similar problems on your own. Sample quantitative problems are presented in chart form, providing a 'general strategy' that you can extrapolate to other problems of the same type. This is followed by the specific solution for the particular example. You should study each example carefully. The problems included in the text as examples represent the fundamental types of problems you will find on the AP exam. Remember, practice makes perfect!
> The chapters of this review book that cover subject matter that is heavily tested on the AP exam, such as Chapter 16 on equilibrium, are more extensive.
> At the end of each chapter there are *multiple choice and comprehension questions with detailed answers* to give you more practice.
> The multiple choice questions at the end of each chapter are designed to be done without a calculator because calculators are not allowed on the Multiple Choice section of the AP exam.
> *Vocabulary words* are bolded and definitions are italicized to enable you to find key concepts and definitions quickly.
> *Two full-scale sample AP exams* with multiple choice questions, free response questions and detailed answers are included at the end of the book. The more problems you practice, the better you will perform on the exam. *Learn from your mistakes as you practice, and you will be less likely to make those mistakes on the actual exam.*

HOW TO USE THIS BOOK

How you use this book depends, in part, on how familiar you are with the chemistry topic being reviewed. As you go through this book, you may take different approaches for different chapters. It is best to start reviewing for the exam earlier rather than later. The night before the exam is definitely too late; a good night's sleep will do you far more good at that point.

For subject areas with which you feel very comfortable, you might choose to skim the key concepts. The important terms are bolded to help you do this. However, we recommend that you go over the sample problems to make sure that you can *apply* your knowledge. For content areas with which you are less confident, read the summary very carefully and work out all the sample problems. Then do all the multiple choice and comprehension questions at the end of the chapter. You may have to refer back to the textbook for a more comprehensive explanation. This book is meant to help you *review* material to which you have already been exposed, *not* to serve as a stand-alone text.

ABOUT THE AP CHEMISTRY TEST SUBJECT CONTENT

The College Board has defined the AP chemistry course content. You can access their website at *www.collegeboard.com/apstudents* to get more details.

This review book parallels the 25 chapters in *Chemistry* by Raymond Chang. However, the vast majority of the subject matter tested on the AP exam is presented in 18 chapters of the Chang textbook. These 18 chapters are listed in normal type in the list below. The titles of the chapters that cover material that is tested very little on the AP exam appear in italics.

As you review for the exam, you should focus mainly on the 18 chapters that contain the material that is most heavily tested on the exam. The chapters of this review book vary considerably in length; chapters that cover frequently tested or complex content are longer.

Chapter 1   Chemistry: The Study of Change
Chapter 2   Atoms, Molecules, and Ions
Chapter 3   Mass Relationships in Chemical Reactions
Chapter 4   Reactions in Aqueous Solutions
Chapter 5   Gases
Chapter 6   Thermochemistry
Chapter 7   Quantum Theory and the Electronic Structure of Atoms
Chapter 8   Periodic Relationships Among the Elements
Chapter 9   Chemical Bonding I: Basic Concepts
Chapter 10   Chemical Bonding II: Molecular Geometry and Hybridization of Atomic
                        Orbitals
Chapter 11   Intermolecular Forces and Liquids and Solids
Chapter 12   Physical Properties of Solutions
Chapter 13   Chemical Kinetics
Chapter 14   Chemical Equilibrium
Chapter 15   Acids and Bases
Chapter 16   Acid-Base Equilibria and Solubility Equilibria
Chapter 17   Chemistry in the Atmosphere

Chapter 18   Entropy, Free Energy, and Equilibrium
Chapter 19   Electrochemistry
Chapter 20   *Metallurgy and the Chemistry of Metals*
Chapter 21   *Nonmetallic Elements and Their Compounds*
Chapter 22   *Transition Metal Chemistry and Coordination Compounds*
Chapter 23   *Nuclear Chemistry*
Chapter 24   *Organic Chemistry*
Chapter 25   *Synthetic and Natural Organic Polymers*

Table 1 gives you an idea of the relative importance of each content area in the Multiple Choice section of the AP exam. Note that content distribution is different from that in the Free Response section. For example, equilibrium concepts are tested more extensively in the Free Response section than in the Multiple Choice section.

Table 1. Average distribution frequency of content areas in multiple choice section

Percent of questions	Topic	Chapter(s) in the Chang textbook relating to the topic area
1% to 3%	• General vocabulary • Precipitation • Nuclear • Organic and complex ions	1 and throughout textbook 4 23 24 (In this review book, organic naming rules are presented as part of Chapter 2.)
4% to 7%	• Gas laws and kinetic molecular theory • Atomic theory • Rates and equilibrium • Oxidation, reduction, and electrochemistry • Thermodynamics • Descriptive chemistry • Laboratory	5 2 13, 14 4, 19 6, 18 4 and throughout the book Throughout the book
8% to 10%	• Mole relationships and stoichiometry problems	3 and throughout the book
10% to 12%	• Acids, bases, and buffers	15, 16
Over 12%	• Bonding, intermolecular forces, and periodic properties • Solutions and phase diagrams	7, 8, 9, 10 11, 12

ABOUT THE AP CHEMISTRY EXAM FORMAT

The AP Chemistry exam is given in two time blocks: Multiple Choice and Free Response. You are given a 10-minute break between the two sections of the exam. The exam description given here is based on the *May 2007 exam.*

Section I. Multiple Choice

The Multiple Choice section is composed of *75 multiple choice questions* to be completed in *90 minutes.* You may *not* use a calculator. The only reference provided to you is a periodic table that contains the element symbols, atomic numbers, atomic masses, and nothing more.

Section II. Free Response

The Free Response section is divided into **Part A** and **Part B** to be completed in *95 minutes.* You may use a calculator for Part A only. There are *four pages of reference material* provided to you: the same periodic table that you had for the Multiple Choice section, plus a list of formulas and the Standard Reduction Potentials Table. *You should be familiar with these pages before* you enter the exam so that if you need to look up something, you'll know exactly where to look. You have only a limited time to complete the exam, so you need to be very efficient about how you use your time.

Part A is generally quantitative. *For Part A, you may use a calculator.* You have 55 minutes to do three problems.
> Question 1 is always an equilibrium problem.
> Question 2 can be in any content area.
> Question 3 can be in any content area.

Note: All quantitative answers must be supported with detailed work in order to receive credit. Be sure you use the proper number of significant figures in your answer. Be sure that your units are in agreement with each other throughout the problem.

Part B is qualitative and *you may not use a calculator.* You have 40 minutes to answer three questions.
> Question 4 asks you to write three balanced net ionic equations. You must write the reactants and predict the products. Equations must be balanced. You will also be asked to answer one question based on each of the three reactions.
> Question 5 can be in any content area.
> Question 6 can be in any content area.

Note: If neither Question 2 nor Question 3 is based on laboratory data and techniques, then Question 5 or 6 will address this topic.

Table 2. Exam format May 2007

Summary of Exam Format

Section I: **Multiple Choice**   **No calculator use**
75 questions   **allowed.**
Time: 90 minutes

───────────────────────── 10-minute break ─────────────────────────

Section II: **Free Response**
Part A. Quantitative Questions 1, 2, and 3   **Calculator allowed.**
Time: 55 minutes.
Question 1. Equilibrium problem
Question 2. Any topic
Question 3. Any topic

Part B. Qualitative Questions 4, 5, and 6   **No calculator use**
Time: 40 minutes.   **allowed.**
Question 4. Equations (three reactions)
Question 5. Any topic
Question 6. Any topic

If you are studying old AP exams, you will notice that prior to 2007 the exam format was slightly different. In the old exams, you were given some choice of which problem to solve (Question 2 or 3, Question 7 or 8). You were also given a choice when doing the equations in Question 4 and you were not required to balance the equations. These rules have changed for 2007. Nonetheless, we still strongly recommend that you use old AP exams as part of your review process. They are available on the Web at *www.collegeboard.com/apstudents*. When you take an old exam, do *all parts of every* problem. If you do this, you will be mimicking the new exam format. If you do every part of the Free Response section of an old exam, you will need to allow yourself more than 95 minutes. We believe 140 minutes would be a realistic amount of time.

ABOUT THE AP GRADING

Multiple Choice Questions. Because a quarter of a point is subtracted from your raw score for each incorrect answer, you should not guess unless you can eliminate some of the choices in the answers. If you have no idea how to answer a question and can eliminate none of the choices, *do not guess the answer* to the question. However, if you can eliminate one or two choices, then guessing from among the remaining answers does make statistical sense.

Free Response Questions. You will be given credit for the parts of the problem done correctly. When answering the quantitative questions, be sure to document your work. Final answers with no work shown are usually assigned *zero* credit.

For example, if 110 g of iron are given in a problem and you convert to moles as part of your solution process, you must show what numbers you used. For example,

$$110. \text{ g Fe} \times \frac{1 \text{ mol Fe}}{55.0 \text{ g Fe}} = 2 \text{ mol Fe}$$

You could condense the above to $\dfrac{110. \text{ g}}{55.0 \text{ g/mol}} = 2 \text{ mol Fe}$

You may NOT just say 2 mol Fe, with no work shown. You must indicate the division process in some way and give the atomic mass of iron.

Be sure to use the correct number of significant figures (addressed in Chapter 1 of this review book). If you are off by more than one significant figure, a point will be deducted from your score.

When answering an essay question, organize your thoughts *before* you begin writing. When you are no longer sure of what you are saying, stop writing! You can begin answering a question correctly; if you continue writing when you are no longer sure of what you are saying, errors tend to creep into your explanation and then you begin to lose points.

Table 3. Overall grade distribution on the AP chemistry exam

Grade distribution		
Multiple Choice:	50% of entire exam grade	
Free Response:	50% of entire exam grade, broken down as follows:	
	Question 1	20%
	Question 2	20%
	Question 3	20%
	Question 4	10%
	Question 5	15%
	Question 6	15%

HOW TO PREPARE FOR THE AP EXAM

Start reviewing for the exam early! You basically have 18 chapters to review. The AP exams are normally given in the first two weeks of May. *Make a timetable for yourself* and pace out the chapters you need to review. Some chapters will take more time than others. For example, Chapters 14 to 16 deal with equilibrium. This is the most important topic on the AP exam and many students find it challenging. You may want to spend extra time on these chapters. Chapters 1, 2, and 3 are very much a review of first-year high school chemistry and you may be able to zip through them pretty quickly.

You will need to devise a plan to suit your own needs as you begin the review process. *Good planning is important when facing so much content to review.* You should focus on the areas that you find difficult and that also have high representation on the AP exam. Table 4 is one example for a study plan.

Table 4. Sample review calendar

Date	Main chapter(s) to be reviewed
March—week 1	1, 2, 3
March—week 2	4, 5
March—week 3	6, 7, 8
March—week 4	9, 10, 11
April—week 1	12, 13
April—week 2	14, 15
April—week 3	16
April—week 4	18, 19, and parts of chapter 23

TIPS FOR TAKING THE EXAM

Before the exam

- Be familiar with the format of the test.
- Be familiar with the four pages of reference materials (periodic table, Standard Reduction Potentials Table, and list of formulas) that are given to you as part of the Free Response section of the exam.
- Take as many practice AP tests as you have time for.
- Get enough sleep the night before (at least 8 hours).
- Wear comfortable clothing to the exam. Dress in layers so you can adjust to the temperature of the room.
- Bring a snack and water for break time.
- Bring a calculator, pencils, and an eraser to the exam. Do not bring extra stuff like bookbags and jackets.

During the exam

Multiple Choice Questions

- Read the questions carefully.
- There are 75 questions to be completed in 90 minutes, which means you have approximately 75 seconds per question. Do NOT spend too much time on any one question. You can go back to the time-consuming questions after you have gone through the exam once.
- If you can eliminate one or two of the five choices from the answers, making an educated guess from among the remaining choices makes sense. If you have no idea, skip the question. For every wrong answer, a quarter of a point is subtracted from your raw score.

Free Response Questions

- Read the questions carefully. Be sure to answer the *entire* question.
- Stop writing when you begin to be unsure about your facts.
- Watch significant figures.
- In the quantitative questions (Questions 1, 2, and 3), do all the quantitative parts first. After 55 minutes you are required to put your calculators away. You can go back to these questions later in the exam and finish off any unanswered parts such as the qualitative parts of the question. However, you *cannot* use your calculator when you go back.
- When answering Question 4, you can use the Standard Reduction Potentials Table given at the beginning of the exam to obtain an oxidation number of an element if necessary.

COMMON MISTAKES TO AVOID

Multiple Choice Questions

> Time yourself. You have 75 seconds per multiple choice question. If a question is taking an inordinate amount of time, go on with the rest of the exam and then go back if you have time. The difference of a single question is unlikely to affect your score. After 90 minutes, the multiple choice part of the exam is over and you cannot go back to it.
> Read the questions carefully and answer the question asked!
> Do not mix up *increasing trend* (highest value is last in the series) with *decreasing trend* (highest value is first in the series).

Free Response Questions

> Time yourself. You have 55 minutes for the three quantitative problems in Part A. That works out to about 18 minutes per question. After 55 minutes you must put your calculator away. You have 40 minutes to do the three qualitative essays in Part B. That works out to about 13 minutes per question. You can go back to Part A questions during this time but won't have your calculator available.
> Read each question carefully and *answer the question asked*. For example, if the question asks:
>
>   1. *Agree or disagree with a statement. Justify your answer.* In your answer you need to say *I agree or I disagree* and then explain your answer. If you only explain your answer (even if you do so correctly), and never answer the "agree vs. disagree" part of the question, points will be deducted from your answer.
>   2. *Increases, decreases, or stays the same. Justify your answer.* In your answer, first state your choice: increases, decreases, or stays the same. Then you must explain your choice. This type of question often occurs with periodic trends and laboratory topics.
>
> *Read each question carefully. Answer the question asked completely.* Think a minute to compose your answer before beginning to write. For example, if you are asked to distinguish between solid $CaCO_3$ and solid $NaCl$, your answer should address *both* chemicals. A complete answer would include what happens with both solids when, for example, HCl (*aq*) is added to each. With $CaCO_3$, CO_2 gas is produced, but with $NaCl$ no gas is produced. An incomplete answer that addresses only the positive result with $CaCO_3$ and omits any mention of what happens with $NaCl$ is an incomplete answer. Credit will be deducted.
> Stop writing when you become unsure of what you are saying. When you answer an essay question, put down what you know for sure. Then stop writing. If you keep going, errors tend to enter your answer and you usually begin to lose credit.
> If you are unsure of the charge of an ion, you can use the Standard Reduction Potentials Table as a reference.
> In Part A, make sure that all of your quantitative work is shown.

➢ Be sure that each of your quantitative answers has the correct number of significant figures.

➢ Be careful about which R value you use. The value 8.31 J/mol • K is used in most thermodynamics work. But 0.0821 L • atm/mol • K is used with the ideal gas law.

➢ Be sure that all of your quantitative answers have units.

➢ Be sure that your units are in agreement when you are doing quantitative work. For example, when using the equation: $\Delta G = \Delta H - T\Delta S$, note that ΔG and ΔH are usually given in kJ/mole and ΔS is given in J/mole. Remember, to convert kJ to J, multiply by 1000.

➢ Remember to use the Kelvin scale with the gas laws and thermodynamics work.

➢ Be careful about the vocabulary you choose to use when answering essay questions. For example, do not confuse:

Volume with mass (point mass). Intermolecular forces usually are involved with volume or surface area issues, not mass.

Intramolecular forces (bonds) with intermolecular attractions. When talking about boiling point and melting point, the issue to focus on usually is intermolecular attractions.

Lab measurements with calculations. When asked what measurements need to be made, the initial temperature and final temperature are measurements. ΔT is a calculation, not a measurement. If you give a calculated value when asked for a measurement, you receive zero credit.

CHAPTER 1
CHEMISTRY: THE STUDY OF CHANGE

This chapter reviews two topics basic to the study of chemistry that most of you learned in first year chemistry:

- The classification of matter and its changes
- Measurement and significant figures

The classification of matter and its changes

The simplest form of matter is an **element**. **Compounds** are combinations of elements that have a definite composition. Elements and compounds can exist in three states of matter: solids, liquids, and gases. The state in which matter exists depends on the temperature.

Matter can undergo physical and chemical changes. **Physical changes** *are those that do not alter the identity of that substance; for example, melting or freezing.* **Chemical changes** *do alter the identity of the substance; for example, rusting or burning.* To determine whether a physical or chemical change has occurred, observations must be made and measurements must be taken. Color, odor, and state are some of the common observations that can be made. Mass, volume, and temperature are the most frequently taken measurements. It is important to distinguish between **extensive properties**, *properties such as mass that depend on how much matter is being considered*, and **intensive properties**, *properties such as density that do not depend on how much matter is being considered.*

Measurement and significant figures

An understanding of **precision**, *how closely two or more measurements of the same quantity agree with one another*, and **accuracy**, *how close a measurement is to the true value of the quantity that was measured*, is important to laboratory measurement and data assessment. You should be very familiar with **SI units, scientific notation**, and **significant figures** in order to handle the calculations associated with chemical change.

> **Take Note:** *All calculations on the AP exam require that the rules for significant figures be obeyed. Credit will be given on the AP exam for the correct number of significant figures and for one more or one less than the correct number of significant figures. If a numerical answer is outside this range in the Free Responses section, a point will be deducted from your score.*

Significant figures are the *meaningful digits in a measured or calculated quantity*. A simplified version of significant figure rules is:

- **All nonzero digits are significant.** Thus <u>845</u> has three significant figures and <u>1.234</u> has four.
- **All zeroes to the right of nonzero digits are significant.** Thus <u>2.0</u> has two significant figures and 0.0000<u>320</u> has three significant figures.*
- **All zeroes between nonzero digits are significant.** Thus <u>606</u> has three significant figures and <u>40,501</u> has five significant figures.
- **Exact numbers obtained from definitions or counting have an infinite number of significant figures.** One meter contains 1000 millimeters and thus one and 1000 have an infinite number of significant figures.

  *   There is one exception: for numbers that do not contain decimal points, the trailing zeroes (i.e., zeroes after the last nonzero digit) may or may not be significant. If there is no decimal, as in the case of the number 400, this implies that the quantity is approximately, but not exactly, <u>4</u>00; in this case, there is only one significant figure.

The mathematical operations with significant figures are the same for addition and subtraction but different than operations involving multiplication and division. In addition and subtraction, the answer is controlled by the number with the <u>fewest</u> decimal places and the answer should contain that number of decimal places (see Example 1). In a multiplication or division operation, the total number of significant figures in the answer is governed by the number with the <u>least</u> number of significant figures regardless of where the decimal point is (see Example 2). Answers that do not contain the correct number of significant figures must be rounded off.

Example 1.   Subtraction with significant figures.

Consider the following experimental data:

mass of precipitate and filter paper	14.1 g
mass of filter paper	0.1983 g

Determine the mass of the precipitate.

General Strategy	Solution to Example 1
Perform the subtraction.	14.**1** g precipitate + filter paper − 0.**1983** g filter paper 13.9017 g precipitate
The 14.**1** g has the fewest decimals. Subtraction answers are governed by the number with the fewest decimal points. The answer must be rounded to one decimal place.	13.9017 g = 13.9 g final answer

Example 2. Division with significant figures.

Laboratory measurements were performed to determine the density of an unknown liquid. The following data were obtained in the lab:

mass of the graduated cylinder empty	10.0500 g
mass of the graduated cylinder + the unknown liquid	91.5900 g
volume of the liquid in the graduated cylinder	88.3 mL

Determine the density of the unknown liquid.

General Strategy	Solution to Example 2
The mass and volume of the liquid must be known in order to calculate the density. The volume is given as 88.3 mL (three significant figures).	Determination of the mass of the liquid: $$\begin{array}{r} 91.5900 \text{ g} \\ -10.0500 \text{ g} \\ \hline 81.5400 \text{ g} \end{array} \text{ (six significant figures)}$$
Calculate the density of the unknown liquid being the mass and volume of the liquid are known.	$$D = \frac{m}{V}$$ $$D = \frac{81.5400 \text{ g}}{88.3 \text{ mL}} = .9234428086 \ \frac{\text{g}}{\text{mL}}$$
There are six significant figures in **81.5400** g and three significant figures in **88.3** mL. The number with fewest significant figures governs the answer in division. The answer should have three significant figures and must be rounded off.	$$0.923 \ \frac{g}{mL}$$

SAMPLE MULTIPLE CHOICE QUESTIONS

1. When the following calculation is performed, the number of significant figures in the answer is:

$$(74.0/23.60) + 15.000$$

   A. 6
   B. 3
   C. 2
   D. 4
   E. 7

2. What is the volume of a sample of gold having a mass of 20.0 g? The density of gold is 19.3 g/mL.

   A. 0.731 mL
   B. 0.893 mL
   C. 1.04 mL
   D. 0.751 mL
   E. 280. mL

3. The term that describes the ability of an instrument to produce the same measurement on repetition is:

   A. quantity
   B. qualitative
   C. accuracy
   D. precision
   E. property

4. Which is an example of an extensive property?

   A. odor
   B. color
   C. melting point
   D. density
   E. mass

5. A temperature of 245 K corresponds to which of the following Celsius temperatures?

   A. $-28\,^{\circ}\text{C}$
   B. $-100\,^{\circ}\text{C}$
   C. $0\,^{\circ}\text{C}$
   D. $273\,^{\circ}\text{C}$
   E. $-73\,^{\circ}\text{C}$

6. Copper is a trace element and nutrient required in the diet of newborn babies, who require 80.0 µg per kg of body weight per day. Formula contains 0.48 µg/mL. How many mL of formula does a 6.60 lb baby require per day? (1 kg = 2.2 lb)

   A. 5.00×10^2 mL
   B. 5.0×10^2 mL
   C. 50. mL
   D. 50.0 mL
   E. 5.00 mL

7. Which of the following is a chemical property?

   A. copper sulfate crystals are blue
   B. water boils at 100 °C
   C. chlorine is a gas at room temperature
   D. sodium reacts with oxygen in the air to form a white oxide
   E. magnesium is malleable

8. In which of the following does a chemical change occur?

   A. $2Ag\ (s)\ +\ O_2\ (g)\ \rightarrow\ 2Ag_2O\ (s)$
   B. $NaCl\ (s) \rightarrow\ NaCl\ (aq)$
   C. $I_2\ (s)\ \rightarrow\ I_2\ (g)$
   D. $Mg\ (s)\ \rightarrow\ Mg\ (l)$
   E. $H_2O\ (g)\ \rightarrow\ H_2O\ (l)$

9. The number of significant figures in 0.00000640301 is:

   A. 6
   B. 4
   C. 11
   D. 2
   E. 5

10. 453 mm expressed in meters is:

   A. 0.0453
   B. 0.453
   C. 4.53
   D. 45.3
   E. 0.00453

Comprehension Questions

1) Classify the following as either physical or chemical changes, and provide a brief explanation for each choice:

        a) A tire's tread wears down with use.
        b) An iron poker in a fire begins to glow red.
        c) A mixture of amino acids is combined to create a protein.
        d) White, mineral-like deposits form at the end of a faucet.
        e) Antacid tablets "fizz" when dropped into a glass of water.

2) Human core body temperature is 98.6 °F. Convert this temperature to both Kelvin and Celsius. Record your answers to the correct number of significant figures.

3) 39.95 g of argon gas, 1 mol, contains approximately 6.022×10^{23} atoms of argon. This quantity of argon will occupy a volume of 22.4 L at 0 °C and 1.00 atm of pressure. Assuming the atoms to be roughly spherical with a radius of 98 pm, calculate the percentage of the volume occupied by the gas that is occupied by the atoms themselves, and the percentage of the total volume that is empty space (i.e., the amount of space between the atoms).

4) A student wishes to determine the density of an irregularly shaped fishing weight to determine if it is made from a pure metal, lead in this case, or an alloy. Describe a practical process, utilizing common lab equipment, by which this measurement could be made. The student discovers that the density of the object is 11.30 g/cm^3, and concludes that it is indeed made from pure lead (lead's density is reported to be 11.34 g/cm^3 at room temperature). Discuss the validity of the claim and possible sources of error that would invalidate the conclusions made.

5) Explain why the Kelvin temperature is considered an absolute temperature scale whereas Celsius is not.

ANSWERS TO MULTIPLE CHOICE QUESTIONS

1. D

        Division requires that the least number of significant figures govern the number of significant figures in the answer. Since 74.0 has only three significant figures, then the answer to the division will be 3.14 (three significant figures).

Significant figures in addition (or subtraction) are governed by the number of decimal places, with the number containing the least number of decimal places controlling the number of significant figures in the answer. The addition of 3.14 (two decimal places) and 15.000 (three decimal places) yields 18.14, which contains two decimal places.

2.   C

$$20.0 \text{ g} \times \frac{1 \text{ mL}}{19.3 \text{ g}} = 1.036 \text{ mL} = 1.04 \text{ mL} \text{ (three significant figures)}$$

3.   D

Precision refers to how closely two or more measurements of the same quantity agree with one another.

4.   E

An extensive property depends on how much matter is being considered. Since mass is the quantity of matter in a given sample, it is an extensive property.

5.   A

The relationship between the Kelvin temperature scale and the Celsius scale is $K = C + 273$. Consequently $245 = C + 273$ or $C = 245 - 273$. $C = -28°C$.

6.   B

Dimensional analysis yields the following solution:

$$6.60 \text{ lb} \times \frac{1 \text{ kg}}{2.2 \text{ lbs}} \times \frac{80.0 \text{ μg}}{1 \text{ kg}} \times \frac{1 \text{ mL}}{0.48 \text{ μg}} = 500 \text{ mL or } 5.0 \times 10^2 \text{ mL}$$

Since 0.48 μg contains two significant figures, the answer requires two significant figures.

7.   D

Silvery sodium reacting with oxygen gas to form a white powdery solid describes a chemical property because a chemical change must occur to observe this property. The reaction is $4Na \, (s) + O_2 \, (g) \rightarrow 2Na_2O \, (s)$.

8.   A

The formation of silver oxide from silver and oxygen is the only change resulting in a new substance. All other choices involve only a change in state.

9.   A

The answer is *0.00000640301*. All nonzero digits are significant and only zeroes to the right of nonzero digits are significant. Zeroes to the left of nonzero digits are not significant.

10.   B

$$453 \text{ mm} \times \frac{1\,m}{1000\,mm} = 0.453 \text{ meters}$$ (three significant figures since the conversion factor is an absolute number)

Answers to Comprehension Questions

1a)   The wearing down of tread is a physical change that occurs through physical contact and friction with the road surface. Although there is the possibility of chemical change when rubber is subjected to atmospheric oxidants and sunlight, a great deal of this change would be the physical removal of rubber by abrasion.

b)   This also represents a physical change. As the iron is heated, energy is added to its outermost electrons and they move to higher energy levels. When they spontaneously move back to lower energy levels, they emit the excess energy that they possessed. Some of this energy can be seen as visible light in the red portion of the electromagnetic spectrum.

c)   Formation of a protein, either as part of a living organism's metabolic processes or in a lab, represents a chemical transformation. Chemical bonds are both broken and formed in this process, and the chemical and physical characteristics of the protein will be different than those of the amino acids.

d)   This formation of mineral deposits is a physical process involving a change of state. Dissolved minerals, for example, $CaCO_3$, in tap water have limited solubility in water and therefore form a solid precipitate at the end of a faucet as the water evaporates. It is possible for these deposits to react with metal materials in the faucet if they are left in place for an extended period, however, they will form even at the end of a plastic faucet.

e)   The antacids that "fizz" or produce a gas, carbon dioxide, do so by a chemical reaction. The antacids contain a carbonate salt, for example, $CaCO_3$, and a solid acid such as citric and/or acetyl salicylic acid. Upon addition to water, the tablets dissolve and begin to react in an acid-base neutralization reaction as shown. The gas produced is CO_2.

$$C_5H_8O_7 + CaCO_3 \rightarrow CaC_5H_7O_7 + H_2O + CO_2$$

2)   Conversion from Celsius to Kelvin is straightforward, and therefore it is common to convert from Fahrenheit to Celsius and then to Kelvin. $°C = (°F \times 1.8) + 32$. Substituting, $°C = (98.6 \, °F \times 1.8) + 32 = \textbf{37.0 °C}$. Kelvin $= °C + 273.15 = 37.0 \, °C + 273.15 = \textbf{310.2 K.}$

3)   Since the question states that the total volume occupied by the gas is 22.4 liters, the calculation that remains is to determine the volume of the gas particles themselves. The atoms can be treated as extremely small spheres with a volume equal to $4/3\Pi r^3$. Substituting a radius of 98×10^{-12} m into the equation:

$$4/3\Pi r^3 = 4/3\Pi(98 \times 10^{-12} \, \text{m})^3 = 4/3\pi \, 9.4 \times 10^{-31} \, \text{m}^3 = 3.9 \times 10^{-30} \, \text{m}^3$$

Multiplying by the conversion 1000 L / m^3 gives a volume of 3.9×10^{-27} L . This value is the volume of a single Ar atom, and must be multiplied by the total number of atoms present, 6.022×10^{23}, to arrive at the volume occupied by all of the atoms in the sample.

$$3.9 \times 10^{-27} \, \text{L/atom} \times 6.022 \times 10^{23} \, \text{atoms} = 2.4 \times 10^{-3} \, \text{L}$$

Calculating the percentage of the space occupied by the atoms:

$$(2.4 \times 10^{-3} \, \text{L} / 22.4 \, \text{L}) \times 100\% = \textbf{0.011\% of the space is occupied by the atoms}$$

themselves, or, subtracting from 100%, a sample of argon gas is **99.99% empty space.**

4)   Density is the ratio between mass and volume for a given object or substance, so it is necessary to determine both the mass and volume of the object, a fishing weight in this case. The mass can be determined with a lab balance. In this case, because of the question that is being asked about the object itself, a balance that reads to the nearest $1/100^{th}$ of a gram would be a good choice. The volume of an irregularly shaped object can be easily measured to a reasonable precision by displacement. This procedure usually involves the use of a graduated cylinder that is partially filled with water. The volume of the water is noted before and after the object is submerged into the water, and the difference in volumes is equal to the volume of the object.

The student's claim that the object is made from pure lead based on a density measurement is a reasonable, but not definitive conclusion. It is assumed that other physical characteristics of the object were taken into account, that is, metallic properties such as luster, malleability, etc. However, the object could have been made from an alloy, a mixture of metals, that just happened to have the same density as lead. A chemical analysis would have to be performed to be sure. The student should have conducted several trials of this measurement to get some idea of the precision of the data generated. Sources of error in this type of measurement include bubbles sticking to the outside of the object being submerged into the water, water splashing out of the graduated cylinder as the object is introduced, weighing the object after its volume is determined (sometimes there is residual water left on the object), and other random errors that would not be detected if a single trial were performed.

5)   Commonly used temperature scales such as Fahrenheit and Celsius use arbitrary values for their zero point such as the freezing point of water, etc. Therefore, these scales have negative temperature values. These negative values correlate with situations in which there is still molecular motion, that is, a positive amount of "temperature" (temperature is proportional to motion of particles). The Kelvin scale is the only one that sets its zero point at the temperature where all molecular motion has ceased, that is, that situation in which there is no "temperature." This is the only case in which a doubling of the measured temperature corresponds with a doubling of molecular motion.

CHAPTER 2
ATOMS, MOLECULES, AND IONS

This chapter is a review of first year chemistry concepts and terminology used to describe atoms and molecules, atomic theory, the periodic table, and the naming of compounds.

> **Take Note:** *You need to know and use correctly the basic vocabulary presented in this chapter. A detailed, accurate knowledge of naming rules for compounds is essential for communicating chemical information, both in writing equations and solving problems. Clear and precise communication is essential for successful performance on the AP exam.*

You can organize Chapter 2 material into categories to help you master the information. The main categories are:

- *Key terms pertaining to atoms and compounds.* These terms include: allotrope, atomic number, diatomic elements, molecule, compound, isotope, ion, mass number, law of definite composition, law of multiple proportions, and law of conservation of matter (Table 1).
- *Key terms pertaining to the periodic table.* These terms include: alkali metals, alkaline earth metals, halogens, metals, metalloids, nonmetals, noble gases, family, and period (Table 2).
- *Key scientists and their contributions to atomic theory.* These include: John Dalton, J. J. Thomson, Marie Curie, R. A. Millikan, E. Rutherford, and J. Chadwick (Table 3).
- Rules for naming inorganic compounds.
- Names of common polyatomic ions (Table 4).
- Rules for naming common organic compounds and some examples (Tables 5–9).

Table 1. Key terms pertaining to atoms and compounds

Term	*Definition* **and some pertinent details**	Examples
Allotropes	*Different forms of the same element in the same physical state.*	O_2 and O_3 C diamond and C graphite
Atomic number (Z)	The *number of protons (Z)* in the nucleus of an element. The periodic table is organized by increasing atomic number.	Hydrogen $Z = 1$ Helium $Z = 2$
Diatomic elements	Diatomic elements exist as *two-atom molecules.* There are seven elements that exist as diatomic molecules at room conditions.	$H_2, O_2, N_2, F_2, Cl_2, Br_2, I_2$

Molecule	*Two or more atoms* form a molecule.	H_2O, O_2
Compound	*Two or more different atoms* form a compound. A given compound always has a definite composition.	H_2O, $NaCl$, $C_6H_{12}O_6$ O_2 is a molecule but *not* a compound because it only has one type of atom.
Isotope	Isotopes are atoms of the same element that have *different numbers of neutrons*. Because of the different numbers of neutrons, isotopes of the same element have different mass numbers and some different properties. For example, the isotope C-12 has a mass number of 12 and is not radioactive. The isotope C-14 has a mass number of 14 and is radioactive.	C-12 and C-14
Ion	An ion is a *charged species* in which the number of protons does not equal the number of electrons. A **cation** is a positive ion; an **anion** is a negative ion. **Monatomic ions** contain one atom and **polyatomic ions** contain more than one atom.	Monatomic cation: Na^+ Monatomic anion: Cl^- Polyatomic cation: NH_4^- Polyatomic anion: OH^-
Mass number (A)	The mass number is equal to the *number of protons plus neutrons* in the nucleus. The mass number minus the number of protons gives the number of neutrons.	U-235 refers to the uranium isotope with 92 protons and a mass number of 235.
Law of Definite Composition	*A given compound always has the same composition.*	Water is always made up of two hydrogen atoms and one oxygen atom. Table salt is always made up of one sodium atom and one chloride atom.
Law of Multiple Proportions	Sometimes *two elements can combine in more than one way* to form two or more different compounds. Subscripts are simple whole number multiples of each other.	H_2O and H_2O_2 In H_2O, 2 g H combine with 16 g O (1 to 8 ratio). In H_2O_2, 2 g H combine with 32 g O (1 to 16 ratio).
Law of Conservation of Matter	*Matter can neither be created nor destroyed* in a nonnuclear chemical reaction.	Mass of reactant molecules = mass of product molecules. $2H_2 + O_2 \rightarrow 2H_2O$ $2(2\text{ g}) + 32\text{ g} \rightarrow 2(18\text{ g})$ $36\text{ g} \rightarrow 36\text{ g}$

Table 2. Key terms pertaining to the periodic table

Term	Definition and some pertinent details
Alkali metals	Group 1A elements: Li, Na, K, Rb, Cs, and Fr. They are the most active metals on the periodic table. They lose one electron and form 1+ ions.
Alkaline Earth Metals	Group 2A elements: Be, Mg, Ca, Sr, Ba, and Ra. They form 2+ ions.
Halogens	Group 7A elements: F, Cl, Br, I, and At. In binary compounds they form 1– ions.
Metals	The metals are on the left side and center of the periodic table. Metals are shiny, malleable, ductile, and are good conductors of heat and electricity.
Metalloids	These elements (B, Si, Ge, As, Sb, Te, Po, At) have properties intermediate between a metal and a nonmetal.
Nonmetals	Nonmetals are on the right side of the periodic table and are poor conductors of heat and electricity.
Noble gases	Group 8A elements: He, Ne, Ar, Kr, Xe, and Rn. He, Ne, and Ar are not chemically active under any conditions.
Family	A vertical column in the periodic table; also known as a **group**. Members of a group have the same number of valence electrons (electrons in the outermost energy shell) and somewhat similar chemical properties.
Period	A horizontal row in the periodic table; also known as a **series**.

Table 3. Key scientists and their contributions to atomic theory

Year	Scientist	Scientific contribution
1808	**John Dalton**	Dalton's atomic theory: • All matter is composed of atoms. • Atoms of one element are all identical in mass and chemical behavior. • Atoms are neither created nor destroyed in chemical reactions. • Atoms combine in small whole number ratios to form compounds. • A given compound has a constant composition (number and type of atoms are the same).
1897	**J. J. Thomson**	Using a cathode ray tube, Thomson determined that cathode rays were composed of negatively charged particles (later called electrons). Thomson calculated the charge-to-mass ratio for electrons. Thomson's model of the atom (the 'plum pudding' model) was later disproved by Rutherford.

1900	**Marie Curie**	Curie worked on the nature of radioactivity and discovered polonium and radium.
1909	**R. A. Millikan**	Millikan performed the oil drop experiment, which enabled him to calculate the charge of an electron.
1911	**E. Rutherford**	Rutherford performed the gold foil experiment in which gold foil was bombarded with alpha particles (helium nuclei). Most alpha particles went through the foil undeflected, leading to the conclusion that the atom was mostly empty space with a very tiny, dense, and positive center (the nuclear model of the atom). Rutherford also studied the nature of radioactivity and described the three types of radiation: alpha, beta, and gamma radiation.
1932	**J. Chadwick**	Chadwick discovered the neutron.

Rules for naming inorganic compounds

1. **Naming a binary compound with a metal and a nonmetal.** By convention, the symbol for the metal is written first. The total charge of a compound is zero. A **binary compound**, a compound *composed of two elements*, ends with the suffix *-ide*. An example is NaCl, sodium chloride.

2. **Naming binary metal/nonmetal compounds in which the metal has more than one oxidation state.** Common metal ions with multiple valences are:
   a. Cu^+, Cu^{2+} named copper(I) and copper(II) or cuprous and cupric ion
   b. Fe^{2+}, Fe^{3+} named iron(II) and iron(III) or ferrous and ferric ion
   c. Sn^{2+}, Sn^{4+} named tin(II) and tin(IV) or stannous and stannic ion
   d. Hg_2^{2+}, Hg^{2+} named mercury(I) and mercury(II) or mercurous and mercuric ion. Note that the mercury(I) ion occurs as the dimer, Hg_2^{2+}.

Some helpful hints to help you with your memorization of oxidation numbers:
   a. The elemental state is zero (Na^0, K^0).
   b. Family 1A metal ions have a +1 oxidation state (Na^+, K^+).
   c. Family 2A metal ions have a +2 oxidation state (Mg^{2+}, Ba^{2+}).
   d. Family 7A (halogen) ions have a −1 oxidation state in binary compounds (KCl, HF).
   e. Common cations that have only one oxidation state are: Al^{3+}, Zn^{2+}, and Ag^+.
   f. Hydrogen usually has a +1 oxidation state (H^+) unless it is a hydride in which hydrogen has a −1 oxidation state (H^-). Hydrides form between hydrogen and a very active metal; an example is sodium hydride, NaH.
   g. Oxygen usually has an oxidation state of −2 in oxides (O^{2-}) unless it is a peroxide in which oxygen has a −1 oxidation state (O_2^{2-}). For example, Na_2O and H_2O_2 are called sodium oxide and hydrogen peroxide, respectively.
   h. See Figure 2.11 for some examples of common monatomic ions.

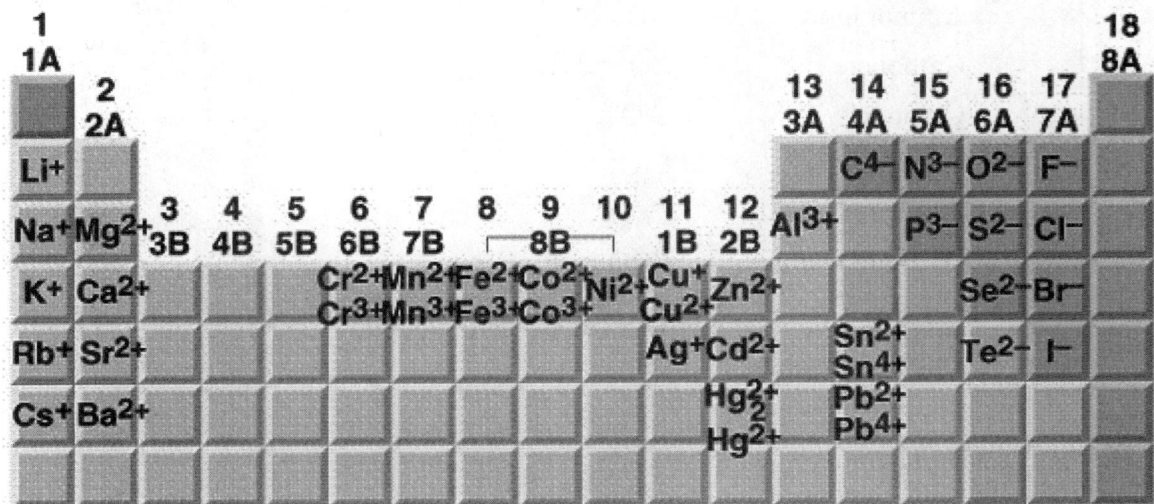

Figure 2.11

3. **Naming binary compounds with two nonmetals (molecular compounds).**
   Prefixes of mono-, di-, tri-, etc., are necessary when one pair of elements can form several different molecular compounds. Elements such as C, N, and S have multiple valences and the name must identify which oxidation number is present. Common examples are:
   a. CO and CO_2 are named carbon monoxide and carbon dioxide, respectively. Carbon oxide is an incorrect name because it does not distinguish between the two compounds.
   b. SO_2 and SO_3 are named sulfur dioxide and sulfur trioxide, respectively.
   c. CF_4 is named carbon tetrafluoride. C usually has an oxidation state of $+4$ or -4. Refer to Tables 5 through 9 for more organic molecule naming guidelines.
   d. NO and N_2O_4 are named nitrogen monoxide and dinitrogen tetraoxide, respectively.

4. **Naming compounds that contain a metal and a polyatomic ion**. When compounds contain polyatomic ions, simply name the metal and name the polyatomic ion. You do not make any changes to the name of the metal ion or the polyatomic ion. For example, $NaNO_3$ is simply sodium nitrate.

Table 4. Most frequently used polyatomic ions

Acetate	CH_3COO^-
Ammonium	NH_4^+
Carbonate	CO_3^{2-}
Chlorate*	ClO_3^-
Chromate	CrO_4^{2-}
Dichromate	$Cr_2O_7^{2-}$
Hydrogen sulfate	HSO_4^-
Nitrate*	NO_3^-
Permanganate	MnO_4^-
Phosphate*	PO_4^{3-}
Sulfate*	SO_4^{2-}

* Refers to the most common form of the polyatomic ion. The number of oxygen atoms can vary with these polyatomic ions but the charge of the polyatomic ion remains the same. Check the naming rules below for the less common forms of these polyatomic ions.

Parentheses are used when the polyatomic ion is used more than once in a compound. For example, barium chlorate is written as $Ba(ClO_3)_2$. If the polyatomic ion is only used once in a compound, the parentheses are not used. For example, sodium chlorate is written as $NaClO_3$.

Some polyatomic ions exist in more than one form:

    a. The name for the most common form of a polyatomic ion ends in *-ate*. Examples are: ClO_3^- chlorate, NO_3^- nitrate, PO_4^{3-} phosphate, and SO_4^{2-} sulfate.

    b. If one oxygen is added to the most common form of the polyatomic ion a prefix of *per-* and suffix of *-ate* are used. For example, ClO_4^- is called perchlorate. Not all polyatomic ions can add an extra oxygen.

    c. If one oxygen is removed from the most common form the name ends in *-ite*. For example, ClO_2^- is called chlorite.

    d. If more than one oxygen is removed from the most common form a prefix of *hypo-* and a suffix of *-ite* are used. For example, ClO^- is called hypochlorite.

Take Note: *If you learn the most common form of the polyatomic ion and the rules above in a to d, you can deduce the names of most polyatomic ions that will appear on the AP exam.*

5. **Naming acids.** (Most acids have H listed as the first element in the compound.)

   a. **Binary acids** (*made of two elements*). The name is composed of prefix *hydro-*, stem and suffix of *-ic*. Examples are HCl and HI named *hydro*<u>chlor</u>*ic* acid and *hydro*<u>iod</u>*ic* acid, respectively.

   b. **Oxoacids** (made of hydrogen and a polyatomic ion that contains oxygen). If the polyatomic ion ends in *-ate*, replace the *-ate* ending with *-ic acid*. Examples are:

   ClO_3^- chlorate becomes chloric acid ($HClO_3$)

   ClO_4^- perchlorate becomes perchloric acid ($HClO_4$)

   ClO_2^- chlorite becomes chlorous acid ($HClO_2$)

   ClO^- hypochlorite becomes hypochlorous acid ($HClO$)

   Note that prefix and suffix rules mirror the polyatomic ion rules given in 4.

6. **Naming bases.** For simple bases containing the –OH group the name is composed of the metal name plus the word *hydroxide*. Examples are NaOH and Al(OH)$_3$, named sodium hydroxide and aluminum hydroxide, respectively.

7. **Naming hydrates**. The hydrate name is made up of several components listed in the following order: metal name, anion name, prefix to indicate the number of water molecules, followed by the word *hydrate*. An example is $CuSO_4 \cdot 5H_2O$ named copper(II) sulfate pentahydrate.

8. **Common and systematic names of some compounds** are given in Table 2.7.

TABLE 2.7 Common and Systematic Names of Some Compounds

Formula	Common Name	Systematic Name
H_2O	Water	Dihydrogen monoxide
NH_3	Ammonia	Trihydrogen nitride
CO_2	Dry ice	Solid carbon dioxide
NaCl	Table salt	Sodium chloride
N_2O	Laughing gas	Dinitrogen monoxide
$CaCO_3$	Marble, chalk, limestone	Calcium carbonate
CaO	Quicklime	Calcium oxide
$Ca(OH)_2$	Slaked lime	Calcium hydroxide
$NaHCO_3$	Baking soda	Sodium hydrogen carbonate
$Na_2CO_3 \cdot 10H_2O$	Washing soda	Sodium carbonate decahydrate
$MgSO_4 \cdot 7H_2O$	Epsom salt	Magnesium sulfate heptahydrate
$Mg(OH)_2$	Milk of magnesia	Magnesium hydroxide
$CaSO_4 \cdot 2H_2O$	Gypsum	Calcium sulfate dihydrate

Rules for naming organic compounds and some common examples.

> **Take Note:** *You will be required to know the names of simple organic compounds and recognize common functional groups (Tables 5 to 9) on the AP exam.*

Table 5. Names of the first ten alkanes (saturated hydrocarbons, single bonds only, name ends in *-ane*, homologous formula: C_nH_{2n+2})

Formula	Name
CH_4	Methane
C_2H_6	Ethane
C_3H_8	Propane
C_4H_{10}	Butane
C_5H_{12}	Pentane
C_6H_{14}	Hexane
C_7H_{16}	Heptane
C_8H_{18}	Octane
C_9H_{20}	Nonane
$C_{10}H_{22}$	Decane

Table 6. Names of alkenes (unsaturated hydrocarbons, contain a double bond, name ends in *-ene*, homologous formula: C_nH_{2n})

C_2H_4	Ethene (common name is ethylene)
C_3H_6	Propene
C_4H_8	Butene

Table 7. Names of alkynes (unsaturated hydrocarbons, contain a triple bond, name ends in *-yne*, homologous formula: C_nH_{2n-2})

C_2H_2	Ethyne (common name is acetylene)
C_3H_4	Propyne
C_4H_6	Butyne

Table 8. Common functional groups

Functional group		Example	Name
R – OH	Alcohol	C_2H_5OH	Ethyl alcohol (ethanol)
R – COOH	Carboxylic Acid	CH_3COOH	Acetic acid (ethanoic acid)
R – COOR	Ester	CH_3COOCH_3	Methyl acetate
R – CHO	Aldehyde	CH_3CH_2CHO	Propanal
R – O – R	Ether	CH_3OCH_3	Dimethyl ether
$\begin{matrix} O \\ \| \\ R - C - R \end{matrix}$	Ketone	$\begin{matrix} O \\ \| \\ CH_3 - C - CH_3 \end{matrix}$	Dimethyl ketone (acetone)

Table 9. Simple organic compounds that are often mentioned

Molecular fomula	Name
C_6H_6	Benzene
$C_6H_{12}O_6$	Glucose
C_2H_2	Acetylene
C_2H_4	Ethylene

SAMPLE MULTIPLE CHOICE QUESTIONS

Questions 1 – 4 refer to the following elements.

    A. Br
    B. C
    C. P
    D. Zn
    E. Rn

1. The element is a member of the noble gas family.
2. The element exists as a diatomic molecule at room conditions.
3. The element forms Na_2XO_3 compound.
4. The element has a monatomic ion with a charge of 2+.

5. Which of the following pairs are isotopes?

    A. O_2 and O_3
    B. Cu^+ and Cu^{2+}
    C. U-235 and U-238
    D. Hydrogen and helium
    E. C_2H_2 and CH

6. Which of the following compounds is incorrectly named?

    A. $NaHSO_4$ sodium hydrogen sulfate
    B. $K_2Cr_2O_7$ potassium dichromate
    C. Fe_2O_3 iron(II) oxide
    D. CO carbon monoxide
    E. $Ba(NO_2)_2$ barium nitrite

7. Which family represents the most active metals?

    A.   Alkali metals
    B.   Alkaline earth metals
    C.   Transitional metals
    D.   Family 7A, the halogens
    E.   Family 8A, the noble gases

8. The scientist who developed the nuclear model of the atom is

    A. John Dalton
    B. J. J. Thomson
    C. Ernest Rutherford
    D. R. A. Millikan
    E. Albert Einstein

9. Which polyatomic ion is incorrectly named?

    A. PO_4^{3-} phosphate
    B. SO_4^{2-} sulfite
    C. CrO_4^{2-} chromate
    D. CO_3^{2-} carbonate
    E. NH_4^+ ammonium

10. The empirical formula for glucose, $C_6H_{12}O_6$, is

    A. CH_2
    B. CH_2O
    C. $C_3H_6O_3$
    D.   CHO
    E.   None of these

Comprehension Questions

1)   Differentiate among atoms, ions, molecules, and compounds using the following substances: Ne, F_2, PCl_5, $KClO_3$, and SO_4^{2-}. Be sure to comment on the number and type of atom(s) involved, electrical charge, types of elements (metal or nonmetal) involved, and characteristics of the formula that allow for identification and differentiation between species.

2)   Define the term *empirical*. How does its meaning relate to the term *empirical formula?* Discuss the differences and similarities between a molecular formula and empirical formula.

3)   Provide names for the following compounds and ions: H_2SO_4, SO_2, NO_2^-, Li_3PO_4, KBr, CO_2, and I^-.

4)   Provide formulae for the following compounds: hypochlorite, strontium ion, phosphorus trichloride, hydrobromic acid, and potassium hydrogen carbonate.

ANSWERS TO SAMPLE MULTIPLE CHOICE QUESTIONS

1. E

    The noble gas family is Family 8A (last column) on the periodic table. Other common groups to know: Family 1A alkali metals, Family 2A alkaline earth metals, Family 7A halogens.

2. A

    Bromine, Br_2, is a diatomic element. The other elements that exist as diatomic molecules at room conditions are: H_2, N_2, O_2, F_2, Cl_2, I_2.

3. B

    Na_2CO_3. This is the formula for sodium carbonate. It is important to know all the polyatomic ions listed in Table 4.

4. D

    Zinc forms 2+ ion. This is the only oxidation state for zinc.

5. C

U-235 and U-238 are isotopes. They have the same number of protons (92) but have a different number of neutrons (143 and 146, respectively).

6. C

The correct name for Fe_2OH_3 is iron(III) oxide. The iron has a +3 oxidation state in this compound. The Roman numeral is used to characterize the oxidation state of the metal ion when the metal has numerous oxidation states. Some common metal ions where using the Roman numeral system is applicable are : $Cu^{+, 2+}$; Hg_2^{2+} / Hg^{2+}; $Fe^{2+, 3+}$; $Sn^{2+, 4+}$.

7. A

The alkali metals are the most active metals. When forming compounds they always have an oxidation state of +1.

8. C

Rutherford's gold foil experiment led to the model of the nuclear atom. In this experiment alpha particles (helium nuclei) were beamed at gold foil. Most of the alpha particles passed through the foil undeflected and very few bounced back. This led to the conclusion that the atom was mostly empty space with a tiny, dense, positive center.

9. B

SO_4^{2-} is the sulfate polyatomic ion. Sulfite is SO_3^{2-}.

10. B

The empirical formula is the most reduced form of the molecular formula. $C_6H_{12}O_6$ is the molecular formula and the empirical formula is CH_2O.

Answers to Comprehension Questions

1)   An atom can be considered a singular particle of matter, and is the smallest amount of an element that retains the properties of that element. There are about 118 different kinds of atoms known to date, and Ne is an example of a substance that exists as isolated atoms. F_2 is a molecular substance, that is, the particles of this substance consist of two atoms bonded together as opposed to the isolated atoms found in the case of Ne.

Ions are charged particles. They can consist of a single atom or groups of differing or similar atoms. SO_4^{2-} is an example of a polyatomic ion as it is a charged particle consisting of several atoms of different elements bonded together. A compound is a pure substance composed of atoms of different elements. Compounds differ from mixtures in that the elements present in a compound always appear in the same ratios, whereas the elements contained in a mixture have no fixed ratios. PCl_5 and $KClO_3$ are both examples of compounds with fixed ratios of elements. PCl_5 is a molecular compound and $KClO_3$ is an ionic compound. The distinction between these two has to do with the types of bonds holding the individual atoms together to make up the compound.

2)   Empirical can be defined as *determined from actual observations or experimental results*. A process known as elemental analysis can be used to determine the mass ratio of elements in a compound. From these mass ratios and atomic masses of the elements a molar ratio of elements can be determined for a compound. This molar ratio is known as an empirical formula as it is derived from experimental results. This formula may or may not represent the molecular formula for a compound depending on how many atoms actually constitute a single molecule or formula unit of the compound. A separate determination of the molecular mass is needed to derive the molecular formula.

3) H_2SO_4, **sulfuric acid**; SO_2, **sulfur dioxide**; NO_2^-, **nitrite**; Li_3PO_4, **lithium phosphate**; KBr, **potassium bromide**; CO_2, **carbon dioxide**; and I^-, **iodide**.

4)   Hypochlorite, **ClO^-**; strontium ion, **Sr^{2+}**; phosphorus trichloride, **PCl_3**; hydrobromic acid, **HBr**; and potassium hydrogen carbonate, **$KHCO_3$**.

CHAPTER 3
MASS RELATIONSHIPS IN CHEMICAL REACTIONS

This chapter reviews the mole concept, balancing chemical equations, and stoichiometry. The topics covered in this chapter are:

- Atomic mass and average atomic mass
- Avogadro's number, mole, and molar mass
- Percent composition calculations
- Empirical and molecular formula determinations
- Chemical equations, amount of reactant and product calculations
- Limiting reagents and reaction yield calculations

> **Take Note:** *It is absolutely essential that you master the mole concept to do well on the quantitative aspects of AP Chemistry!!*
> *When solving quantitative problems on the Free Response section of the AP exam, supporting work must be shown to receive credit. Using dimensional analysis is a very powerful technique in solving problems.*
> *Be sure to report your answer to the correct number of significant figures (see Chapter 1 in this review book).*

Atomic mass and average atomic mass

Atomic mass is the *mass of an atom in atomic mass units (amu)*. **One amu** is defined as *1/12 of one C-12 atom*. The C-12 isotope has a mass of exactly 12.000 amu. The C-12 isotope provides the relative scale for the masses of the other elements.

Average atomic mass is the *value reported on the periodic table, which takes into account the various isotopes of an element and their respective frequencies.* To calculate the average atomic mass of an element, add up the different masses of the isotopes (using amu) multiplied by each isotope's abundance (percent occurrence in nature divided by 100).

Average atomic mass of element x = (isotope$_1$ in amu × abundance$_1$) + (isotope$_2$ × abundance$_2$) + (isotope$_n$ × abundance$_n$)

Example 1. Determining average atomic mass.

The natural abundance of the C-12 isotope in nature is 98.90% and the C-13 isotope is 1.100%. What is the average atomic mass reported on the periodic table for the element carbon? (The atomic masses of C-12 and C-13 are 12.00000 amu and 13.00335 amu, respectively.)

Solution to Example 1.

$$(12.00 \text{ amu} \times 0.9890) + (13.00335 \text{ amu} \times 0.0110) = 12.01 = \text{average atomic mass of carbon}$$

Molecular mass refers to the *mass of a molecule expressed in atomic mass units.* The molecular mass is the sum of the atomic masses of the atoms in the molecular formula.

Example 2. Determining molecular mass.

Determine the molecular mass of barium hydroxide, $Ba(OH)_2$.

Solution to Example 2.

1 Ba atom   + 2 O atoms   + 2 H atoms
137.33 amu + 2(16.00 amu) + 2(1.0079 amu) = *171.34 amu*

Avogadro's number, mole, and molar mass

Avogadro's number refers to *6.022 $\times 10^{23}$ particles.* The *quantity of an element that contains Avogadro's number of particles* is called a **mole**. The **molar mass** of an element or compound is the *mass of one mole of its atoms or molecules expressed in grams.*

Some examples of mole, atom, ion, molecule, and mass equivalencies are:

- 1 mole of C atoms = 6.022×10^{23} atoms of C = 12.01 g of C

- 1 mole of O_2 molecules = 6.022×10^{23} molecules of O_2 = 32.00 g O_2

- 1 mole of of O_2 *molecules* = 2 moles of oxygen *atom*s = 12.044×10^{23} atoms of O

- 1 mole O^{2-} ions = 6.022×10^{23} O^{2-} ions = 16.00 g O^{2-} ions

Percent composition calculations

The **percent composition** of an element in a compound =

$$\frac{n \times \textbf{molar mass of element}}{\textbf{molar mass of compound}} \times \textbf{100\%} \text{ where}$$

n is the number of moles of the element in the compound

Example 3. Determining the percent composition.

Determine the percent composition of chloride in barium chloride, $BaCl_2$.

$$\frac{\text{Mass of } 2\ Cl^-}{\text{Mass of } BaCl_2} \times 100 = \frac{2(35.45\ g\ Cl^-)}{2(35.45\ g\ Cl^-) + 137.33\ g\ Ba} \times 100\% = 34.0\ \%\ Cl^-$$

Empirical and molecular formula determinations

The **empirical formula** is the *most reduced form of the molecular formula*. For instance, sugar, which has the molecular formula of $C_6H_{12}O_6$, has an empirical formula of CH_2O. Empirical formulas can be determined from percent composition data. If the molar mass is also known, the molecular formula can be determined as well (see Example 4).

Example 4. Determining empirical and molecular formulas.

An unknown compound is known to contain 30.43% N and 60.56% O and has a molar mass of 92.00 g. What are the empirical formula and the molecular formula for the compound?

General Strategy	Solution to Example 4
Assume a 100 g sample.	30.43 g of N 60.56 g O
Convert gram quantities to moles by dividing by the molar mass of the atom.	$\dfrac{30.43\ g\ N}{14.01\ g\ N/mol\ N} = 2.17\ mol\ N$ $\dfrac{60.56\ g\ O}{16.00\ g\ O/\ mol\ O} = 4.35\ mol$
Divide by the smallest number of moles to obtain whole number ratios. The whole number ratios are the subscripts in the empirical formula.	$\dfrac{2.17\ mol\ N}{2.17\ mol\ N} = 1\ mol\ N$ $\dfrac{4.35\ mol\ O}{2.17\ mol\ N} = 2\ mol$ *Empirical formula* $= NO_2$
To determine the molecular formula divide the molar mass by the empirical mass. Multiply the subscripts of the empirical formula by this factor.	Empirical mass of NO_2 = 14.01 g N + 2(16.00) g O = 46.01 g NO_2 $\dfrac{\text{Molar mass}}{\text{Empirical mass}} = \dfrac{92.00\ g}{46.01\ g} = 2$ Molecular formula is twice the empirical formula: *Molecular formula* $= N_2O_4$

Chemical equations, amount of reactant and product calculations

Chemical equations are a shorthand method for *representing chemical reactions.* *Starting materials* are called **reactants** and are on the left side of the equation. The arrow indicates a reaction has taken place in the forward direction. A double-ended arrow `substances formed in a chemical reaction,* are on the right side of the equation. The numbers in front of the reactant and product molecules or atoms are called **coefficients**. They refer to the *molar relationships between substances* in a chemical reaction. The coefficients are used to balance the *atoms* in an equation. Atoms and mass are conserved in a reaction, molecules may not be. See Table 3.1.

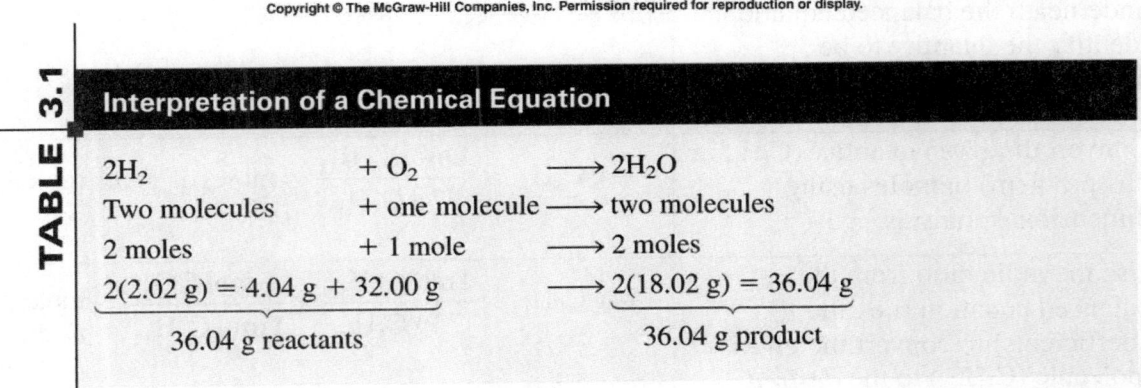

When balancing a chemical equation by inspection, it is usually best to balance the elements of hydrogen and oxygen last. Subscripts in molecules *cannot* be changed to balance a chemical equation. Coefficients are used to balance the equation.

Example 5. Balancing a chemical equation.

Consider the reaction: $C_5H_{12}(l) + 8O_2(g) \rightarrow 5CO_2(g) + 6H_2O(g)$
State what quantities are conserved and which are not conserved.

Solution to Example 5.
There are 5 atoms of C, 12 atoms of H, and 16 atoms of O on each side of the equation. The atoms are conserved as the reaction proceeds forward. The mass of reactants is equal to the mass of products. Mass is conserved. There are 9 reactant molecules and 11 product molecules. The number of molecules is not conserved.

The balanced chemical equation can be used to calculate the amount of reactants and products used or produced in a reaction. Examples 6, 7, and 8 represent some common types of stoichiometry problems. Example 6 describes a general strategy that can be applied when solving stoichiometry problems.

Example 6. Stoichiometry problem determining amount of product produced.

Pentane is burned in an excess of oxygen to produce carbon dioxide and water. If 144 g of pentane are burned, how many grams of carbon dioxide are produced?

General Strategy	Solution to Example 6
Write and balance the equation.	C_5H_{12} (l) + $8O_2$ (g) → $5CO_2$ (g) + $6H_2O$ (g)
Place the data from the problem underneath the balanced equation. Identify the quantity to be calculated (the *desired* molecule, CO_2 in this problem).	C_5H_{12} (l) + $8O_2$ (g) → $5CO_2$ (g) + $6H_2O$ (g) 144 g excess ?g
Convert the *given* quantity (C_5H_{12} in this problem) to moles using dimensional analysis.	$144\,g\,C_5H_{12} \times \dfrac{1\,mol\,C_5H_{12}}{72\,g\,C_5H_{12}} = $ moles of *given* molecule C_5H_{12}
Use the mole ratio from the balanced equation (i.e., the coefficients) to convert the *given* molecule (C_5H_{12}) to the *desired* molecule (CO_2).	$144\,g\,C_5H_{12} \times \dfrac{1\,mol\,C_5H_{12}}{72\,g\,C_5H_{12}} \times \dfrac{5\,mol\,CO_2}{1\,mol\,C_5H_{12}} = $ moles of *desired* molecule, CO_2
Convert the *desired* molecule to the appropriate units (grams of CO_2).	$144\,g\,C_5H_{12} \times \dfrac{1\,mol\,C_5H_{12}}{72\,g\,C_5H_{12}} \times \dfrac{5\,mol\,CO_2}{1\,mol\,C_5H_{12}} \times \dfrac{44\,g\,CO_2}{1\,mol\,CO_2}$ $= 110.\,g\,CO_2\,produced$

Example 7. Stoichiometry problem determining the amount of reactant consumed.

The Haber process is a reaction in which ammonia is produced by combining elemental hydrogen and nitrogen gas. How many grams of nitrogen are required to completely react with 6.0 grams of hydrogen?

General Strategy	Solution to Example 7
Write and balance the equation.	$3H_2\,(g) + N_2\,(g) \Leftrightarrow 2NH_3\,(g)$
Place the data from the problem underneath the balanced equation. Identify the quantity to be calculated (the *desired* molecule, N_2 in this problem).	$3H_2\,(g) + N_2\,(g) \Leftrightarrow 2NH_3\,(g)$ 6.0 g $\underline{?g}$
Convert the *given* quantity (H_2 in this problem) to moles using dimensional analysis.	$6.0\,g\,H_2 \times \dfrac{1\,mol\,H_2}{2.0\,g\,H_2} = \text{moles of } given \text{ molecule, } H_2$
Use the mole ratio from the balanced equation (i.e., the coefficients) to convert the *given* molecule (H_2) to the *desired* molecule (N_2).	$6.0\,g\,H_2 \times \dfrac{1\,mol\,H_2}{2.0\,g\,H_2} \times \dfrac{1\,mol\,N_2}{3\,mol\,H_2}$ moles of *desired* molecule, N_2
Convert *desired* molecule to the appropriate units (grams of N_2).	$6.0\,g\,H_2 \times \dfrac{1\,mol\,H_2}{2.0\,g\,H_2} \times \dfrac{1\,mol\,N_2}{3\,mol\,H_2} \times \dfrac{28.0\,g\,N_2}{1\,mol\,N_2}$ $= 28\,g\,N_2\ required$

> **Take Note:** *When doing stoichiometry problems on the Free Response section of the AP exam, you must show supporting work describing the solution process to receive credit for the problem.*
> *It is critical that equations be written and balanced correctly. It is essential that you be familiar with polyatomic ions, correct oxidation numbers for all species, common organic molecules, and the diatomic elements. You must also be proficient at balancing equations.*

Limiting reagents and reaction yield calculations

Reactants in a chemical reaction may not be present in stoichiometric amounts. One of the reactants, known as the **limiting reactant,** is *consumed completely*; the other reactant, known as the **excess reactant,** is only *partially consumed* in the reaction. The *limiting reactant determines how much product is formed.*

For example, if 1 mol of iodine is reacted with 0.50 mol of hydrogen according to the following reaction, only 1.0 mol of hydrogen iodide is produced.

$$I_2\,(g) \;+\; H_2\,(g) \Leftrightarrow 2HI\,(g)$$
     1.0 mol   0.5 mol
excess reactant   limiting reactant

H_2 is the limiting reactant in this reaction. When all the H_2 is consumed, 0.50 mol of I_2 are left over. H_2 determines the amount of HI produced.

$$0.50\,\text{mol}\,H_2 \times \frac{2\,\text{mol}\,HI}{1\,\text{mol}\,H_2} = 1.0\,\text{mol}\,HI\ \text{produced}$$

When more than one reactant quantity is given in a problem, it is likely that one of the reactants will be consumed completely (the limiting reactant) while the other reactant is not (the excess reactant). To determine which reactant is limiting, determine the amount of mole product produced by each reactant. The reactant that produces the least amount of product is the limiting reactant.

Example 8. Stoichiometry problem involving limiting reactants.

When 56 g of silicon are combined with 35 g of chlorine gas in a reaction vessel:
        a. How many moles of $SiCl_4$ are formed?
        b. What is the limiting reactant?
        c. How many moles of the excess reactant are left?

General Strategy	Solution to Example 8
Write and balance the equation.	$Si\,(s) + 2Cl_2\,(g) \rightarrow SiCl_4\,(l)$
Place the data from the problem underneath the balanced equation. Identify the quantity to be calculated (the *desired* molecule, $SiCl_4$ in this problem).	$Si\,(s) + 2\,Cl_2\,(g) \rightarrow SiCl_4\,(l)$ 56 g 35 g ?g
Convert the *given* quantities to moles using dimensional analysis.	$56\,g\,Si \times \dfrac{1\,\text{mol}\,Si}{28.1\,g\,Si} = 2.0\,\text{mol}\,Si$ $35\,g\,Cl_2 \times \dfrac{1\,\text{mol}\,Cl_2}{71.0\,g\,Cl_2} = 0.50\,\text{mol}\,Cl_2$
Use the mole ratio from the balanced equation (i.e., the coefficients) to convert the *given* molecules to the *desired* product molecule and compare. *The reactant molecule that produces the least product molecule is the limiting reactant.*	$2.0\,\text{mol}\,Si \times \dfrac{1\,\text{mol}\,SiCl_4}{1\,\text{mol}\,Si} = 2.0$ mol $SiCl_4$ are formed *if* all Si reacted. $0.50\,mol\,Cl_2 \times \dfrac{1\,\text{mol}\,SiCl_4}{2\,\text{mol}\,Cl_2} = 0.25$ mol $SiCl_4$ are formed *if* all of Cl_2 reacted. Cl_2 produces less product, so *Cl_2 is the limiting reactant.*

Determine how many moles of excess reactant are left.	$0.50 \, \text{mol Cl}_2 \times \dfrac{1 \, \text{mol Si}}{2 \, \text{mol Cl}_2} = 0.25$ mol of Si are consumed in the reaction. 2.0 mol Si available – 0.25 mol Si reacted = 1.75 mol of Si are left over = *1.8 mol of Si* (correct number of significant figures).
Answers to Example 8.	*a. 0.25 mol SiCl₄ are produced* *b. Cl₂ is the limiting reactant* *c. 1.8 mol Si remain unreacted*

Percent yield (sometimes called **reaction yield**) refers to the *actual yield of a reaction*. In Example 8 above the theoretical yield was calculated by *assuming* that all the reactant had been converted to product. In real life, all the reactant may not react and less than theoretical yield is produced. The formula to calculate percent yield is:

$$\% \, \text{yield} = \frac{\textbf{Actual yield}}{\textbf{Theortical yield}} \times \textbf{100}$$

In Example 8, the theoretical yield is 0.25 mol of SiCl₄. If in the lab only 0.20 mol of SiCl₄ were produced, then the percent yield would be:

$$\% \, yield = \frac{0.20}{0.25} \times 100\% = 80.\%$$

SAMPLE MULTIPLE CHOICE QUESTIONS

1. How many atoms of helium are there in a balloon that contains 40. 0 g of helium gas?

    A. 3.01×10^{23}
    B. 6.02×10^{23}
    C. 12.04×10^{23}
    D. 6.02×10^{24}
    E. 12.04×10^{24}

2. What is the mass percent of nitrogen in ammonium carbonate, $(NH_4)_2CO_3$?

    A. 29%
    B. 14.5%
    C. 42%
    D. 50%
    E. 58%

3. Octane fuel is burned in air. When the equation is balanced with the lowest whole-number coefficients, the coefficient for the water molecule is:

    A. 2
    B. 9
    C. 12
    D. 16
    E. 18

4. Iron rusts readily in air according to the reaction:

$$4Fe\ (s) + 3O_2\ (g) \rightarrow 2Fe_2O_3\ (s)$$

    When 112 g of iron rust, how much iron(III) oxide is formed?

    A. 160 g
    B. 320 g
    C. 676 g
    D. 722 g
    E. 1280 g

5. When ammonia gas reacts with hydrogen chloride gas, a white solid, ammonium chloride, forms.

$$NH_3\ (g)\quad +\quad HCl\ (g)\quad \rightarrow\quad NH_4Cl\ (s)$$

    If 6.02×10^{23} molecules of ammonia react with 12.04×10^{23} molecules of hydrogen chloride, how many molecules are in the reaction vessel when the reaction is complete?

    A. 1.00×10^{23}
    B. 3.01×10^{23}
    C. 6.02×10^{23}
    D. 12.04×10^{23}
    E. None of these

6. Which of the following nitrogen oxide compounds is the empirical formula for a compound that is analyzed as 47% nitrogen and 53% oxygen?

   A. NO
   B. N_2O_3
   C. N_2O_4
   D. NO_2
   E. N_2O_2

7. A 100.0 g sample of impure calcium carbonate was heated. It decomposed to form carbon dioxide gas and calcium oxide. After heating, the solid residue weighed 78 grams. What was the percent of calcium carbonate by mass in the original sample?

   A. 10.%
   B. 15%
   C. 25%
   D. 50.%
   E. 75%

8. ___KOH + ___H_3PO_4 → ___K_3PO_4 +___ H_2O

   When 1 mol of KOH neutralizes H_3PO_4 according to this equation, how many moles of water are formed?

   A. 1
   B. 2
   C. 3
   D. 4
   E. 5

9. A solution contains 0.10 mol of $Pb(NO_3)_2$ and 0.050 mol of BaI_2. How many moles of PbI_2 will precipitate?

   A. 0.050
   B. 0.10
   C. 0.15
   D. 0.20
   E. None of these

10. What mass of oxygen gas is produced when 0.10 mol of water is electrolyzed?

   A. 0.32 g
   B. 3.2 g
   C. 1.6 g
   D. 16 g
   E. 32 g

Comprehension Questions

1)   The main group element gallium is one of the very few metallic elements that can exist in the liquid state at room temperature, that is, providing that it's a warm summer day. Gallium's melting point is 29.8 °C or about 86 °F. Compounds of gallium have unique electrical properties and have therefore found use in products such as light emitting diodes (LEDs). This element has two stable isotopes, Ga = 69 (atomic mass = 68.926 amu) and Ga = 71 (atomic mass = 70.925), and an average atomic mass of 69.723 amu. Calculate the percent abundance of each isotope.

2)   Twelve-gauge copper wire, like the kind commonly used in residential electrical systems, is roughly cylindrical and has a diameter of approximately 0.1040 in. Copper's density is 8.92 g/cm^3 and copper atoms have an approximate atomic radius of 135 pm.

   a)   Calculate the number of atoms it would take to span the thickness, that is, diameter, of one of these wires. Express this value as a number of atoms and a number of moles of atoms.

   b)   Calculate the mass, in grams, of a 100-ft piece of copper wire.

   c)   How many moles of copper atoms would be found in a piece of this wire that is exactly 100 ft long? How many atoms?

3)   Oxidation of carbon-containing compounds can take place not only through reaction with molecular oxygen, as in common combustion, but also by reaction with a variety of other oxidizing agents. Sugar-containing candies, which we will represent with the formula $C_{12}H_{22}O_{11}$, react violently at elevated temperatures with the strong oxidant potassium chlorate, $KClO_3$, according to the following reaction, which closely resembles combustion:

$$C_{12}H_{22}O_{11} + KClO_3 \rightarrow CO_2 + H_2O + KCl$$

   a)   Provide coefficients to balance the equation.

   b)   What quantity of carbon dioxide could be produced from the reaction of 1.50 mol of sugar with an excess of potassium chlorate? Express your answer in grams and moles.

   c)   What minimum amount of $KClO_3$, in grams, would need to be reacted to produce 3.25 g of water?

d)   If a reaction is set up in which 16.1 g of $KClO_3$ is combined with 3.42 g of candy, what quantity of H_2O will be formed? Which reactant is in excess and by how much? Give both answers in grams.

4)   The arthritis drug Celebrex is a *selective* inhibitor of the enzyme that causes inflammation in humans and consequently has very few, if any, of the side effects associated with traditional nonsteroidal anti-inflammatory drugs, NSAIDs. (These compounds also inhibit enzymes responsible for noninflammatory processes.) It has therefore found widespread use in patients suffering from many inflammatory disorders. Celebrex's molecular formula is $C_{17}H_{14}N_3SO_2F_3$.

a)   Calculate the molecular mass and percent composition of Celebrex.

b)   This anti-inflammatory agent is synthesized from the condensation of 4-sulphonamidophenyl hydrazine, $C_6H_9N_3SO_2$, and the Claisen condensation product of 4-methyl acetophenone and ethyl trifluoroacetate, $C_{11}H_9O_2F_3$ according to the following reaction:

$$C_6H_9N_3SO_2 + C_{11}H_9O_2F_3 \rightarrow C_{17}H_{14}N_3SO_2F_3 + 2H_2O$$

Suppose that a chemist sets up a reaction to prepare Celebrex by combining 20.0 g of each of the above reactants. How much of the anti-inflammatory compound could be synthesized from this reaction? Express your answer in grams and moles.

c)   Which reactant is in excess? Which is limiting?

d)   By what amount, expressed in grams and moles, is the excess reactant in excess?

e)   Suppose the chemist isolates 20.8 g of the purified drug from this reaction. What is the percent yield for this process?

ANSWERS TO SAMPLE MULTIPLE CHOICE QUESTIONS

1. D

Dimensional analysis yields the following solution:

$$40.0\,g\,He \times \frac{1\,mol\,He}{4.00\,g\,He} \times \frac{6.022 \times 10^{23}}{1\,mol\,He} = 6.02 \times 10^{24}\ atoms\ He$$

2. A

Percent nitrogen is the mass of nitrogen divided by molar mass times 100:

$$\frac{2N}{(NH_4)_2CO_3} \times 100\% = \frac{2(14)}{2(14)+8(1)+12+3(16)} \times 100\% = 29\% \, N$$

3. E

The first step is to write the equation correctly. This is a combustion reaction involving the burning of a hydrocarbon in which CO_2 and H_2O are produced:
$$C_8H_{18} + O_2 \rightarrow CO_2 + H_2O$$
Using coefficients, balance the equation by inspection. Begin by balancing the carbons (step 1), then hydrogen (step 2), and then oxygen (step 3), and then multiply by 2 to obtain whole-number coefficients (step 4).

(step 1)	$C_8H_{18} + O_2 \rightarrow 8CO_2 + H_2O$
(step 2)	$C_8H_{18} + O_2 \rightarrow 8CO_2 + 9H_2O$
(step 3)	$C_8H_{18} + 25/2\,O_2 \rightarrow 8CO_2 + 9H_2O$
(step 4)	$2C_8H_{18} + 25\,O_2 \rightarrow 16CO_2 + 18H_2O$

4. A

Dimensional analysis yields the following solution:

$$112\,g\,Fe \times \frac{1\,mol\,Fe}{56\,g\,Fe} \times \frac{2\,mol\,Fe_2O_3}{4\,mol\,Fe} \times \frac{160\,g\,Fe_2O_3}{1\,mol\,Fe_2O_3} = 160.\,g\,Fe_2O_3$$

5. D

This is a limiting reactant type of problem. The general strategy is to convert all quantities to moles and then determine which reactant is totally consumed. The limiting reagent (also called limiting reactant) determines the amount of product produced.

NH_3	+	HCl	$\rightarrow$	NH_4Cl
6.02×10^{23}		12.04×10^{23}		
1 mol		2 mol		
limiting reactant		excess reactant		

Since there is a 1 to 1 ratio between NH_3 and HCl, all of the NH_3 is consumed and 1 mol of HCl is left unreacted. One mole of NH_4Cl is formed.

Net result:
$$1\,mol\,HCl\,\text{left unreacted} + 1\,mol\,NH_4Cl\,\text{produced} = 2\,mol\,\text{of molecules}$$
$$= 12 \times 10^{23}\,molecules$$

6. A

This is a traditional empirical formula problem. Begin by assuming a 100-g
sample, convert all elements to moles by dividing by atomic weight, divide by the
smallest number of moles to get whole-number ratios that serve as the subscripts
in the empirical formula.

General Strategy	*Example 6* worked out
Assume a 100-g sample and convert the percentages of elements to grams.	46.7 g N 53.3 g O
Convert gram quantities to moles by dividing by the molar mass of the atom.	$46.7\,\text{g N} \times \dfrac{1\,\text{mol N}}{14.0\,\text{g}} = 3.33\,\text{mol N}$ $53.3\,\text{g O} \times \dfrac{1\,\text{mol O}}{16.0\,\text{g O}} = 3.33\,\text{mol O}$
Divide by the smallest number of moles to get whole-number ratios. The whole number ratios are the subscripts in the empirical formula.	$N_{\frac{3.33}{3.33}}\;O_{\frac{3.33}{3.33}} = NO$ *Empirical formula = NO*

7. D

The 100.0 g is made of $CaCO_3(s)$ and some other material. The weight 'loss' as
the reaction proceeds is due to carbon dioxide gas being released.

$$CaCO_3\,(s) \rightarrow CaO\,(s)\ + CO_2\,(g)$$

By converting the mass of CO_2 to moles of CO_2 it is possible to calculate the
moles of $CaCO_3(s)$ in the original sample, since every mole of CO_2 came from a
mole of $CaCO_3$.

100.0 g solid reactant − 78 g solid product residue = 22 g CO_2 product gas

$$22\,\text{g}\,CO_2 \times \frac{1\,\text{mol}\,CO_2}{44\,\text{g}\,CO_2} \times \frac{1\ \text{mol}\,CaCO_3}{1\,\text{mol}\,CO_2} \times \frac{100.0\,\text{g}\,CaCO_3}{1\,\text{mol}\,CaCO_3} = 50.\,\text{g}\,CaCO_3$$

$$\%\,CaCO_3\ \text{in sample} = \frac{\text{part}}{\text{whole}} \times 100\% = \frac{50.\,\text{g}\,CaCO_3}{100\,\text{g sample}} \times 100 = 50.\%\,CaCO_3$$

8. A

When dealing with a stoichiometry problem, the first step is always to balance the equation and then use the coefficients of the balanced equation to relate one molecule to another using dimensional analysis.

$$3KOH + H_3PO_4 \rightarrow K_3PO_4 + 3H_2O$$
1 mol   ? mol

$$1\,mol\,KOH \times \frac{3\,mol\,H_2O}{3\,mol\,KOH} = 1\,mol\,H_2O$$

9. A

This problem is based on the mole relationships between molecules and ions.

0.10 mol of $Pb(NO_3)_2$ contains 0.10 mol Pb^{2+} ion since there is a 1 to 1 ratio between $Pb(NO_3)_2$ and Pb^{2+}.

0. 050 mol BaI_2 contains 0.10 mol I^- ion since there is a 1 to 2 ratio between BaI_2 and I^- as illustrated in the following step:

$$0.050\,mol\,BaI_2 \times \frac{2\,mol\,I^-}{1\,mol\,BaI_2} = 0.10\,mol\,I^-$$

This is a limiting reactant type problem (refer to Example 8). You must determine which reactant ion limits the amount of precipitate, PbI_2, produced. The ion that forms the lesser amount limits the reaction to that amount.

$$0.10\,mol\,of\,Pb^{2+} \times \frac{1\,mol\,PbI_2}{1\,mol\,Pb^{2+}} = 0.10\,mol\,PbI_2$$

$$0.10\,mol\,I^- \times \frac{1\,mol\,PbI_2}{1\,mol\,I^-} = 0.050\,mol\,PbI_2 = amount\,of\,precipitate\,formed$$

10. C

Write a balanced equation for the reaction and then use dimensional analysis to obtain the answer.

$$2\,H_2O \rightarrow 2H_2 + O_2$$
0.10 mol   ? g

$$0.10\,mol\,H_2O \times \frac{1\,mol\,O_2}{2\,mol\,H_2O} \times \frac{32.0\,g\,O_2}{1\,mol\,O_2} = 1.6\,g\,O_2$$

Answers to Comprehension Questions

1)   The average atomic mass of an element is the weighted average, that is, the sum of the masses of the various isotopes multiplied by their percent abundance. In this case, you are not given the percent abundance of either isotope, that is, two variables and only one equation. Solution of the problem only becomes possible with the recognition that the sum of the percents abundance is 100 and the relative abundance of each isotope can then be represented as X and $100 - X$ (or $1 - X$ if the percents are represented as decimals). Therefore:

$69.723 = 68.926X + 70.925(1 - X)$

$69.723 = 68.926X + 70.925 - 70.925\,X$

$-10.202 = -10.999X$

$X = 0.92754$ or **92.754%**; in this case X represents the percent abundance of the isotope with a mass of 68.926 amu. The percent abundance of the isotope weighing 70.925 amu is $100\% - 92.754\%$ or **7.246%**.

2)   a)   Solution to this part of the problem simply involves conversion to a common unit for length and then division of the wire's diameter by the diameter of a single atom. Therefore:

diameter of wire = 0.1040 in. × (2.54 cm / 1 in.) = 0.2642 cm

0.2642 cm × (1 m / 100 cm) × (10^{12} pm / 1 m) = 2.642×10^{9} pm / 135 pm/atom = **1.96×10^{7} atoms per diameter of wire**

        b)   The mass of 100 ft of this wire will be determined by multiplying the volume of that quantity of wire by the density of copper. Since the wire is cylindrical, its volume can be calculated using the formula $\pi r^{2} l$ (r = radius, l = length of cylinder)

Substituting:

the radius of the wire equals ½ the diameter, so 0.2642 cm × 0.5 = 0.1321 cm

the volume of the wire = π × (0.1321 cm)^{2} × (100 ft × (12 in. / ft) × (2.54 cm / in)) = 167.1 cm^{3}

the mass of the wire = volume × density = 167.1 cm^{3} × 8.92 g/cm^{3} = **1491 g**

c)   The number of moles of copper found in 100 ft or 1491 g of this wire can be determined by dividing the mass of the wire by the molar mass of copper, 63.546 g / mol. The number of atoms can be determined by multiplying the moles of copper by Avogadro's number, 6.022×10^{23} particles per mole. Therefore:

1491 g Cu / 63.546 g/mol = **23.46 mol of Cu**

23.46 mol of Cu $\times$ 6.022×10^{23} atoms/mol = $\mathbf{1.412 \times 10^{25}}$ **atoms of Cu**

3)   a) $C_{12}H_{22}O_{11} + 8KClO_3 \rightarrow 12CO_2 + 11H_2O + 8KCl$

b)   The coefficients of a balanced equation represent mole ratios in which the reactants combine and the products are produced. The mass of a product or reactant may be determined by multiplying the moles of that substance by its molar mass. Therefore:

$1.50C_{12}H_{22}O_{11} \times$ (12 mol CO_2/1 mol $C_{12}H_{22}O_{11}$) = **18.0 mol** CO_2

18.0 mol $CO_2 \times$ 44.01 g CO_2/mol = **792 g of** CO_2

c)   Beginning with the quantity of a reagent in grams, you must first convert to moles to use the mole ratios found in the balanced equation. The moles of $KClO_3$ necessary will then have to be re-converted to grams to provide an answer:

3.25 g $H_2O \times$ (1 mol H_2O/18.02 g) $\times$ (8 mol $KClO_3$/11 mol H_2O) $\times$ (122.55 g $KClO_3$/ 1 mol $KClO_3$) = **16.1 g of** $KClO_3$

d)   A convenient way to determine the limiting reactant in this situation is to calculate the maximum amount of product, H_2O in this case, that can be formed from each reactant assuming the other to be in excess. The answer with the smaller amount of product will have started with the limiting reactant. According to part c above, 16.1 g of $KClO_3$ will form **3.25 g of** H_2O.

3.42 g $C_{12}H_{22}O_{11} \times$ (1 mol $C_{12}H_{22}O_{11}$/342.34 g $C_{12}H_{22}O_{11}$) $\times$ (11 mol H_2O/1 mol $C_{12}H_{22}O_{11}$) $\times$ (18.02 g H_2O/1 mol H_2O) = **1.98 g** H_2O

Since there's enough $KClO_3$ to make 3.25 g of H_2O and only enough candy to make 1.98 g H_2O, the candy is the limiting reactant. To determine the amount of $KClO_3$ in excess, the amount necessary to react with 3.42 g of candy is calculated and subtracted from the initial amount, 16.1 g.

3.42 g $C_{12}H_{22}O_{11}$ × (1 mol $C_{12}H_{22}O_{11}$/342.34 g $C_{12}H_{22}O_{11}$) × (8 mol $KClO_3$/1 mol $C_{12}H_{22}O_{11}$) × (122.55 g $KClO_3$/1 mol $KClO_3$) = **9.79 g $KClO_3$** needed to react with 3.42 g of candy

16.1 g $KClO_3$ − 9.79 g $KClO_3$ = **6.3 g $KClO_3$ in excess**

4)   a)   The molar mass of Celebrex, $C_{17}H_{14}N_3SO_2F_3$, is (17 × 12.01) + (14 × 1.008) + (3 × 14.01) + (1 × 32.07) + (2 × 16.00) + (3 × 19.00) = **381.4 g/mol**

Percent composition is determined by dividing the mass of each element in the compound by the total mass of the compound and then multiplying by 100%.

%C = [((17 × 12.01) g/mol) / (381.4 g/mol)] × 100% = **53.53% C**

%H = [((14 × 1.008) g/mol) / (381.4 g/mol)] × 100% = **3.700% H**

%N = [((3 × 14.01) g/mol) / (381.4 g/mol)] × 100% = **11.02% N**

%S = [((1 × 32.07) g/mol) / (381.4 g/mol)] × 100% = **8.41% S**

%O = [((2 × 16.00) g/mol) / (381.4 g/mol)] × 100% = **8.39% O**

%F = [((3 × 19.00) g/mol) / (381.4 g/mol)] × 100% = **14.94% F**

b)   As in question 3d, the maximum amount of product can be calculated by performing two separate calculations each starting with a different reactant.

20.0 g $C_6H_9N_3SO_2$ × (1 mol $C_6H_9N_3SO_2$/187.18 g $C_6H_9N_3SO_2$) × (1 mol $C_{17}H_{14}N_3SO_2F_3$/ 1 mol $C_6H_9N_3SO_2$) = **0.107 mol of Celebrex** × (381.4 g $C_{17}H_{14}N_3SO_2F_3$/1 mol $C_{17}H_{14}N_3SO_2F_3$) = **40.8 g Celebrex**

20.0 g $C_{11}H_9O_2F_3$ × (1 mol $C_{11}H_9O_2F_3$/230.18 g $C_{11}H_9O_2F_3$) × (1 mol $C_{17}H_{14}N_3SO_2F_3$/ 1 mol $C_{11}H_9O_2F_3$) = **0.0869 mol of Celebrex** × (381.4 g $C_{17}H_{14}N_3SO_2F_3$/1 mol $C_{17}H_{14}N_3SO_2F_3$) = **33.1 g Celebrex**

Therefore the maximum amount of Celebrex that can be synthesized is **33.1 g or 0.0869 mol.**

c)   The limiting reactant is the Claisen condensation product, $C_{11}H_9O_2F_3$, and the phenylhydrazine compound, $C_6H_9N_3SO_2$, is in excess.

d)   As in 3d, the amount of excess reactant can be calculated by starting with the limiting reactant and determining the amount of the other reactant needed to completely react with it.

20.0 g $C_{11}H_9O_2F_3$ × (1 mol $C_{11}H_9O_2F_3$/230.18 g $C_{11}H_9O_2F_3$) × (1 mol $C_6H_9N_3SO_2$/ 1 mol $C_{11}H_9O_2F_3$) × (187.18 g $C_6H_9N_3SO_2$/1 mol $C_6H_9N_3SO_2$) = 16.3 g $C_6H_9N_3SO_2$ needed to react with 20.0 g $C_{11}H_9O_2F_3$

20.0 g $C_6H_9N_3SO_2$ – 16.3 g $C_6H_9N_3SO_2$ = **3.7 g $C_6H_9N_3SO_2$ in excess**

Multiplying by the reciprocal of the molar mass of $C_6H_9N_3SO_2$,

3.7 g $C_6H_9N_3SO_2$ × (1 mol $C_6H_9N_3SO_2$/187.18 g $C_6H_9N_3SO_2$) = **0.0198 mol $C_6H_9N_3SO_2$ in excess**

e)   Percent yield is equal to 100% times the ratio of actual yield to theoretical yield (from part b), therefore:

% yield = (20.8 g / 33.1 g) × 100% = **62.8%**

CHAPTER 4
REACTIONS IN AQUEOUS SOLUTIONS

This chapter reviews the vast array of chemical changes that occur in water solution, both qualitatively and quantitatively. The major topics in Chapter 4 are:

- An introduction to aqueous reactions, including electrolytes and solubility rules
- A review of the six basic reaction types
- The quantitative aspects of reactions in solution

An introduction to aqueous reactions

While it may appear that there are many different types of reactions, there are actually only six basic categories of reactions in inorganic chemistry:

- Synthesis or combination
- Decomposition
- Single displacement
- Double displacement
- Oxidation-reduction
- Complex ion formation

For you to be able to identify the reaction type and then to predict the products of the reaction, it is important:

- to understand the basic terminology associated with reactions in solution; and
- to understand the ionic and molecular species that reagents will form in solution

Only after the correct reactants and products are obtained and the equation has been balanced can you deal quantitatively with these reactions. *A firm knowledge of solubility rules, strong electrolytes, weak electrolytes, oxidation numbers, and trends in the periodic table is essential* and will aid immensely in mastering this material. Reading the strong electrolyte and solubility rules over periodically will help you master them (see Tables 4.1 and 4.2).

Take Note: *The ability to correlate oxidation numbers, activities, electrolyte rules, and solubility rules to the periodic table is essential for successful completion of the equation questions on the Multiple Choice section of the AP exam. The periodic table is the only reference table provided on this part of the exam.*

The Free Response section of the AP exam contains a periodic table as a reference, the Standard Reduction Potentials Table, from which the charges of many ions can be determined, and a list of formulas and definitions. The Standard Reduction Potentials Table can also be employed to determine the activity of the elements.

TABLE 4.1

Classification of Solutes in Aqueous Solution

Strong Electrolyte	Weak Electrolyte	Nonelectrolyte
HCl	CH_3COOH	$(NH_2)_2CO$ (urea)
HNO_3	HF	CH_3OH (methanol)
$HClO_4$	HNO_2	C_2H_5OH (ethanol)
H_2SO_4*	NH_3	$C_6H_{12}O_6$ (glucose)
$NaOH$	$H_2O^\dagger$	$C_{12}H_{22}O_{11}$ (sucrose)
$Ba(OH)_2$		
Ionic compounds		

* H_2SO_4 has two ionizable H^+ ions.

† Pure water is an extremely weak electrolyte.

TABLE 4.2

Solubility Rules for Common Ionic Compounds in Water at 25°C	
Soluble Compounds	**Exceptions**
Compounds containing alkali metal ions (Li^+, Na^+, K^+, Rb^+, Cs^+) and the ammonium ion (NH_4^+)	
Nitrates (NO_3^-), bicarbonates (HCO_3^-), and chlorates (ClO_3^-)	
Halides (Cl^-, Br^-, I^-)	Halides of Ag^+, Hg_2^{2+}, and Pb^{2+}
Sulfates (SO_4^{2-})	Sulfates of Ag^+, Ca^{2+}, Sr^{2+}, Ba^{2+}, Hg^{2+}, and Pb^{2+}
Insoluble Compounds	**Exceptions**
Carbonates (CO_3^{2-}), phosphates (PO_4^{3-}), chromates (CrO_4^{2-}), sulfides (S^{2-})	Compounds containing alkali metal ions and the ammonium ion
Hydroxides (OH^-)	Compounds containing alkali metal ions and the Ba^{2+} ion

Notice the patterns for electrolytes that are salts, bases, and acids:

1. All **salts** (*metal/nonmetal combinations*) that are soluble are **strong electrolytes** and therefore *dissociate completely in solution* and exist as ions.

2. All alkali (Group 1) metals form **strong bases**. These bases *dissociate completely* in water to form a *metal ion and OH⁻*. Most other metallic bases are insoluble, with the notable exception of the Group 2 metal hydroxide barium hydroxide, $Ba(OH)_2$, which dissociates completely.

   Ammonia, the most common **weak base**, *ionizes to a limited extent* in aqueous solution, forming only a few NH_4^+ and OH^- ions

   $(NH_3 + H_2O \Leftrightarrow NH_4^- + OH^-)$. Most of the ammonia remains as molecular ammonia, NH_3.

3. There are seven commonly recognized **strong acids** (H_2SO_4, HNO_3, $HClO_4$, $HClO_3$, HCl, HBr, and HI), which *ionize completely and form ions in solution*.

   Essentially, all other acids are **weak acids,** which *ionize to a limited extent*, forming only a few ions in solution. The weak acids are written as molecular substances. HF and CH_3COOH are two common *weak acids whose formulas will be left as molecular species*.

Finally, there are three forms in which a reaction may be written: a **molecular equation,** *which simply shows the reactants and products as molecules*, an **ionic equation**, *which shows how the molecules exist in water solution*, or the **net ionic equation**, *which shows* *only the substances that actually undergo a chemical change.* For example, if solutions of sodium hydroxide and hydrochloric acid are mixed, the reaction can be represented with any of the three equation types:

NaOH (aq) + HCl (aq) → NaCl (aq) + H_2O (l) **molecular equation**

Na^+ (aq) + OH^- (aq) + H^+ (aq) + Cl^- (aq) → Na^+ (aq) + Cl^- (aq) + H_2O (l) **ionic equation**

OH^- (aq) + H+ (aq) → H_2O (l) **net ionic equation**

Spectator ions, *those substances that remain unchanged from reactant to product,* are not included in the net ionic equation.

Take Note: *The AP exam contains questions on equations in both the Multiple Choice section and the Free Response section.*

*The **Multiple Choice** section questions involve many aspects of equations, including balancing, predicting products, picking appropriate reagents, identifying reaction type, calculating oxidation numbers, and stoichiometry.*

***Beginning with the 2007 exam, the Free Response section will contain a required question containing three equations.** *All three equations must be done. The question will consist of three parts for each equation:*

1. *You must write the reactants and predict the products of the reaction.*
2. *You must balance the equation. **The final answer must be the balanced net ionic equation.***
3. *You must answer a question about that particular equation.*

The best strategy for predicting the correct products is to translate the names of the reactants into formulas and then categorize the reaction type before beginning to predict the products. Remember, there are only a limited number of reaction types.

* *The Free Response section of old AP exams contained a required question devoted to the correct prediction of products for a chemical reaction. In the past, you were given eight equations from which you chose five to do. Keep this in mind when you use old AP exams to study.*

A review of the six basic reaction types

Take Note: *Chapter 4 of the Chang textbook covers reactions that occur in aqueous solution. Most reactions occur in aqueous solution, but a few do not. The AP exam states that you should assume that all reactions occur in aqueous solution unless otherwise stated. A few reactions that fall within the six reaction types but do not occur in aqueous solution are included in the following discussion.*

I. **Combination or synthesis reactions** lead to the formation of a new compound by the combining of:

> (1) two elements; or
> (2) two simple molecules.

Examples include:

1. Two elements:

$$3Mg\ (s)\ +\ N_2\ (g)\ \rightarrow\ Mg_3N_2\ (s)$$

2. Two simple molecules:

$$SO_2\ (g)\ +\ CaO\ (s)\ \rightarrow CaSO_3\ (s)$$

$$N_2O_5\ (g) + H_2O\ (l) \rightarrow 2H^+\ (aq) + 2NO_3^-\ (aq)\quad \text{nonmetal oxide = acid anhydride}$$

$$Na_2O\ (s)\ +\ H_2O\ (l)\ \rightarrow\ 2Na^+(aq) + 2OH^-(aq)\ \text{metal oxide = basic anhydride}$$

II. **Decomposition reactions** are easily recognized by the fact that there is usually only one reactant and the decomposition usually requires the application of some form of energy. Decomposition reactions in which compounds come apart to form new materials can result from:

> (1) the simple decomposition of a binary compound into its elements;
> (2) the decomposition of an acid or base into a nonmetal oxide or metal oxide and water; or
> (3) the decomposition of certain polyatomic ions.

Some examples of decomposition reactions are:

1. Simple decomposition:

$$2H_2O\ (l) \rightarrow\ 2H_2\ (g)\ +\ O_2\ (g)$$

2. Decomposition of an acid or base to a nonmetal oxide or metal oxide and water:

H_2SO_3 (aq) → H_2O (l)   + SO_2 (g) (acid decomposition to water and nonmetal oxide)

2M*OH (aq) → H_2O (l) + M_2O* (s)   (base decomposition to water and metal oxide)

*M = metal, +1 oxidation number

Notice that the decompositions of the acids and bases are the reverse of the combination reactions in which nonmetal oxides in water form acids and metal oxides in water form bases.

3. Decomposition of polyatomic ions:

$M*CO_3$ (s) → $MO*$ (s) + CO_2 (g)   *M = metal with +2 oxidation state

2M*ClO_3 (s) → 2MCl* (s) + $3O_2$ (g)   *M = metal with +1 oxidation state

---

Take Note: *Although there are many polyatomic ions, you will find that carbonate and chlorate are the two most often used on the AP exam to illustrate the decomposition of polyatomic ions. The carbonate decomposition produces a metal oxide and carbon dioxide, but the chlorate releases all of its oxygen as O_2 gas, leaving a solid metal chloride.*

III. **Single displacement reactions** consist of:

(1) simple single displacement; and
(2) complex ion reactions involving displacement.

The classic single displacement occurs when a more active single element displaces a less active element in the compound. There are also single displacement reactions in which a more powerful ligand displaces the ligand in a complex ion. Examples of single displacement reactions follow:

Take Note: *You will have the Standard Reduction Potentials Table available to you on the Free Response section of the AP exam. An element's position on this table is indicative of the element's activity.*

---

1. Simple single displacement:

$$Cu\ (s)\ +\ 2AgNO_3\ (aq)\ \rightarrow\ Cu(NO_3)_2\ (aq)\ +\ 2Ag\ (s)\ \text{molecular equation}$$

or

$$\textbf{Cu}\ +\ \textbf{2Ag}^+\ \rightarrow\ \textbf{Cu}^{2+}\ +\ \textbf{2Ag}\qquad \textbf{net ionic equation}$$

2. Complex ion displacement:

$$AgCl\ (s)\ +\ 2NH_3\ (aq)\ \rightarrow\ Ag(NH_3)_2^+\ +\ Cl^-\quad \textbf{net ionic equation}$$

IV. **Double displacement reactions** are reactions in which two compounds exchange their ions to form two or more new compounds. Double displacement reactions occur if a precipitate forms, a gas forms, or a molecular substance such as water forms. The two most common forms of double displacement reactions are:

> (1) precipitation reactions; and
> (2) acid-base neutralization reactions.

Four examples of double displacement reactions are:

1) Precipitation reaction:

$$3Ba(OH)_2\ (aq)\ +\ Fe_2(SO_4)_3\ (aq) \rightarrow 3BaSO_4\ (s)\ +\ 2Fe(OH)_3\ (s)\ \text{molecular}$$
$$\text{equation}$$

or

$$\textbf{Ba}^{2+} + \textbf{3OH}^- + \textbf{Fe}^{3+} + \textbf{SO}_4{}^{2-} \rightarrow \textbf{BaSO}_4\ (s)\ +\ \textbf{Fe(OH)}_3\ (s)\ \ \textbf{net ionic equation}$$

In this double displacement reaction *two precipitates* form. It is not uncommon for a precipitation reaction to have two precipitates as products.

2) Neutralization reaction:

$$HCl\ (aq)\ +\ NaOH\ (aq)\ \rightarrow\ NaCl\ (aq)\ +\ H_2O\ (l)\quad \text{molecular equation}$$

or

$$\textbf{H}^+\ +\ \textbf{OH}^-\ \rightarrow \textbf{H}_2\textbf{O}\quad \textbf{net ionic equation}$$

The formation of the *molecular substance* water in this neutralization accompanies the salt formation associated with the neutralization process.

3) Neutralization reaction:

NH_4Cl (*aq*)   +   NaOH (*aq*) → NaCl (*aq*) + NH_3 (*aq*) + H_2O (*l*)   molecular equation

or

$$NH_4^+ + OH^- \rightarrow NH_3 \, (aq) + H_2O \text{ net ionic equation}$$

The product of this double displacement is NH_3, a weak base. When the weak acids H_2SO_3 or HNO_2 form as the result of a double displacement reaction, they usually undergo a further decomposition reaction to their nonmetal oxides and water (see the discussion of decomposition reactions).

4) There is a special case of neutralization in which no aqueous environment is present:

CaO (*s*)   + SO_2 (*g*)   → $CaSO_3$ (*s*)
basic anhydride   **acidic anhydride**
metal oxide   *nonmetal oxide*   salt

Notice that this is also a combination (synthesis) reaction.

V. **Oxidation-reduction reactions** are electron transfer reactions that result in new products. In oxidation-reduction (redox) reactions there are two separate steps. One involves *the loss of electrons by an element*, resulting in an increase in the element's oxidation number, or its **oxidation**. The other step involves the *gain of electrons by an element*, resulting in the reduction of the element's oxidation number, or its **reduction**. The *element that is oxidized* is called the **reducing agent**. The *element that is reduced* is referred to as the **oxidizing agent**. A single displacement reaction involving a metal in an elemental state is an oxidation-reduction reaction. If you examine the oxidation number changes in the following simple single displacement reaction, you will see that it is also technically an oxidation-reduction reaction:

```
   0          +1 +5–2        +2 +5–2              0     (oxidation numbers)
  Cu (s)   +   2AgNO₃ (aq)  →  Cu(NO₃)₂ (aq)   +   2Ag (s) molecular equation
```

or, separating the half-reactions:

$$Cu\,(s) \;\longrightarrow\; Cu^{2+}\,(aq) \;+\; 2e^- \qquad \text{(oxidation half-reaction)}$$
$$\underline{2(Ag^+\,(aq) \;+\; 1e^- \;\longrightarrow\; Ag\,(s))} \qquad \text{(reduction half-reaction)}$$
$$2Ag^+\,(aq) \;+\; Cu\,(s) \longrightarrow\; 2Ag\,(s) \;+\; Cu^{2+}\,(aq) \quad \text{net ionic equation}$$

The presence of a substance in its elemental state in the reactants or products indicates that a redox reaction is occurring. Recognition of some of the most powerful oxidizing and reducing agents will also lead to correct identification of an oxidation-reduction reaction. (The colors of some polyatomic ions that are common to the AP exam are provided.)

Table 1. Powerful oxidizing and reducing agents

Powerful Oxidizing Agents in Acid	Powerful Reducing Agents in Acid
$MnO_4^- \;\rightarrow\; Mn^{2+}$ purple colorless $Cr_2O_7^{2-} \;\rightarrow\; Cr^{3+}$ orange	$HSO_3^{2-} \rightarrow SO_4^{2-}$ colorless colorless

Powerful Oxidizing Agents in Base	Powerful Reducing Agents in Base
$MnO_4^- \rightarrow MnO_2$ (neutral to slightly basic) purple black precipitate $MnO_4^- \rightarrow MnO_4^{2-}$ (strongly basic) purple green $CrO_4^{2-} \rightarrow Cr^{3+}$ yellow	$SO_3^{2-} \rightarrow SO_4^{2-}$ colorless colorless

Consider the reaction of hydrochloric acid and potassium permanganate. Recognizing that permanganate, MnO_4^-, is a strong oxidizing agent will lead you to identify this reaction as an oxidation-reduction reaction. Further, MnO_4^- in acid solution results in the formation of Mn^{2+}. Since the Mn in MnO_4^- is reduced, something else must be oxidized. Since H^+ and K^+ are already in their highest oxidation states, Cl^- is the only chemical species capable of losing electrons. Consequently, the following skeleton reaction for the oxidation-reduction reaction begins to emerge (see Chapter 19 of this review book for balancing redox equations):

$$\overset{+1}{H^+} \;+\; \overset{-1}{Cl^-} \;+\; \overset{+1}{K^+} \;+\; \overset{+7\,-2}{MnO_4^-} \;\rightarrow\; \overset{0}{Cl_2} \;+\; \overset{+2}{Mn^{2+}}$$

or

$$5Cl^- \;+\; 8H^+ \;+\; MnO_4^- \;\rightarrow\; Mn^{2+} \;+\; 4H_2O \qquad \textbf{net ionic equation}$$

One final, special case of an oxidation-reduction reaction is a **disproportionation reaction,** *in which one reactant reacts to form two products with different oxidation states.* The elements most frequently involved in disproportionation reactions on the AP exam are chlorine and iodine. Notice the oxidation number changes for chlorine in the following reaction:

$$\overset{0}{Cl_2}\ (g) \ + \ \overset{-2\ +1}{2OH^-}\ (aq) \ \rightarrow \ \overset{+1\ -2}{ClO^-}\ (aq) \ + \ \overset{-1}{Cl^-} \ + \ \overset{+1\ -2}{H_2O}\ (l)$$

or

$$4OH^-\ (aq) \ + \ 2Cl_2\ (g) \ \rightarrow \ 2ClO^-\ (aq) \ + \ 2Cl^-\ (aq) \ + \ 2H_2O\ (l) \ \textbf{net ionic equation}$$

VI. The **formation of complex ions** is the final category of reactions. Complex ions form from the reaction of some transition metal ions and ligands (see Chang, Chapter 22 and Chapter 22 of this review book). Some examples of complex ions are:

Complex Ion	Name of Ion	Color of Ion
$Fe(CN)_6^{3-}$	hexacyanoferrate(III)	red
$Ag(NH_3)_2^+$	diamminesilver(I)	colorless
$CO(NH_3)_6^{3+}$	hexamminecobalt(III)	yellow
MnF_6^{4-}	hexafluoromanganate(II)	pink
$Cu(NH_3)_4^{2+}$	tetrammine copper(II)	blue
$Zn(OH)_4^{2-}$	tetrahydroxozincate(II)	colorless
$Al(H_2O)_6^{3+}$	hexaaquaaluminum(III)	colorless

> **Take Note:** *On the AP exam, if a transition metal ion is in solution and a concentrated solution of a potential ligand is added, it is usually a signal that a complex ion will form.*

A common reaction for complex ions is substitution or ligand exchange (a single displacement reaction). For example, examine the following reactions:

$$Cu(OH)_2\ (s) \ + \ 4NH_3\ (aq) \ \rightarrow \ Cu(NH_3)_4^{2+} \ + \ 2OH^-\ (aq) \ \textbf{net ionic equation}$$

$$AgCl\ (s) \ + \ 2NH_3\ (aq) \ \rightarrow \ Ag(NH_3)_2^+ \ + \ Cl^- \ \ \textbf{net ionic equation}$$

You should notice that in both examples, a precipitate forms a soluble complex ion after the ligand, NH_3, is added.

Quantitative aspects of reactions in solution

Stoichiometry in solution requires that the amount of reactants in solution be known. Concentration units, usually expressed in molarity, are used. Important analytical laboratory techniques such as gravimetric analysis, based on mass, and titration, based on volume, depend on the mole ratios of correctly balanced equations.

Molarity

Concentration can be expressed in many different ways (see Chapter 12 of this review book), but the most commonly used unit is **molarity (M)**, which is defined as *the number of moles of solute per liter of solution*. A 1.46 M solution of $C_6H_{12}O_6$ means that 1.46 mol of sugar is dissolved in enough water to make 1.0 L of solution. It is important to remember that molarity refers only to the amount of solute originally dissolved in solution and does not take into account subsequent processes such as the dissociation of the salt. In a 1 M solution of KCl, the subsequent dissociation of this strong electrolyte yields 1 mol of K^+ ions and 1 mol of Cl^- ions in solution. However, a 1 M solution of $Ba(NO_3)_2$ and its subsequent dissociation yields 1 mol of Ba^{2+} ions and 2 mol of NO_3^- ions in solution: $Ba(NO_3)_2 \rightarrow Ba^{2+} + 2NO_3^-$.

Gravimetric analysis

Gravimetric analysis is an *analytical technique based on the measurement of mass*. One type of gravimetric analysis experiment involves the formation, isolation, and mass determination of a precipitate to determine the composition of an unknown material.

Example 1. A gravimetric determination of composition.

A 0.5662-g sample of an ionic compound containing chloride and an unknown metal is dissolved in water and treated with an excess of $AgNO_3$ solution. If 1.0882 g of AgCl precipitate forms, what is the percent by mass of Cl in the original compound?

General Strategy	Solution to Example 1
Determine the mass percentage of the *desired element* in the precipitate.	$\dfrac{35.45\,g\,Cl}{143.54\,g\,AgCl} \times 100\% = 24.72\%\,Cl^-$
Determine the grams of the *desired element* that have been incorporated into the precipitate.	$0.2472 \times 1.0882\,g = 0.2690\,g\,Cl^-$ in sample of precipitate
Calculate the percentage of the *desired element* in the original sample.	$\dfrac{0.2690\,g\,Cl^-}{0.5662\,g\,sample} = 47.51\%\,Cl^-$ in original sample

Acid-base titration

A quantitative study of acid-base neutralization is usually carried out by a technique called **titration**, in which *a solution of known concentration, the **standard solution**, is gradually added to another solution of unknown concentration until the chemical reaction is complete*. The **equivalence point** is *reached when the acid is completely reacted with the base* and is signaled by a sharp change in color of an indicator (see Chapter 16 of this review guide for a complete discussion of titration).

Example 2. The determination of the molarity of a solution by titration.

> In a titration experiment, a student finds that 35.18 mL of a KOH solution are needed to neutralize 0.5468 g of potassium acid phthalate (KHP), $KHC_8H_4O_4$, a monoprotic acid. What is the molarity of the KOH?

General Strategy	Solution to Example 2
Calculate the moles of the *given* molecule.	$0.5468\,g\,KHP \times \dfrac{1\,mol\,KHP}{204.2\,g\,KHP} = 2.678 \times 10^{-3}\,mol\,KHP$
Use the mole ratio between the *given* molecule and the *desired* molecule to determine the moles of the desired molecule present in solution.	KHP is a monoprotic acid that can be represented generically as HA. $$HA \ + \ KOH \rightarrow KA \ + \ H_2O$$ Since 1 mol HA = 1 mol KOH, there must be 2.678×10^{-3} moles of KOH in the 35.18 mL of KOH.
Place the <u>moles</u> of the *desired* molecule over the <u>volume</u> of the solution to determine the molarity.	$\dfrac{2.678 \times 10^{-3}\,mol\,KOH}{35.18\,mL} \times \dfrac{1000\,mL}{1\,L} = 0.07612\,M\,KOH$

Redox titrations

Titrations can be used to carry out redox reactions quantitatively. Just as an acid can be titrated against a base, an oxidizing agent can be titrated against a reducing agent using a similar procedure. The equivalence point is reached when the reducing agent is completely oxidized by the oxidizing agent. This is signaled by a sharp color change. Redox titrations require the same type of calculations (mole method) as acid-base neutralization. The difference is that the equations and stoichiometry tend to be more complex.

Example 3. Determination of molarity in a redox titration.

A 16.42 mL volume of 0.1327 M $KMnO_4$ solution is needed to oxidize 25.00 mL of a $FeSO_4$ solution in an acidic medium. Determine the molarity of the $FeSO_4$ solution.

General Strategy	Solution to Example 3
Write and balance the net ionic redox equation.	$5Fe^{2+} + MnO_4^- + 8H^+ \rightarrow Mn^{2+} + 5Fe^{3+} + 4H_2O$
Calculate the number of moles of *standard solution* used. $KMnO_4$ in this problem is the **standard solution**; *its molarity and volume are known.*	$16.42 \text{ mL} \times \dfrac{0.1327 \, mol \, MnO_4^-}{1000 \, mL} = 2.179 \times 10^{-3} \text{ mol } MnO_4^-$
From the coefficients of the balanced net equation, the mole ratio between the *standard solution* and *desired* molecule is used to calculate the moles of *desired* molecule present.	$2.179 \times 10^{-3} \text{ mol } MnO_4^- \times \dfrac{5 \, mol \, Fe^{2+}}{1 \, mol \, MnO_4^-} = 1.090 \times 10^{-2} \text{ mol}$ Fe^{2+}
Determine the concentration of the *desired* solution by placing the moles of the *desired* molecule over the volume of the *desired* solution.	$M = \dfrac{mol}{L}$ $= \dfrac{1.090 \times 10^{-2} \, mol \, Fe^{2+}}{25.00 \, mL} \times \dfrac{1000 \, mL}{L} = 0.4360 \, M \, Fe^{2+}$

Take Note: *On the AP exam, it is important for you to have correctly written and balanced equations in order to perform accurate stoichiometric calculations. Remember to have the correct number of significant figures in your final answer.*

SAMPLE MULTIPLE CHOICE QUESTIONS

1. When the following equation is balanced and all the coefficients are the lowest possible whole number, what is the coefficient of the NO_3^- ion?

   $$\_\_Cu\,(s) + \_\_H^+\,(aq) + \_\_NO_3^-\,(aq) \rightarrow \_\_Cu^{2+}\,(aq) + \_\_NO_2\,(g) + \_\_H_2O\,(l)$$

   A. 2   B. 3   C. 4   D. 6   E. 8

2. The correct <u>ionic</u> equation between solutions of ammonium nitrate and sodium hydroxide is:

   A. $NH_4NO_3\,(aq) \ + \ NaOH\,(aq) \rightarrow \ NaNO_3\,(aq) \ + \ NH_4OH\,(l)$

   B. $NH_4^+\,(aq) + NO_3^-\,(aq) + Na^+\,(aq) + OH^-\,(aq) \rightarrow NH_3\,(aq) +$
   $$H_2O\,(l) + Na^+\,(aq) + NO_3^-\,(aq)$$

   C. $NH_4^+\,(aq) + NO_3^-\,(aq) + Na^+\,(aq) + OH^-\,(aq) \rightarrow NH_4OH\,(aq) +$
   $$Na^+\,(aq) + OH^-\,(aq)$$

   D. $NH_4^+\,(aq) + NO_3^-\,(aq) + NaOH\,(aq) \rightarrow NH_3\,(aq) + H_2O\,(l) +$
   $$Na^+\,(aq) + NO_3^-\,(aq)$$

   E. $NH_4NO_3\,(aq) \ + \ Na^+\,(aq) \ + \ OH^-\,(aq) \ \rightarrow \ NH_4OH\,(l) \ + \ Na^+\,(aq) +$
   $$NO_3^-\,(aq)$$

3. What is the <u>net ionic equation</u> for the reaction that would occur between aqueous nickel(II) nitrate solution and concentrated ammonia?

   A. $Ni^{2+} + 2NO_3^- + 2NH_4OH \rightarrow Ni(OH)_2 + 2NH_4^+ + 2NO_3^-$

   B. $Ni^{2+} + 4H_2O \rightarrow Ni(H_2O)_4^{2+}$

   C. $Ni^{2+} + 4NH_3 \rightarrow Ni(NH_3)_4^{2+}$

   D. $Ni^{2+} + 2NH_3 + H_2O \rightarrow NiO + 2NH_4^+$

   E. $Ni^{2+} + 2NO_3^- + 2NH_4^+ + 2OH^- \rightarrow 2\,NH_4NO_3 + Ni(OH)_2$

4.   Consider the following reaction:

$$5H_2O(l) + S_2O_3^{2-} + 4Cl_2(aq) \rightarrow 8Cl^-(aq) + 2SO_4^{2-}(aq) + 10H^+$$

Identify the oxidizing agent in this reaction:

A. Cl^-
B. $S_2O_3^{2-}$
C. H_2O
D. Cl_2
E. H^+

5.   A concentrated solution of sodium hydroxide is added to a solution of potassium permanganate. The permanganate ion reacts to form:

A. Mn^{2+}
B. MnO_4^-
C. Mn
D. Mn_2O_3
E. MnO_4^{2-}

6.   Solutions of calcium chloride and sodium carbonate are mixed. Which is the correct net ionic equation for the reaction?

A. $Ca^{2+} + CO_3^{2-} \rightarrow CaCO_3$
B. $Na^+ + Cl^- \rightarrow NaCl$
C. $CaCl_2 + Na_2CO_3 \rightarrow CaCO_3 + 2NaCl$
D. $Ca^{2+} + 2Cl^- + 2Na^+ + CO_3^{2-} \rightarrow CaCO_3 + 2Na^+ + 2Cl^-$
E. $Ca^{2+} + 2Cl^- + 2Na^+ + CO_3^{2-} \rightarrow CaCO_3 + 2NaCl$

7.   Consider the reaction:

$$Ba(OH)_2\ (aq) + H_2SO_4(aq) \rightarrow BaSO_4\ (s) + 2H_2O\ (l)$$

Which reaction type(s) best identify the chemical change shown here:

            I.   Double displacement
            II. Precipitation
            III. Single displacement
            IV. Redox
             V. Neutralization

A. I only
B. II and V
C. I, II, and IV
D. V only
E. I, II, and V

8.   Determine the oxidation number of the <u>chromium</u> in $Al_2(Cr_2O_7)_3$.

A. +3
B. +4
C. +5
D. +6
E. +7

9.   Which of the following compounds would NOT be soluble to a large extent in water, but would be soluble in hydrochloric acid?

A. $NaCl$ (s)
B. $MgCO_3$ (s)
C. $CuSO_4$ (s)
D. $Ba(NO_3)_2$ (s)
E. $KClO_4$ (s)

10. Identify the substance that is reduced in the following reaction:

$$ClO_3^- + 6Br^- + 6H^+ \rightarrow 3Br_2 + 3H_2O + Cl^-$$

A. ClO_3^-
B. Br^-
C. H^+
D. Br_2
E. Cl^-

11. Predict the products of a reaction in which an acidified solution of $KMnO_4$ is mixed with a solution of $FeCl_2$.

A. H_2O, Mn, Cl_2
B. Fe^{3+}, Mn^{2+}, H_2O
C. H_2, MnO_2, Fe^{3+}
D. H_2O, K, Cl_2
E. Mn, H_2O, Fe

12. A total of 50.0 mL of 0.100 M H_2SO_4 are used to titrate 10.0 mL of NaOH to its endpoint. What is the molarity of the NaOH?

   A. 0.50 M
   B. 0.025 M
   C. 0.100 M
   D. 1.00 M
   E. 1.0 M

13. Silver nitrate solution ($AgNO_3$) is used to precipitate sulfur from an alkali metal sulfide. The original sample of alkali metal sulfide had a mass of 0.12 g. The precipitate silver sulfide, Ag_2S, when dried, had a mass of 0.50 g. What is the mass percent of sulfur in the original alkali metal sulfide (Ag_2S = 250 g/mol)?

   A. 25%
   B. 10%
   C. 76%
   D. 53%
   E. 5.0%

14. 20.0 mL of a 0.100 M solution of calcium chloride, $CaCl_2$, are mixed with 10.0 mL of a 0.100 M solution of silver nitrate, $AgNO_3$. What mass of silver chloride precipitate will form?

   A. 1.29 g
   B. 0.144 g
   C. 0.646 g
   D. 0.108 g
   E. 0.530 g

15. Enough $Al_2(SO_4)_3$ solid is added to water to prepare 100.0 mL of a 0.150 M solution of aluminum sulfate. What is the concentration of sulfate ion in solution?

   A. 0.150 M
   B. 0.200 M
   C. 0.350 M
   D. 0.450 M
   E. 0.750 M

Comprehension Questions

1)   Listed from a) to h) are one or more reactants for a chemical reaction. In every case a reaction will occur under the stated conditions. Unless otherwise stated, the temperature is assumed to be 25 °C and the pressure of gases, including the atmosphere, to be 1.0 atm. In each case, predict the product(s) and provide a balanced (with lowest whole-number coefficients), net ionic equation. It is not necessary to provide states of matter (e.g., solid, liquid, etc.). Answer the follow-up question for each reaction based on your prediction of products.

   a)   solid calcium oxide is added to water

           i)   Would the pH of the resulting solution be above, below, or equal to 7?

   b)   a mixture of ethane and oxygen gases are exposed to an open flame

           i)   What are the oxidation states of carbon and oxygen before and after the reaction?

   c)   an acidified solution of potassium permanganate is added to a solution of lithium chloride

           i)   What substance is the oxidizing agent in the above reaction?

   d)   equal volumes of equimolar solutions of nitrous acid and lithium hydroxide are combined

           i)   What is the chemical driving force for the above reaction, that is, why does this reaction proceed to products rather than remain a mixture of aqueous solutions?

   e)   solutions of barium nitrate and potassium sulfide are combined

           i)   What physical state would you expect a pure sample of barium nitrate to have at 25 °C and 1 atm of pressure?

   f)   a strip of magnesium metal is immersed in a solution of copper(II) sulfate

           i)   What color would you expect the copper(II) sulfate solution to be?

   g)   sulfur dioxide gas is bubbled into distilled water

           i)   What effects would raising the temperature have on the above process?

   h)   solid calcium carbonate is heated

i)   What would be the effect of performing this reaction in a closed vessel?

2)   A group I metal chlorate, 0.7562 g, and manganese(IV) oxide, 10 mg, are heated in a crucible to approximately 350 °C at which point the mixture has liquified and begins to bubble. Heating is continued until bubbling ceases, and the mixture is allowed to cool to room temperature and solidify. The residue in the crucible has a mass of 0.4621 g.

   a)   Write a balanced chemical equation for the above process (use the symbol M to represent the group I metal). What is the chemical formula of the gas given off during the process?

   b)   Using the data provided, determine the percent of oxygen in the original chlorate salt.

   c)   What is the identity of the group I metal?

3)   50.00 mL of a 0.1514 M sulfurous acid solution is titrated to a phenolphthalein endpoint with 0.09857 M potassium hydroxide. What volume of potassium hydroxide will be needed to reach the endpoint?

ANSWERS TO MULTIPLE CHOICE QUESTIONS

1.   A

   Examination of the oxidation numbers for this reaction yields the following results.

   $$\overset{0}{Cu}(s) + \overset{+1}{H^+}(aq) + \overset{+5\,-2}{NO_3^-}(aq) \rightarrow \overset{+2}{Cu^{2+}}(aq) + \overset{+4\,-2}{NO_2}(g) + \overset{+1\,-2}{H_2O}(l)$$

   Examination of the changes in the oxidation numbers shows that Cu has been oxidized to Cu^{2+} and N in NO_3^- has been reduced from +5 to +4 in NO_2.

   The nitrate half-reaction must be multiplied by two so that the electrons lost in the oxidation will equal the electrons gained in the reduction. (See Chapter 19 of this review book for details on balancing redox equations.) Adding the two half-reactions yields the overall equation for the redox reaction:

   $$Cu \rightarrow Cu^{2+} + 2e^-$$
   $$+ \quad (1e^- + 2H^+ + NO_3^- \rightarrow NO_2 + H_2O)2$$
   $$\overline{4H^+ + 2NO_3^- + Cu \rightarrow Cu^{2+} + 2NO_2 + 2H_2O}$$

2.   B

Ammonium nitrate is NH_4NO_3 and sodium hydroxide is $NaOH$. Both of these reactants are strong electrolytes and should appear in solution as ions: $NH_4^+ + NO_3^-$ and Na^+ and OH^-. Since these are two compounds, the reaction type they exhibit is a double displacement. The products of the reaction are $NaNO_3$, which is a strong electrolyte in solution, Na^+ and NO_3^-, and NH_4OH, which exists as NH_3 and H_2O. The ionic equation is:

$$NH_4^+\,(aq) + NO_3^-\,(aq) + Na^+\,(aq) + OH^-\,(aq) \rightarrow NH_3\,(aq) + H_2O\,(l) + Na^+\,(aq) + NO_3^-\,(aq)$$

3.   C

Nickel(II) nitrate is soluble in solution and exists as Ni^{2+} and NO_3^-. The addition of concentrated NH_3 to a solution containing the transition metal nickel usually signals the formation of a complex ion. Since the oxidation number of the nickel is +2, the most likely number of ligands is four: $Ni(NH_3)_4^{2+}$

4.   D

The oxidizing agent in a reaction is reduced. A check of the oxidation numbers in this reaction will show that Cl_2 is reduced from 0 to -1:

$$\begin{array}{cccccc}
+1\ -2 & +2\ -2 & 0 & -1 & +6\ -2 & +1 \\
5H_2O\ (l)\ + & S_2O_3^{2-}\ + & 4Cl_2\ (aq) & \rightarrow\ 8Cl^-\ (aq)\ + & 2SO_4^{2-}\ (aq)\ + & 10H^+
\end{array}$$

5.   E

Permanganate (MnO_4^{2-}) is a powerful oxidizing agent that is purple. In a strongly basic environment, permanganate is reduced to MnO_4^{2-}, which is dark green.

6.   A

Calcium chloride $(CaCl_2)$ and sodium carbonate (Na_2CO_3) are both soluble strong electrolytes and exist as ions in solution: Ca^{2+} and $2Cl^-$, and $2Na^+$ and CO_3^{2-}. Since these are two compounds, they undergo a double displacement reaction. The products of the double displacement are $NaCl$, which is a strong electrolyte and remains in solution, and $CaCO_3$, which is insoluble and forms a precipitate in solution. Removal of the spectator ions Na^+ and Cl^- from the equation yields the net ionic equation:

$$Ca^{2+} + CO_3^{2-} \rightarrow CaCO_3$$

7.   E

The reactants are two ionic compounds and this is a <u>double displacement</u> reaction. Examination of the products and the solubility rules reveals that $BaSO_4$ is insoluble, making this a <u>precipitation reaction</u>. The $Ba(OH)_2$ is a base and H_2SO_4 is an acid, which also makes this reaction a <u>neutralization</u> reaction.

8.   D

The calculation of the oxidation number yields *+6*

$$Al_2(Cr_2O_7)_3$$

Each aluminum ion has an oxidation number of +3 for a total on the two aluminum ions of +6. $2Al^{3+}$ = +6

Therefore, the total oxidation number on the three dichromate ions is –6. $3(Cr_2O_7)$ = –6 because the total charge on the compound is 0. Hence, each dichromate ion has an oxidation number of –2.

Since oxygen usually has an oxidation number of –2,

$2Cr + 7(-2) = -2$ (the oxidation number on the $Cr_2O_7^{2-}$)
$2Cr = +12$
Therefore, each *Cr = +6.*

9.   B

Checking the solubility rules shows that all compounds except $MgCO_3$ are soluble in water. The $MgCO_3$ does not dissolve in water, but it will undergo the following reaction with hydrochloric acid:

$$MgCO_3\ (s)\ +\quad HCl\ (aq)\ \rightarrow\quad MgCl_2\ (aq)\ +\quad CO_2\ (g)\ +\quad H_2O\ (l)$$

The net equation for this reaction is:

$$MgCO_3\ (s)\ +\ H^+\ \rightarrow\ Mg^{2+}\ (aq) +\ CO_2\ (g)\ +\ H_2O\ (l)$$

10. A

The substance that is reduced has a reduction in its oxidation number. Calculating the oxidation numbers of the reactants and products shows the Cl in ClO_3^- has an oxidation state of $+5$ and the Cl^- has an oxidation state of -1.

$$\overset{+5\;-2}{ClO_3^-} + \overset{-1}{6Br^-} + \overset{+1}{6H^+} \rightarrow \overset{0}{3Br_2} + \overset{+1\;-2}{3H_2O} + \overset{-1}{Cl^-}$$

11. B

Acidified $KMnO_4$ is a powerful oxidizing agent. In acid solution, MnO_4^- will be reduced to Mn^{2+} (see Table 1). Iron, Fe^{2+}, is a metal and is more likely to lose an electron and be oxidized to its higher state, Fe^{3+}, than Cl^- is to be oxidized to Cl_2. Water is also a product in an acidified solution of $KMnO_4$. The water helps to conserve O atoms from the reactant MnO_4^- ions.

12. D

This titration problem involves a diprotic acid. The correctly balanced equation for the reaction is:
$$H_2SO_4 + 2NaOH \rightarrow Na_2SO_4 + 2H_2O$$

$$50.0 \text{ mL } H_2SO_4 \times \frac{0.100\, mol\, H_2SO_4}{1000\, mL} \times \frac{2\, mol\, NaOH}{1\, mol\, H_2SO_4} = 0.0100 \text{ mol NaOH}$$

$$\frac{0.0100\, mol\, NaOH}{10.0\, mL} \times \frac{1000\, ml}{1L} = 1.00 \text{ M NaOH}$$

13. D

All of the sulfur contained in the alkali metal sulfide is precipitated by silver and is combined in the silver sulfide.

$$0.50 \text{ g } Ag_2S \times \frac{32.0\, g\, S}{250\, g\, Ag_2S} = 0.064 \text{ g S (from original alkali metal sample)}$$

$$\%S = \frac{0.064\, g\, S}{0.12\, g\, sample} \times 100\% = 53\%\ sulfur$$

14. B

Since both reactants are in solution, the net ionic equation should be found first, and then checked for a limiting reagent.

$$Ca^{2+} + 2Cl^- + 2Ag^+ + 2NO_3^- \rightarrow AgCl + Ca^{2+} + 2NO_3^-$$

The net equation is $Ag^+ + Cl^- \rightarrow AgCl$

CaCl₂ 20.0 mL $\times \dfrac{0.100\,mol\,CaCl_2}{1000\,mL}$ = .00200 mol CaCl₂ = **2**(.00200) mol Cl⁻

$= .00400$ mol Cl⁻ when CaCl₂ dissociates in solution

AgNO₃ 10.0 mL $\times \dfrac{0.100\,mol\,AgNO_3}{1000\,mL}$ = .00100 mol AgNO₃ = .00100 mol Ag⁺

when AgNO₃ dissociates

Since Ag^+ and Cl^- react in a 1 mol : 1 mol ratio, Ag^+ is the limiting reagent.

0.00100 mol $Ag^+ \times \dfrac{1\,mol\,AgCl}{1\,mol\,Ag^+} \times \dfrac{143.5\,g\,AgCl}{1\,mol\,AgCl} = 0.144\,g\,AgCl$

15. D

Aluminum sulfate, Al₂(SO₄)₃, is a strong electrolyte. It dissociates completely. For every 1 mol of aluminum sulfate, 2 mol of aluminum ion, Al^{3+}, and 3 mol of sulfate ion, SO_4^{2-}, will be released into solution, $Al_2(SO_4)_3 \rightarrow 2Al^{3+} + 3SO_4^{2-}$. Therefore, if 0.150 mol of Al₂(SO₄)₃ are placed in solution, three times that amount, or 0.450 mol of sulfate, will be released. The molarity of the sulfate is *0.450 M.*

Answers to Comprehension Questions

1) a) **CaO + H₂O → Ca(OH)₂**

Calcium oxide is considered a metallic oxide or basic anhydride. Addition of water to this class of compounds yields a metallic hydroxide or base. Since the product is a base, the pH of the solution would be above 7.

b) **2C₂H₆ + 5O₂ → 2CO₂ + 3H₂O**

This is a common combustion reaction of a hydrocarbon yielding carbon dioxide and water. The oxidation states of carbon before and after combination with oxygen are +3 and +4, respectively. Those for oxygen are 0 and –2, respectively.

c) $8H^+ + MnO_4^- + 2Cl^- \rightarrow Mn^{2+} + Cl_2 + 4H_2O$

Permanganate is a strong oxidizing agent and produces manganese(II) ions in acidic solution.

d) $HNO_2 + OH^- \rightarrow NO_2^- + H_2O$

This is an acid-base neutralization reaction between a weak acid and strong base. The acid, because it is only partially ionized, must be represented as a molecular species. The driving force for this reaction, as in many double displacement reactions, is the formation of a primarily molecular species, water in this case. Once the bonds between hydrogen and oxygen form, they are not easily broken under the conditions of this reaction. This product mixture represents the low energy combinations of atoms for this reaction.

e) $Ba^{2+} + S^{2-} \rightarrow BaS$

This is an example of a common double displacement reaction in which the cation of one soluble compound, barium nitrate, combines with the anion of another soluble substance, potassium sulfide, to form an insoluble precipitate, barium sulfide, upon mixing. The starting compound barium nitrate would be a crystalline solid in its pure form. Most if not all pure ionic substances are solids at room temperature owing to the nature of the bonds holding all of the particles together.

f) $Mg + Cu^{2+} \rightarrow Mg^{2+} + Cu$

This is the single displacement version of a redox reaction in which a more active metal, magnesium, replaces a less active one, copper, in a compound. The element that existed as an ion is reduced to its elemental form, and the element present as neutral atoms is oxidized to its stable cationic form, +2 for a group II element in this case. The copper(II) sulfate solution would be a deep blue. Aqueous solutions of copper(II) ions are blue.

g) $SO_2 + H_2O \rightarrow H_2SO_3$

Sulfur dioxide can be thought of as a nonmetal oxide or acidic anhydride. This class of compounds, when combined with water, yields an acid as a product. Raising the temperature would have a couple of effects on this reaction. One, raising the temperature generally increases reaction rates, and two, for this specific case, raising the temperature might inhibit product formation due to decreased solubility of the starting material. The solubility of gases generally decreases as the temperature of an aqueous solution increases.

h) $CaCO_3 \rightarrow CaO + CO_2$

The clue to predicting products for this decomposition reaction is that there is only one reactant. In general, carbonates decompose to form carbon dioxide gas and metallic oxides, calcium oxide in this case. If this reaction were run open to the atmosphere, it would proceed until there was essentially no carbonate left. If it were run in a closed vessel, one that could withstand the pressure of the gas produced, the reaction would eventually reach an equilibrium in which measurable amounts of all three species would be present.

2)   a)   Metallic chlorates decompose upon heating to form metallic chlorides and oxygen gas. The manganese(IV) oxide is added as a catalyst. Therefore:

$$2MClO_3 \rightarrow 2MCl + 3O_2$$

Since the metal is known to be from group I, the formula of the chloride must be $MClO_3$. The residue in the crucible is the solid metallic chloride left from the reaction plus the catalyst. **The gas given off during the reaction is oxygen, O_2.**

b)   Subtracting the mass of the catalyst from the remaining mass of solid, 0.4621 g – 0.010 g = 0.4521 g, gives a mass of 0.4521 g of remaining metal chloride. Subtracting this mass from the original mass of metal chlorate, 0.7562 g – 0.4521 g = 0.3041 g, gives the mass of the oxygen driven off. The percent oxygen in the original sample can be determined by dividing the mass of oxygen driven off by the original mass of the sample,

(0.3041 g oxygen / 0.7562 g $MClO_3$) × 100% = **40.2% oxygen by mass**

c)   Calculating the percent by mass of oxygen for each of the group I metal yields only one result that correlates with the data presented. **The metal is potassium, K.**

molar mass of $KClO_3$ = (1 × 39.098) + (1 × 35.453) + (3 × 15.999) = 122.55 g/mol

calculated % oxygen in $KClO_3$ = ((3 × 15.999) / 122.55) × 100% = 39.2% oxygen.

3)   Sulfurous acid is a weak, diprotic acid and therefore reacts in a 1:2 mole ratio with potassium hydroxide as shown.

$$H_2SO_3 + 2KOH \rightarrow K_2SO_3 + 2H_2O$$

If the volume and molarity of a solution are known, then moles of the substance can easily be calculated by multiplying these two quantities. The balanced chemical equation, as always, represents mole ratios that combine or react.

0.05000 L H_2SO_3 × (0.1514 mol H_2SO_3 / liter of solution) × (2 mol KOH / 1 mol H_2SO_3) = 0.01514 mol of KOH needed to react with this quantity of acid.

Dividing the moles of base by the concentration of the base in moles/liter will give the volume of base needed to reach the endpoint,

0.01514 mol of KOH / 0.09857 M = **0.1536 L or 153.6 mL**

CHAPTER 5
GASES

This chapter reviews the properties of gases, including:

- General properties of gases and the basic gas laws, including Boyle's law, Charles' law, Gay-Lussac's law, Avogadro's law, and the general gas law.
- The ideal gas law
- Stoichiometry of gaseous reactions
- Dalton's law of partial pressures
- Kinetic molecular theory
- Graham's law
- Nonideal gases and the van der Waals equation

General properties of gases and the basic gas laws

Matter exists as a solid, a liquid, or a gas. Atoms and molecules in gases have the lowest density of all the states of matter and the weakest intermolecular forces (see Chapter 11 of this review book).

Gases expand to fill the space available to them. They take the shape of their container and are evenly distributed within it. They mix completely and uniformly with other gases when confined in the same space, providing no chemical reaction occurs.

There are four variables that characterize a gas:

- pressure (P),
- volume (V),
- temperature (T), and
- the number of moles (n).

The basic gas laws show the mathematical relationships that exist among these variables. The basic gas laws include Boyle's law, Charles' law, Gay-Lussac's law, and Avogadro's law. Their mathematical relationships are summarized in Table (note that the constant k is different for each of the laws):

Table 1. Summary of the basic gas laws

Relationship	Law	Equation	Mathematical relationship
pressure-volume relationship **(temperature and moles are constant)**	Boyle's	$P = \dfrac{k}{V}$ or $PV = k$ Equation for a change in pressure with corresponding change in volume on fixed sample of gas: $P_1 V_1 = P_2 V_2$	inverse
volume-temperature relationship **(pressure and moles are constant)**	Charles'	$V = kT$ or $\dfrac{V}{T} = k$ Equation for a change in temperature with corresponding change in volume on fixed sample of gas: $\dfrac{V_1}{V_2} = \dfrac{T_1}{T_2}$	direct
pressure-temperature relationship **(volume and moles are constant)**	Gay-Lussac's	$P = kT$ or $\dfrac{P}{T} = k$ Equation for change in temperature with corresponding change in pressure: $\dfrac{P_1}{P_2} = \dfrac{T_1}{T_2}$	direct
volume-amount relationship **(pressure and temperature are constant)**	Avogadro's	$V = kn$ or $\dfrac{V}{n} = k$	direct

A useful form basic gas laws in which the number of moles of the gas is constant is referred to as the **general gas law:**

$$\frac{V_1 P_1}{T_1} = \frac{V_2 P_2}{T_2}$$

This equation allows for the calculation of any one of the variables (V, T, P) for a *fixed* sample of gas.

The ideal gas law

A combination of all four basic gas relationships yields a single equation called the **ideal gas law:**

$$PV = nRT \text{ where}$$

V is volume of the gas in liters
P is pressure of the gas in atmospheres
R is the ideal gas constant, 0.0821 L·atm·mol^{-1}·K^{-1}
T is temperature of the gas in Kelvin

The ideal gas law assumes gas molecules have no intermolecular interactions and take up no volume. It is important for you to realize that the volumes mentioned in gas law problems refer to the volume of the container.

Standard conditions, STP, for a gas *are 1 atm and 273 K.* The pressure of a gas can be expressed in a variety of units: *1 atm = 760 torr = 760 mm Hg = 101.3 kPa.* Gas temperatures must always be in **Kelvin:** $K = °C + 273$. Remember that the volume occupied by *1 mol of ideal gas at standard conditions is 22.4 L.*

The ideal gas law is extremely useful. Knowing three of the four variables (P, V, n, T), allows for a straightforward calculation of the fourth. The equation can also be used to determine the molar mass and the density of a gas by substituting and rearranging the ideal gas law:

1. **Molar mass from gas density:**

    $$PV = nRT$$

    Since n (moles) $= \dfrac{m(mass)}{M(molar\ mass)}$, then $\dfrac{m}{M}$ can be substituted for n (moles):

$$PV = \frac{m}{M}RT$$

Rearranging this equation to solve for M (molar mass) yields the following:

$$M = \frac{mRT}{VP} \quad \text{or,} \quad \text{since } \frac{m(mass)}{V(volume)} \text{ is density } (d)$$

$$M = \frac{dRT}{P}$$

2. **Calculation of gas density**:

You can also rearrange the above equation to calculate the gas density:

$$d = \frac{m}{V} = \frac{PM}{RT}$$

Stoichiometry of gaseous reactions

The ideal gas law relates gas volumes and moles of gas in a reaction. These relationships can be used to solve stoichiometry problems.

Example 1. Stoichiometry with gaseous reactants.

Consider the following combustion reaction where all reactants and products are at the same temperature:

$$2C_2H_6(g) + 7O_2(g) \rightarrow 4CO_2(g) + 6H_2O(l)$$

If 10.0 L of C_2H_6 are burned at STP, how many liters of oxygen would be required for a complete reaction at the same conditions?

General Strategy	Solution to Example 1
Write and balance the equation.	$2C_2H_6(g) + 7O_2(g) \rightarrow 4CO_2(g) + 6H_2O(l)$
Place the data from the problem underneath the balanced equation.	$2C_2H_6(g) + 7O_2(g) \rightarrow 4CO_2(g) + 6H_2O(l)$ 10.0 L ? L
Convert the *given* quantity to moles.	$10.0\,L\,C_2H_6 \times \dfrac{1\,mol\,C_2H_6}{22.4\,L_{STP}} = \text{moles of } C_2H_6$
Use the mole ratio from the balanced equation to go from the *given* molecule to the *desired* molecule.	$10.0\,L\,C_2H_6 \times \dfrac{1\,mol\,C_2H_6}{22.4\,L} \times \dfrac{7\,mol\,O_2}{2\,mole\,C_2H_6} =$ moles of desired molecule

Convert the *desired* molecule to the appropriate units.	$10.0\,L\,C_2H_6 \times \dfrac{1\,mol\,C_2H_6}{22.4\,L\,C_2H_6} \times \dfrac{7\,mol\,O_2}{2\,mol\,C_2H_6} \times \dfrac{\mathbf{22.4\,L\,O_2}}{\mathbf{1\,mol\,O_2}} =$ $35.0\,L_{STP}\,O_2$
An *alternative solution* to the problem is to use Avogadro's law, which shows that at the same temperature and pressure, the moles of gas are directly related to their volume. In other words, *the coefficients in the balanced equation are mole ratios as well as volume ratios for gaseous molecules.*	$10.0\,L_{STP}\,C_2H_6 \times \dfrac{7\,L\,O_2}{2\,L\,C_2H_6} = 35.0\,L_{STP}\,O_2$

Dalton's law of partial pressures

Another consequence of the ideal gas law is **Dalton's law of partial pressures:** *when two or more gases are mixed in a container, the total pressure in the container is the sum of the pressures of the individual components. Each gas exerts a pressure,* called the **partial pressure,** equal to the pressure the gas would exert if it were the only gas in the container.

$$P_{total} = p_1 + p_2 + p_3 + p_n$$

The partial pressure of each gas is directly related to its mole fraction in the mixture. A common misconception is that if another gas is added to a container with a mixture of gases, the individual partial pressures change. Only the total pressure of the container will change. Each partial pressure remains the same.

> **Take Note:** *One of the most common applications of Dalton's law of partial pressures that you will find on the AP exam is calculating the pressure of a gas collected over water. Since water evaporates and mixes with the gas, the total pressure in the collection vessel is the sum of the partial pressures of the gas collected and the water vapor:* $P_{total} = P_{gas} + P_{water\;vapor}$ *or* $P_{gas} = P_{total} - P_{water\;vapor}$

Kinetic molecular theory

Gas laws predict gas behavior but do not explain gas behavior. The kinetic molecular theory (KMT) of gases provides explanations at the molecular level for the observed macroscopic behavior of gases.   The **kinetic molecular theory has four basic tenets:**

1. The distances between the molecules of a gas are so large that the molecules can be considered points; they possess mass but have no volume.
2. Gas molecules are in constant random motion. They have perfectly elastic collisions, i.e., no energy is lost when the molecules collide.
3. Gas molecules do not attract or repel one another.
4. The *average* kinetic energy of the molecules is proportional to the temperature of the gas. The temperature is measured in Kelvin units.

Graham's law

Effusion is *the escape of a gas through a pinhole into a vacuum.* The rates of effusion of gases are inversely related to the square roots of their molar masses. Comparing the rates of effusion of two different gases at the same temperature and pressure gives Graham's law:

$$\frac{r_2}{r_1} = \sqrt{\frac{M_1}{M_2}}$$

M is the molar mass in grams
r is the rate of effusion

Notice that the *molecular mass (M) of the gas and its rate (r) of effusion are inversely related*, resulting in light molecules traveling quickly and heavy molecules slowly. An extension of Graham's law shows that the *time (t) it takes a gas to effuse is directly proportional to the molecular mass* of that gas. To reflect this relationship, Graham's law can also be written as:

$$\frac{t_2}{t_1} = \sqrt{\frac{M_2}{M_1}} \text{ where}$$

t is the effusion time for the gases

> **Take Note:** *A common question on the Multiple Choice section of the AP exam requires you to extend Graham's law and address the time it takes for a gas to effuse.*

Nonideal gases and the van der Waals equation

An ideal gas molecule has no volume and no intermolecular attractions. Ideal gases obey the **ideal gas law:** $P_{ideal}V_{ideal} = nRT$. However, real gas molecules do take up space and do experience intermolecular attractions. *Real gases approach ideal behavior at high temperature and low pressure.* At high temperatures, the effect of the intermolecular attractions that exist among real gas molecules is negligible. At low pressures, the molecular volume of the real gas is negligible when compared to the volume of the container.

The **van der Waals equation** is a modification of the ideal gas law to reflect the measurements made in lab on real gases:

$$(\mathbf{P}_{measured} + a\frac{n^2}{V^2}) \ (\mathbf{V}_{measured} - nb) = n\mathbf{R}T$$

> $P_{measured}$ is the gas pressure measured in lab
> a is the proportionality constant related to the strength of the
>   intermolecular attractive forces and is unique to each substance
> $V_{measured}$ is the volume of the gas container measured in lab
> b is the proportionality constant related to the volume of a gas atom or
>   molecule and is unique to each substance
> n is the number of moles of gas

The term a $\frac{n^2}{V^2}$ is the correction factor added to the measured pressure of a gas so that $(P_{measured} + a\frac{n^2}{V^2}) = P_{ideal}$. The correction factor is added because the intermolecular attractions cause the gas particles to hit the sides of the container with less momentum. Therefore, the measured pressure is less than ideal pressure.

The term nb is the correction factor subtracted from the measured volume of a gas container so that $V_{measured} - nb = V_{ideal}$. This correction factor is subtracted because the entire volume of the container is not available to the gas particles since the gas particles themselves occupy some of that space. In general, the larger the gas particle, the greater the value of factor b.

Example 2. Predicting the deviation from ideal behavior of a gas.

> Which gas will show the greater deviation from ideal behavior, helium, He, or sulfur dioxide, SO_2?

General Strategy	Solution to Example 2
Evaluate the pressure deviation.	SO_2 is a polar molecule with strong intermolecular forces while He exhibits only dispersion forces (see Chapter 11 of this review book). SO_2 will cause a greater deviation from ideal pressure.
Evaluate the volume deviation.	SO_2 is a much larger molecule than He and has a greater deviation from ideal volume.
Conclusion	*SO_2 will deviate from ideal behavior to a greater extent than He.*

Take Note: *The AP exam rarely asks you for calculations using the van der Waals equation. The AP exam is more likely to ask for comparisons of molecules and how they would deviate from ideality based on their size, molecular architecture, and the type of intermolecular forces present (see Chapters 9, 10, and 11 of this review book).*

Take Note: *On the Free Response section of the AP exam, all of the equations associated with gases will be provided for you. However, there is only a list of equations. You must know how to apply them.*

SAMPLE MULTIPLE CHOICE QUESTIONS

1. A sample of an ideal gas is placed in a sealed container of fixed volume. The gas is heated from 25.0 °C to 45.0 ° C. Which of the following factors will increase?

    I.   The pressure of the gas
    II. The average kinetic energy of the gas
    III. The density of the gas
    IV. The intermolecular forces

    A. I and II
    B. I, II, and III
    C. III and IV
    D. I and IV
    E. I only

2. Consider the following chemical reaction:

    $$NH_4HS(s) \rightarrow NH_3(g) + H_2S(g)$$

    0.020 mol of NH_4HS are placed in a 2.0 L container and heated to 127 °C. All of the solid decomposes. Determine the pressure inside the container.

    A. 0.020 atm
    B. 0.16 atm
    C. 0.33 atm
    D. 1.0 atm
    E. 0.66 atm

3. Two gaseous elements, X and Y, are mixed in a container of fixed volume. The initial partial pressure of X is 2.5 atm and of Y is 9.00 atm. A solid product forms in the container. At the end of the reaction, only Y remains and exerts a pressure of 4.00 atm. The temperature is constant throughout the process. Determine the empirical formula of the compound X_nY_m that forms.

    A. XY
    B. XY_2
    C. X_2Y_9
    D. XY_3
    E. X_2Y_5

4.   The effusion rate of gas X is four times faster than that of sulfur dioxide (SO_2). Determine the identity of gas X.

    A. O_2
    B.   He
    C. CH_4
    D.   NO
    E.   Ne

5.   The empirical formula of a compound is CH. The vapor density of this compound is 2.90 g/L at STP. Determine the molecular formula of this compound.

    A. C_6H_6
    B. C_5H_5
    C. C_4H_4
    D. C_3H_3
    E. C_2H_2

6.   A 60.0-g sample of gas occupies 10.0 L at 27.0 °C and 3.00 atm. Which answer represents the calculation of the molar mass of the gas? ($R = 0.0821$ L·atm · $mol^{-1}·K^{-1}$)

    A. $\dfrac{(60.0)(0.0821)(300)}{(3.00)(10.0)}$

    B. $\dfrac{(3.00)(10.0)}{(0.0821)(60.0)(300)}$

    C. $\dfrac{(60.0)(10.0)(300)}{(3.00)(0.0821)}$

    D. $\dfrac{(10.0)(0.0821)(60.0)}{(3.00)(300)}$

    E.   none of the above

7.   The following gases are all at STP. Which gas has the highest average kinetic energy?

    A.   All the gases have the same kinetic energy.
    B. N_2
    C. O_2
    D. SO_2
    E.   Ne

8.   Which of the following gases will exhibit the greatest deviation from ideal gas behavior?

A. CO_2
B. CH_4
C. CCl_4
D. O_2
E. SO_2

9.   Two containers of gas have the same volume, temperature, and pressure. One container has a sample of oxygen that has a mass of 3.95 g. The second container contains a gas that has a mass of 2.99 g. Determine the molar mass of the gas in the second container.

A.   69 g/mol
B.   24 g/mol
C.   90 g/mol
D.   44 g/mol
E.   11 g/mol

10.   10.0 g of gas A, 10.0 g of gas B, 10.0 g of gas C, and 10.0 g of gas D are added to a closed container of fixed volume. The total pressure in the container is 760 torr. The partial pressure of A:

A.   is 25% of the total pressure.
B.   depends on the size of the container.
C.   is a function of its molar mass.
D.   is directly related to its mole fraction.
E.   none of the answers is correct.

Comprehension Questions

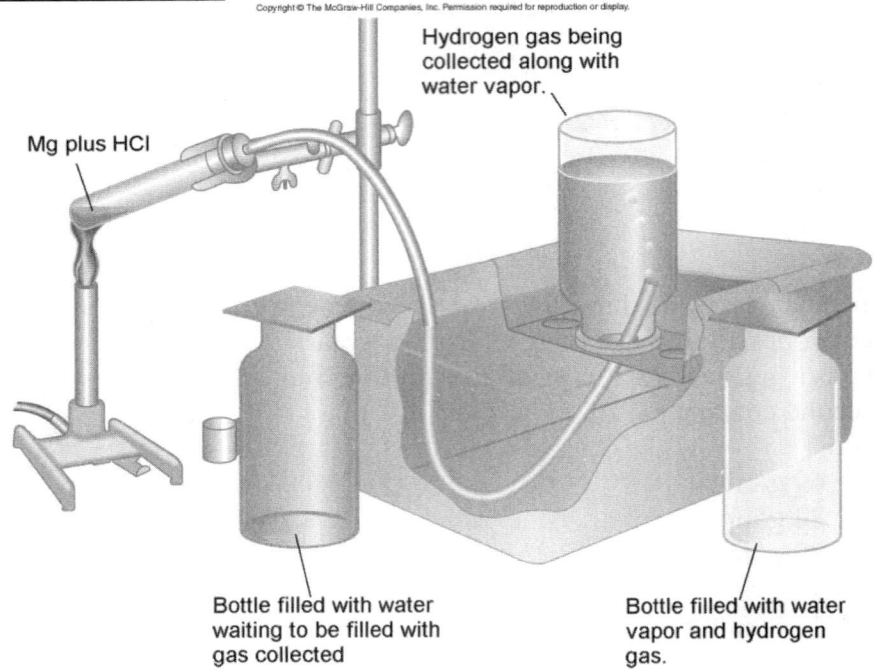

Mg plus HCl

Hydrogen gas being collected along with water vapor.

Bottle filled with water waiting to be filled with gas collected

Bottle filled with water vapor and hydrogen gas.

1)   A sample of magnesium metal was allowed to react with excess hydrochloric acid solution. The gas liberated was collected over water at 23 °C in the setup shown. The volume of the gas collected was 47.1 mL, and the atmospheric pressure in the lab was 744 mm Hg. The vapor pressure of water at 23 °C is 21.1 mm Hg.

   a) Write a balanced equation for the reaction described above.
   b) What volume, in mL, would the gas collected occupy if it was dried and stored at STP?
   c) Determine the mass of magnesium metal, in grams, that would need to be reacted to liberate the quantity of gas collected above.
   d) What assumptions have to be made for the results of the calculation in c) to be valid?

2)   Use the kinetic molecular theory to explain the following observations:

        a)   The plunger of a syringe containing 20.0 mL of air at 1 atm of pressure is pulled out so that the volume inside the syringe is increased to 40.0 mL. The total pressure of the gases inside the syringe decreases to approximately 0.5 atm.

b)   A steel cylinder containing nitrogen gas initially pressurized to 2000 psi at 24 °C was sitting in the sun on a cylinder delivery truck in central Florida. When the driver looked at the gauge on the cylinder after finishing his lunch it indicated a pressure of slightly more than 2200 psi.

c)   One mole of hydrogen gas, molar mass = 2.02 g, and one mole of krypton gas, molar mass = 83.80g, both occupy a volume of 22.4 L at STP.

3)   Consider the apparatus pictured. Calculate the partial pressures of all gases and the total pressure of the system after the stopcocks are opened and the gases are allowed to mix. What is the mass of the gas contained in bulb C after the mixing occurs? The temperature of the entire system remains constant at 22 °C. The volume of the tubing connecting the flasks is negligible.

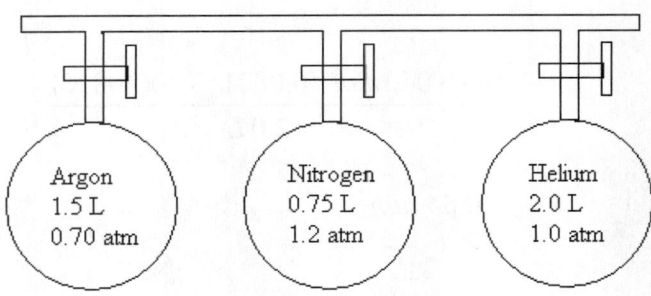

4)   Water vapor near its boiling point deviates significantly from ideal behavior in terms of pressure and volume measurements made at a given temperature. Explain this observation in terms of false assumptions made in the kinetic molecular theory, and describe how the van der Waals equation compensates for the differences observed.

ANSWERS TO SAMPLE MULTIPLE CHOICE QUESTIONS

1.   A

A rise in temperature will increase the average kinetic energy of a gas sample. The increase in kinetic energy will cause the molecules to strike the walls of the container with more momentum and more often, thereby increasing the pressure. Since the volume remains the same, the gas density is unaffected. The increase in kinetic energy will reduce the effect of the intermolecular forces.

2.   E

From the simple stoichiometry of the reaction, 0.020 mol of solid NH_4HS decompose, forming 0.020 mol of gaseous NH_3 and 0.020 mol of gaseous H_2S. Therefore, the total number of moles of gaseous product is 0.040 mol.

Using the ideal gas law: $PV = nRT$ (T in Kelvin, K = 273 +127 °C = 400 K)

$$P = \frac{nRT}{V}$$

$$= \frac{(0.040\,mol)(0.0821^{L \cdot ATM}_{mol \cdot K})(400\,K)}{2.0\,L}$$

$$= 0.66\,atm$$

3.   B

The volume and temperature are constant in the reaction. Therefore, the partial pressure ratio reflects the mole ratio. All of X is used (2.5 atm). Since 4.00 atm of Y remain, 5.00 atm reflects the amount of Y incorporated into the compound. The simple mole ratio of X to Y is 1:2. The empirical formula is XY_2.

4.   B

Comparing the rates of effusion and molecular mass requires the application of Graham's law:

$$\frac{r_2}{r_1} = \sqrt{\frac{M_1}{M_2}}$$

Therefore,

$$\frac{1}{4} = \sqrt{\frac{x\,g/mol}{64\,g/mol}} \qquad x = 4\ g/mol = \mathbf{He}$$

5.   B

The mass of the empirical formula CH is 13 g/mol. The density of the gas at STP
can be used to determine the molar mass of the gas.

$$\frac{2.90\,g}{1\,L} \times \frac{22.4\,L}{1\,mol}STP = \frac{65.0\,g}{1\,mol}$$

Since 65.0/13 = 5, the molecular formula is five times greater than CH, or C_5H_5.

6.   A

The ideal gas law will be needed to determine the number of moles present.
Volume (10.0 L), pressure (3 atm), and temperature (K = °C + 273 or 300 K) are
given.

$$PV = nRT$$

$$(3\text{ atm})(10.0\text{L}) = \frac{60\,g}{M}\left(\frac{.0821\,L \cdot atm}{mol\,K}\right)(300\text{ K})$$

$$M = \frac{(60\,g)(.0821\,L \cdot atm)(300\,K)}{(3\,atm)(10.0\,L)}$$

7.   A

Average kinetic energy is directly related to temperature. Since all the gases are
at the same temperature, they all have the same kinetic energy.

8.   E

SO_2 is a bent molecule and is polar as a result of its bonding and geometry. All of
the other gas molecules exhibit only dispersion forces as a result of their
bonding and geometry.

9.   B

Both containers are the same volume, temperature and pressure and must
therefore contain the same number of moles of gas. A simple ratio derived from
Avogadro's law will give the molar mass of the unknown gas:

$$\frac{32\,g\,/\,mol}{3.95\,g} = \frac{x\,g\,/\,mol}{2.99\,g}$$

$$x = 24 \text{ g/mol}$$

10.   D

Dalton's law of partial pressures demonstrates that the partial pressure of a gas is directly related to its mole fraction in the mixture.

Answers to Comprehension Questions

1a)   Active metals react with hydrochloric acid to form chloride salts of the metals along with hydrogen gas. Therefore the balanced equation is **Mg + 2HCl → H$_2$ + MgCl$_2$**.

1b)   The key to these kinds of problems is to remember that gases collected over water are not pure samples of gas; they contain water vapor. The gas of interest, hydrogen in this case, will occupy the total volume specified, but its pressure will only be part of the total pressure given for the mixture (Dalton's law). In this case you need to first use the ideal gas law to determine the number of moles of hydrogen collected, and then use the combined gas law to determine its volume at STP. Alternatively you can use the molar volume of a gas at STP.

P_{H_2} = (744 mmHg – 21.1 mm Hg) / 760 mm Hg = 0.951 atm

$n_{H_2} = PV/RT$ = (0.951atm × 0.0471 L) / (0.08206 L·atm/mol·K × 296.15 K) = 0.00184 mol

Volume of H$_2$ at STP = 0.00184 mol × 22.4 L/mol = 0.0413 L or **41.3 mL**

1c)   This is actually a stoichiometry problem involving volumes of gases and masses of metal. The balanced equation tells us that there will be 1 mol of H$_2$ formed for each mole of magnesium reacted, therefore

mass of Mg reacted = (0.00184 mol of H$_2$) × (1 mol of Mg / 1 mol of H$_2$) × (24.31 g Mg / 1 mol Mg) = **0.0448 g Mg**

1d)   This calculation is only valid assuming a 100% yield for the reaction, 100% purity for the starting magnesium, negligible solubility of H$_2$ in water at this temperature, and ideal behavior of all the gases at this temperature.

2a)   The kinetic molecular theory of gases states that the measurable pressure that gases exert on their container is due to collisions between the gas particles and the walls of the container. The total pressure is directly proportional to the frequency of collisions or number of collisions per unit area. By increasing the volume of the syringe, without changing the number of gas particles or their temperature, the particle density of the gas was decreased, thereby reducing the number of collisions with the container per unit area of container surface. That is, the same number of particles are spread out over a greater interior volume, thereby reducing the frequency with which they collide with the walls.

2b)   The pressure of the gas in the cylinder, as described in a), is due to collisions with the interior surface of the cylinder. In this case, the frequency of collisions per unit area remained the same, but the force with which each particle hit was increased. The kinetic energy of the particles, $KE = \frac{1}{2} mv^2$, is directly proportional to the velocity, v, of the particles (or, more appropriately, μ, their root mean square speed), and the velocity of the particles is directly proportional to the absolute temperature (for a mole of ideal gas, $KE = 3/2\ RT$). As the temperature of the gas in the cylinder increased while sitting in the sun, to approximately 130 °F or 55 °C in this case, the particles' average speed and kinetic energy increased. This resulted in more frequent and forceful collisions, resulting in increased pressure (in proportion to the absolute temperature change of the cylinder).

2c)   The kinetic molecular theory describes gases as mostly empty space (about 99.99%). The actual volume of the gas particles is considered to be insignificant compared with the overall volume of the gas. Therefore, even though krypton gas atoms are larger than those of hydrogen, the overall volume of samples of these gases containing the same number of particles at the same temperature and pressure will be the same.

3)   The first part of this problem is basically an application of Boyle's law. The initial pressure and volume of each gas are known, and the final volume of each gas, the entire volume of the system in this case, is also known. The problem then becomes one of solving for the individual partial pressures of each of the gases and then adding them together, Dalton's law, to get the final total pressure.

$V_{total} = 1.5\ L + 0.75\ L + 2.0\ L = 4.25\ L$

$P_1 \times V_1 = P_2 \times V_2 \quad \text{or} \quad P_2 = (P_1 \times V_1) / V_2$

$P_{final\ Ar} = (0.70\ \text{atm} \times 1.5\ L) / 4.25\ L = 0.247\ \text{atm}$

$P_{final\ N2} = (1.20\ \text{atm} \times 0.75\ L) / 4.25\ L = 0.212\ \text{atm}$

$P_{final\ He} = (1.0\ \text{atm} \times 2.0\ L) / 4.25\ L = 0.471\ \text{atm}$

$$\boldsymbol{P}_{\textbf{total}} = P_{\text{final Ar}} + P_{\text{final N2}} + P_{\text{final He}} = 0.247 \text{ atm} + 0.212 \text{ atm} + 0.471 \text{ atm} = \textbf{0.930 atm}$$

The second part of the problem involves solving for the number of moles of each gas that occupies the 2.0 L flask on the right, and multiplying each by the molar mass of that gas.

Mass of Ar in 2.0 L bulb = moles of Ar $\times M_{Ar} = [(P_{Ar} \times V_{Ar}) / (R \times T)] \times M_{Ar}$

Mass of Ar in 2.0 L bulb = $[(0.247 \text{ atm} \times 2.0 \text{ L}) / 0.0821 \text{ L·atm·mol}^{-1}\text{·K}^{-1} \times 295)] \times 39.95$ g/mol = 0.815 g

Mass of N_2 in 2.0 L bulb = $[(0.212 \text{ atm} \times 2.0 \text{ L}) / 0.0821 \text{ L·atm·mol}^{-1}\text{·K}^{-1} \times 295)] \times 28.01$ g/mol = 0.490 g

Mass of He in 2.0 L bulb = $[(0.471 \text{ atm} \times 2.0 \text{ L}) / 0.0821 \text{ L·atm·mol}^{-1}\text{·K}^{-1} \times 295)] \times 4.00$ g/mol = 0.156 g

Total mass of gas in bulb C = 0.815 g + 0.490 g + 0.156 g = **1.46 g**

4)   The kinetic molecular theory makes two simplifying but false assertions about the behavior of gases. One is that the actual volume of the particles of a gas is negligible compared with the volume occupied by the gas, and the other is that there are no inter-particle interactions or forces in the gas state. When gases are subjected to either high pressures or low temperatures, they begin to deviate from ideal behavior. As gases approach their boiling point, they begin to exhibit significant, attractive interparticle interactions (van der Waals forces) that cause the volume to deviate negatively from that predicted by the ideal gas equation. Also, as the particles get closer to each other, their actual volume begins to be significant compared with the overall volume that the sample occupies. The net result for something like water is that the quantity PV/RT for 1 mol of water vapor near its boiling point will be less that 1.

The van der Waals equation contains terms, including gas specific constants a and b, that correct for the inaccuracies in pressure and volume, respectively. The measured, pressure of real gases near their boiling point is usually lower than the ideally predicted pressure, and the volume is usually slightly larger.

CHAPTER 6
THERMOCHEMISTRY

This chapter reviews basic thermodynamics. The topics include:

- The basic terms associated with energy changes
- The first law of thermodynamics
- A comparison of ΔE and ΔH
- Calorimeter calculations
- Calculations using Hess' law
- Calculations using heats of formation

The basic terms associated with energy changes

Energy exists in many forms, but heat is the form of energy most commonly associated with chemical change. *Reactions that release energy to the surroundings* are referred to as **exothermic** reactions. Reactions that *absorb energy from the surroundings* are referred to as **endothermic** reactions.

The study of the energy associated with a change in the state of a system is governed by the **Law of Conservation of Energy**, which states that *energy can be converted from one form to another but cannot be created or destroyed.*

All the macroscopic properties that describe a system, such as temperature, volume, and pressure, are collectively referred to as the *state* of the system. The value of **state functions** such as ΔT, ΔE, and ΔH are *independent of the path* taken from initial state to final state. For example, in the following diagram the difference in potential energy between the bottom and the top of the mountain is the same regardless of the path taken.

The first law of thermodynamics

A system has an **internal energy, E,** that results from the *kinetic and potential energies in the system.* The reactants have their own sum of energies, E_1, and the products have their own sum of energies, E_2. Therefore, if the system undergoes a change from reactants to products, *the change in internal energy for the system,* ΔE_{system}, can be expressed as

$$\Delta E_{system} = E_2 - E_1$$

If the ΔE_{system} is exothermic, a quantity of energy is lost to the surroundings and $E_{surroundings}$ will gain that quantity of energy. *The energy in the universe is conserved*:

$$\Delta E_{system} + \Delta E_{surrounding} = 0$$

The first law of thermodynamics can be stated in equation form:

$$\Delta E = q + w \quad \text{where}$$

ΔE is the change in internal energy of the system in kJ/mol
q is the heat given off or absorbed by the system
w is the work done *by* the system or *on* the system

It is important to keep the correct sign conventions for q and w. If energy is ADDED to the system, q is positive (+) and if the system LOSES energy, q is negative (–). If work is done BY the system on the surroundings, then w is negative (–), and if work is done ON the system, w is positive (+). Work that involves the expansion or contraction of a gas is expressed as $w = -P\Delta V = -P(V_f - V_i)$.

ΔE is a state function, but q and w are not. For example, if a fully charged battery is attached to a motor, the motor will run until the battery is dead. While the motor runs, useful work and heat are produced and $\Delta E = q + w$. If the same fully charged battery has a crowbar tossed across its electrodes, it quickly discharges and goes from fully charged to dead, but produces only heat, q. In both cases, the battery has gone from fully charged to dead, and therefore the change in energy (ΔE) is the same. In the case of the motor, however, heat and work are produced whereas in the case of the crowbar, heat is produced, but no useful work is done.

A Comparison of ΔE and ΔH

A quantity closely related to ΔE is ΔH, enthalpy. The change in **enthalpy** is defined as:

$$\Delta H = \Delta E + P\Delta V$$

The form of this equation often used to calculate the change in enthalpy for a reaction is:

$$\Delta H = \Delta E + \Delta n R T \quad \text{where}$$

ΔH is the enthalpy change in kJ/mol
ΔE is the change in energy in kJ/mol
Δn is the change in the number of moles of gas
   ($\Delta n = \text{moles}_{\text{gaseous product}} - \text{moles}_{\text{gaseous reactant}}$)
R is 8.31 J/mol·K
T is the temperature in Kelvin

If the reactants and products of a reaction are in solution or there are no changes in the number of moles of gas during the reaction, the term $\Delta n R T$ drops out and $\Delta H = \Delta E$.

Depending on the experimental conditions, the calculated heat leads to either ΔE or ΔH. If the experiment is carried out in a **bomb calorimeter**, which is a *reaction vessel with a constant volume*, all energy changes are in the form of heat. When the volume is held constant, $\Delta V = 0$, the value of w is necessarily 0 (because $w = -P\Delta V$). The end result is:

$$\Delta E = q_v \quad \text{where}$$

q_v is the heat absorbed or released when the experiment is carried out at constant volume

A **Styrofoam cup calorimeter** is open and *subject to atmospheric pressure*. The heat determination in this type of calorimeter leads to ΔH values:

$$\Delta H = q_p \quad \text{where}$$

q_p is the heat absorbed or released when the experiment is carried out at constant pressure

---

Take Note: *The AP exam contains a list of constants and their values. It is important that you watch the units. R is in joules,* $\dfrac{8.31\,J}{mol \cdot K}$, *but ΔE and ΔH are usually in kilojoules/mol.*

Calorimeter calculations

Calorimeter calculations (bomb or styrofoam cup) can be presented in two ways. One calculation presents the individual components of the calorimeter from which the heat change for the reaction can be calculated: the amount of water or water solution, the specific heat of water, and the initial and final temperature measurements of the water.

$q = m$ **(specific heat)** ΔT where

q is the heat, in joules or kilojoules, absorbed or released by the reaction
m is the amount of water or water solution, usually in grams
specific heat for water (and for dilute aqueous solutions) is 4.184 J/g·°C
ΔT is the temperature change in °C or K

The second calculation presents a **calorimeter constant**, *the heat capacity*, for the entire calorimeter and the initial and final temperature:

$q = $ **(calorimeter constant*)** ΔT where

calorimeter constant*
q is the heat in joules or kilojoules absorbed or released by the reaction
ΔT is the change in temperature

*The calorimeter constant units can be expressed in $\dfrac{J}{°C}$, $\dfrac{J}{K}$, $\dfrac{kJ}{°C}$, or $\dfrac{kJ}{K}$.

Take Note: *If AP exam questions present the calorimeter constant as* $\dfrac{J}{K}$, *but give you the initial and final temperature units in Celsius, the ΔT in Celsius corresponds to the ΔT in Kelvin. The energy change involved in a 1°C temperature change is the same as for a 1° temperature change on the Kelvin scale.*

Example 1. Determination of heat of reaction in a bomb calorimeter.

A 1.435g sample of naphthalene, $C_{10}H_8$, is burned in a bomb calorimeter. The initial temperature of the water in the calorimeter was 20.28 °C. The final temperature of the water was 25.95 °C. If the heat capacity of the bomb calorimeter is 10.17 $\dfrac{kJ}{°C}$, what is the molar heat of combustion of naphthalene?

General Strategy	Solution to Example 1
Determine the temperature change in the calorimeter and determine whether the reaction is exothermic or endothermic.	$\Delta T = T_{final} - T_{initial}$ $\quad = 5.67\ °C$ Since the temperature in the calorimeter rose, this is an exothermic reaction.
Determine the heat change in the calorimeter. Assume that no heat is lost to the surroundings since this is a bomb calorimeter. $q = $ *(calorimeter constant)* ΔT	$q = $ *(calorimeter constant)* ΔT $\quad = (10.17\ \dfrac{kJ}{°C})\,(5.67\ °\,C)$ $\quad = 57.66\ kJ$ $\quad = -57.66\ kJ$ since the reaction is exothermic
The heat determined above is the heat released by the 1.435 g sample. Using the molar mass of naphthalene, calculate the molar heat of combustion.	molar mass of $C_{10}H_8 = 128\ \dfrac{g}{mol}\ C_{10}H_8$ molar heat of combustion $= -\dfrac{57.66\,kJ}{1.435\,g\ C_{10}H_8} \times \dfrac{128\,g\ C_{10}H_8}{1\,mol\ C_{10}H_8}$ $\Delta E_{combusiton} = -5.15 \times 10^3\ \dfrac{kJ}{mol}$

Take Note: *It is important to remember that for calorimeter calculations on the AP exam (especially for laboratory questions), ΔT is not a measurement, but rather a calculated value. $T_{initial}$ and T_{final} are the measurements.*

Calculations using Hess' law

Some reaction heats cannot be directly determined in a calorimeter because the reactions proceed too slowly or side reactions produce substances other than the desired compound. In this case, Hess' law can be used to determine the enthalpy change for the reaction. **Hess' law** states *that when reactants are converted to products, the change in enthalpy is the same whether the reaction takes place in one step or a series of steps.* Hess' law is based on the fact that ΔH is a state function and depends only on the initial and final state of the system. Example 2 is an illustration of the application of Hess' law:

Example 2. An application of Hess' law.

Calculate the *enthalpy change* for the reaction:

$$N_2\,(g) \;+\; O_2\,(g) \;\rightarrow\; 2NO\,(g) \qquad\qquad \Delta H = \;?$$

given the following thermochemical equations:

1) $N_2\,(g) \;+\; 2O_2\,(g) \;\rightarrow\; 2NO_2\,(g) \qquad \Delta H = \;+67.6\ kJ$

2) $2NO\,(g) \;+\; O_2\,(g) \;\rightarrow\; 2NO_2\,(g) \qquad \Delta H = \;-113.2\ kJ$

General Strategy	Solution to Example 2		
The algebraic solution requires that the equations added together must equal the net equation. In this case, the second reaction is reversed to obtain the net equation. You change the *sign* of ΔH of a reaction when you *reverse* its direction.	$N_2\,(g) \;+\; 2O_2\,(g) \;\rightarrow\; \cancel{2NO_2\,(g)}$ $+ \underline{\cancel{2NO_2\,(g)} \;\rightarrow 2NO\,(g) \;+\; O_2\,(g)}$ $N_2\,(g) \;+\; O_2\,(g) \;\rightarrow\; 2NO\,(g)$	$\Delta H = \;+67.6\ kJ$ $\Delta H = \;+113.2\ kJ$ $\Delta H = \;+180.8\ kJ$	

Calculations using heats of formation

The **Heat of Formation Table** provides yet another way in which the enthalpy of a reaction can be determined. The **heat of formation** refers to *the energy absorbed or released when 1 mol of a compound is formed from its elements at 1 atm and 25 °C. The heats of formation of elements are defined as zero at these conditions.* Since the enthalpy change (ΔH) is a state function, its value depends only on the initial and final states. It can be calculated by:

$$\Delta H^0_{reaction} = \Sigma\,\Delta H^0_{f\,(products)} - \Sigma\,\Delta H^0_{f\,(reactants)} \quad \text{where}$$

$\Sigma\,\Delta H^0_{f\,(products)}$ is the sum of the standard heats of formation of the products in kJ/mol

$\Sigma\,\Delta H^0_{f\,(reactants)}$ is the sum of the standard heats of formation of the reactants in kJ/mol

When performing calculations using the Heat of Formation Table, be sure you watch the units. The Heat of Formation Table units are $\dfrac{kJ}{mol}$. If more than 1 mole of reactant or product is involved, then the heat of formation must be multiplied by the number of moles, or **mol** $\times \dfrac{kJ}{mol}$.

Example 3. Using heats of formation to determine the enthalpy change of a reaction.

Pentaborane-9, B_5H_9, is a colorless, highly reactive liquid that explodes when exposed to oxygen. The equation for this combustion is:

$$2B_5H_9\,(l) \;+\; 12O_2\,(g) \;\rightarrow\; 5B_2O_3\,(s) \;+\; 9H_2O\,(l)$$

Using the following standard heats of formation, ΔH_f^0, determine the kilojoules of heat produced by this reaction.

$$B_2O_3(s) \;=\; -1263.6\;\dfrac{kJ}{mol}$$

$$H_2O\,(l) \;=\; -285.5\;\dfrac{kJ}{mol}$$

$$B_5H_9\,(l) \;=\; 73.2\;\dfrac{kJ}{mol}$$

General Strategy	Solution to Example 3
The data provided is heats of formation, ΔH_f. The appropriate equation is: $$\Delta H^0_{reaction} = \Sigma\,\Delta H^0_{f\,(products)} - \Sigma\,\Delta H^0_{f\,(reactants)}$$	$$\Delta H^0_{reaction} = \Sigma\,\Delta H^0_{f\,(products)} - \Sigma\,\Delta H^0_{f\,(reactants)}$$ $$\Delta H^0_{rxn} = [5\Delta H^0_{f\,(B_2O_3)} + 9\Delta H^0_{f\,(H_2O)}] - [2\Delta H^0_{f\,(B_5H_9)} + 12\Delta H^0_{f\,(O_2)}]$$
The units are $\dfrac{kJ}{1\,mol}$. Remember to multiply by the number of moles of each reactant or product.	$\Delta H_{reaction} = [\,5(-1263.6\,\dfrac{kJ}{mol}) + 9(-285.5\,\dfrac{kJ}{mol})\,] -$ $[2(73.2\,\dfrac{kJ}{mol}) + 12(0\,\dfrac{kJ}{mol})]$ $= -9040\;kJ$ *are released when $\underline{2}$ mol of B_5H_9 burn*

Take Note: *The problems on the AP exam will provide either the values for the heats of formation in the problem or sufficient thermodynamic information for you to calculate their values. Remember that elements at 25 °C and 1 atm are defined as having a heat of formation,* ΔH_f^0, *of zero.*

SAMPLE MULTIPLE CHOICE QUESTIONS

1. The specific heat of an unknown metal is $\dfrac{0.230 J}{g\,^\circ C}$. A 45.0 g sample of the metal is heated from 25.0 °C to 58.0 °C. How much heat is required to heat the metal?

    A. 10.4 J
    B. 13.3 J
    C. 600. J
    D. 259 J
    E. 342 J

2. A 5.30 g sample of gold is heated from 20.0 °C to 35.0 °C and absorbs 2.07 J of heat. Determine the specific heat of gold.

    A. .077 $\dfrac{J}{g\,^\circ C}$

    B. 10.6 $\dfrac{J}{g\,^\circ C}$

    C. 0.932 $\dfrac{J}{g\,^\circ C}$

    D. .0260 $\dfrac{J}{g\,^\circ C}$

    E. 30.1 $\dfrac{J}{g\,^\circ C}$

3. A reaction is carried out in a constant pressure calorimeter. Appropriate measurements are taken and the energy for the change is calculated. The quantity of heat exchanged under these conditions leads to the calculation of which state function?

    A. Δq
    B. ΔH
    C. ΔE
    D. ΔT
    E. Δw

4. A balloon expands isothermally from 1.0 L to 10.0 L against a constant external pressure of 7.0 atm. The work done by the system is:

A. -63 L·atm
B. $+63$ L·atm
C. -70 L·atm
D. $+70$ L·atm
E. cannot be calculated

5. The following reaction is carried out in a bomb calorimeter:

$$4NH_3(g) \ + \ 5O_2(g) \ \rightarrow \ 4NO(g) \ + \ 6H_2O(g)$$

The calorimeter constant is $\dfrac{45.0\,kJ}{K}$. When 34.0 g of NH_3 are reacted in the calorimeter, the temperature rises by 10.0 °C. What is the change in energy, ΔE, for the reaction?

A. -90.0 kJ
B. -450 kJ
C. -900 kJ
D. -625 kJ
E. -560 kJ

6. Consider the following heats of combustion:

$$CH_3OH\,(l) \ + \ \frac{3}{2}O_2(g) \ \rightarrow \ CO_2(g) \ + \ 2H_2O\,(l) \qquad \Delta H° \ = \ -730\,\frac{kJ}{mol}$$

$$C\,(graphite) \ + \ O_2(g) \ \rightarrow \ CO_2(g) \qquad\qquad\qquad \Delta H° \ = \ -390\,\frac{kJ}{mol}$$

$$H_2(g) \ + \ \tfrac{1}{2}O_2(g) \ \rightarrow \ H_2O\,(l) \qquad\qquad\qquad\qquad \Delta H° \ = \ -290\,\frac{kJ}{mol}$$

Determine the enthalpy of formation of methanol, $CH_3OH\,(l)$.

A. -240 kJ/mol
B. $+730$ kJ/mol
C. $+50$ kJ/mol
D. -1400 kJ/mol
E. -680 kJ/mol

7.   Two reactants are mixed in a bomb calorimeter and a reaction occurs. The temperature of the calorimeter rises. Which of the following statements regarding the reaction is true?

A. The reaction is absorbing heat.
B. The reaction is releasing heat.
C. The reaction is endothermic.
D. The reaction heat is equal to $q\Delta T$.
E. A and C are correct.

8.   Consider the following reaction:

$$C_3H_8 (g) \;+\; 5O_2 (g) \;\rightarrow\; 3CO_2 (g) \;+\; 4H_2O (g) \qquad \Delta H = -2440 \text{ kJ}$$

The heat of formation (ΔH_f^0) of CO_2 is $-390 \dfrac{kJ}{mol}$ and the heat of formation (ΔH_f^0) of H_2O is $-240 \dfrac{kJ}{mol}$. What is the heat of formation of propane (C_3H_8)?

A. +2440 kJ/mol
B. −2440 kJ/mol
C. 0 kJ/mol
D. +310 kJ/mol
E. −310 kJ/mol

9.   Consider the following reaction:

$$2C_2H_6 (g) \;+\; 7O_2 (g) \;\rightarrow\; 4CO_2 (g) \;+\; 6H_2O (l) \qquad \Delta H_{rxn} = -3120. \text{ kJ}$$

If 3.00 g of ethane are burned, how many kJ of heat are produced?

A. −156 kJ
B. −312 kJ
C. −1560 kJ
D. −624 kJ
E. −752 kJ

10. Consider the equation for photosynthesis:

$$6CO_2 (g) \quad + \quad 6H_2O (g) \quad \rightarrow \quad C_6H_{12}O_6 (s) \quad + \quad 6O_2 (g) \qquad \Delta H = +43.0 \ \frac{kJ}{mol}$$

If 4.00 kJ of heat are added, how many liters of oxygen at STP are produced?

A. 0.10 L
B. 22.4 L
C. 5.6 L
D. 12.5 L
E. 19.1 L

Comprehension Questions

1) Consider the following reaction representing the combustion of sulfur dioxide gas:

$$2SO_2 (g) + O_2 (g) \rightarrow 2SO_3 (g)$$

a) Calculate ΔH^0_{rxn} for this process given the following information:

Species	$\Delta H°_f$ (kJ/mol)
$SO_2 (g)$	−296.1
$O_2 (g)$	0
$SO_3 (g)$	−395.2

b) Is this reaction exothermic or endothermic?

c) Calculate the amount of heat liberated in the above process when 2.075 g of $SO_3 (g)$ is produced.

d) What volume, in liters, of $SO_3(g)$ would be produced at 20.0 °C and 1.00 atm of pressure in the above combustion reaction involving the transfer of 50.00 kJ of heat?

2) Given the following data:

$$CO (g) + H_2O (g) \rightarrow CO_2 (g) + H_2 (g) \qquad \Delta H^0 = -41.4 \ kJ$$

$$H_2 \, (s) + \tfrac{1}{2} \, O_2 \, (g) \rightarrow H_2O \, (g) \qquad\qquad \Delta H^0 = -\,241.8 \text{ kJ}$$

$$C \, (s) + O_2 \, (g) \rightarrow CO_2 \, (g) \qquad\qquad \Delta H^0 = -\,393.5 \text{ kJ}$$

a) Calculate ΔH^0_{rxn} for the following process used in the formation of water gas (a mixture of H_2 and CO):

$$C \, (s) + H_2O \, (g) \rightarrow CO \, (g) + H_2(g)$$

b) Is the formation of water gas from hydrogen and coke, C, endothermic or exothermic? What does this indicate about the relative amounts of heat energy or enthalpy contained in the reactants and products?

3) The combustion of methane, $CH_4 \, (g)$, the primary component of natural gas, is an exothermic process that is commonly used as a heat source for cooking, drying clothes, heating water, and domestic heat.

$$CH_4 \, (g) + 2O_2 \, (g) \leftrightarrow CO_2 \, (g) + 2H_2O \, (g) \qquad\qquad \Delta H = -\,890.4 \text{ kJ}$$

a) Is the above reaction exothermic or endothermic?

b) What would be the enthalpy change, in joules, associated with the formation of 10.00 L of carbon dioxide gas, measured at 754 mm Hg and 21.0 °C, by the above process?

c) How much heat, in kilojoules, would be evolved or consumed during the formation of 310.0 g of methane from CO_2 and H_2O?

d) How much CH_4, in grams, would be necessary to heat 153 L of water, the amount commonly found in a domestic water heater, from 22 °C to 55 °C? Assume insignificant loss of heat to the surroundings from a well-insulated water heater, and a density of 1.00 g/cm^3 for tap water.

4)   A student was asked to determine the enthalpy change for a chemical reaction using a constant pressure calorimeter, that is, a polystyrene cup, and an ordinary lab thermometer. The reaction under consideration is the neutralization of ammonium chloride by sodium hydroxide.

$$NH_4Cl + NaOH \rightarrow NH_3 + NaCl + H_2O$$

a)   Write the net ionic equation for the above neutralization reaction.

b)   In an effort to collect accurate data, the student decided to first measure the specific heat of the calorimeter. To do this, she added 50.0 g of room temperature water to the foam cup; both were at the same temperature, 21.2 °C. She then added 50.0 g of warm water, 78.5 °C, to the contents of the cup. The highest temperature that the mixture achieved, after thorough mixing, was 48.5 °C. What is the heat capacity of the calorimeter in joules?

c)   The student then added 50.0 mL of 2.00 M NH_4Cl solution, (density = 1.04 g/mL) to the dried calorimeter followed by 50.0 ml of 2.00 M NaOH solution, (density = 1.02 g/mL). The initial temperatures of both solutions and the cup were 22.3 °C. The final maximum temperature of the reaction mix, after stirring, was 23.5 °C. Calculate the molar heat of neutralization of ammonium chloride by sodium hydroxide. Assume that the specific heat of these acid and base solutions is equal to that of water, $4.184 \text{ J·g}^{-1}\text{·°C}^{-1}$.

d)   Was measurement of the cup's heat capacity necessary? Explain.

ANSWERS TO MULTIPLE CHOICE QUESTIONS

1.   E

Since the amount of material, specific heat, and temperatures are given, the amount of heat is calculated according to the equation:

$q = m$ (specific heat) ΔT

$= \quad (45.0 \text{ g}) \, (\dfrac{0.23\,J}{g\cdot°C}) \, (58.0\text{ °C} - 25.0\text{ °C})$

$= \quad 342\ kJ$

2.   D

Since the amount of heat, material, and temperatures are supplied, the specific heat can be calculated using the equation:

$$q = m \text{ (specific heat) } \Delta T$$

$$2.07 \text{ J} = (5.30 \text{ g}) \text{ (specific heat) } (35.0 \text{ °C} - 20.0 \text{ °C})$$

$$\text{specific heat} = \frac{2.07 J}{(5.30 \, g)(15.0 \, °C)}$$

$$= .0260 \frac{J}{g \cdot °C}$$

3.   B

The heat determined from the measurements taken from a constant pressure calorimeter is the enthalpy, ΔH.

4.   A

Since work $(-P\Delta V)$ is done by the system, energy is lost by the system:

$$w = -P\Delta V = -P (V_f - V_i)$$

$$= -(7.0 \text{ atm}) (10.0 \text{ L} - 1.0 \text{ L})$$

$$= -63 \text{ } L \cdot atm$$

This expansion is carried out isothermally, which means that the initial and final temperatures are the same. Since the system has expended energy, $w = -63$ L·atm, it must have absorbed the same amount of heat, $q = +63$ L·atm.

5.   C

The solution to this problem requires two steps: 1) determine the heat exchanged using the calorimeter measurements; and 2) adjust the heat to the stoichiometry of the reaction.

1) q = calorimeter constant × temperature change

$$q = (\frac{45.0\,kJ}{K})\ (10.0\ ^{\circ}C)$$

$$= 450.\ kJ$$

2) $34.0\ g\ NH_3 \times \dfrac{1\,mol}{17.0\,g} = 2.00\ mol\ NH_3$ used in the calorimeter

$\dfrac{450.\,kJ}{2.00\,mol} \times 4.00\ mol\ NH_3$ (coefficient of NH_3 in equation) $=\ 900\ kJ$

$\Delta H_{rxn} = -900\ kJ$ The temperature of the calorimeter rose so the reaction is exothermic.

6.   A

The heat of formation refers to the amount of heat involved when 1 mol of product forms from its elements at standard conditions. As a first step, write the equation for the formation of methanol.

$$C\ (graphite)\ +\ 2H_2\ (g)\ +\ \tfrac{1}{2}O_2(g)\ \rightarrow\ CH_3OH\ (l)$$

Then apply Hess' law to obtain the solution.

$\cancel{CO_2(g)}\ +\ \cancel{2H_2O}\ (l)\ \rightarrow\ CH_3OH\ (l)\ +\ \dfrac{3}{2}O_2\ (g)\quad \Delta H^0\ =\ +730\dfrac{kJ}{mol}$

$C\ (graphite)\ +\ O_2\ (g)\ \rightarrow\ \cancel{CO_2(g)}\qquad\qquad\qquad \Delta H^0\ =\ -390\ \dfrac{kJ}{mol}$

$\underline{*2(H_2\ (g)\ +\ \tfrac{1}{2}O_2\ (g)\ \rightarrow\ \cancel{H_2O\ (l)})\qquad\qquad\qquad \Delta H^0\ =\ 2(-290\dfrac{kJ}{mol})}$

$C\ (graphite)\ +\ 2H_2(g)\ +\ \tfrac{1}{2}O_2(g)\ \rightarrow\ CH_3OH\ (l)\quad \Delta H^0_{f}\ =\ -240\ kJ/mol$

7.   B

Since the temperature of the calorimeter is rising, the reaction is releasing heat. This is an exothermic reaction.

8. D

This problem involves the heats of formation:

$$\Delta H^0_{reaction} = \Sigma \Delta H^0_{f\,(products)} - \Sigma \Delta H^0_{f\,(reactants)}$$

$$\Delta H_{reaction} = [3\Delta H_{f\,(CO_2)} + 4\Delta H_{f\,(H_2O)}] \;-\; [5\Delta H_{f(O_2)} + 1\Delta H_{f(C_3H_8)}]$$

$$-2440 \text{ kJ} = [3(-390\;\frac{kJ}{mol}) + 4(-240\;\frac{kJ}{mol})] \;-\; [5(0) + x]$$

$$x = +310 \text{ kJ}$$

$$\Delta H^0_f = +310\;\frac{kJ}{mol}\,C_3H_8$$

9. A

The solution involves the stoichiometry of the thermochemical equation.

$$3.00 \text{ g } C_2H_6 \;\times\; \frac{1\,mol}{30.0\,g}C_2H_6 \;\times\; \frac{-3120.\,kJ}{2\,mol\,C_2H_6} \;=\; -156\,kJ$$

10. D

The solution involves the stoichiometry of the thermochemical equation

$$4.00 \text{ kJ} \;\times\; \frac{6\,mol\,O_2}{43.0\,kJ\cdot mol^{-1}} \;\times\; \frac{22.4\,L}{1\,mol\,O_2\,STP} \;=\; 12.5\,L\,O_2$$

Answers to Comprehension Questions

1)   a)   The standard enthalpy change for a reaction, ΔH^0_{rxn}, can be calculated for a reaction using the formula $\Delta H^0_{rxn} = \Sigma \Delta H^0_{f\,(products)} - \Sigma \Delta H^0_{f\,(reactants)}$. Substituting:

ΔH^{0}_{rxn} = [2 mol × (−395.2 kJ/mol)] − [(2 mol × (−296.1 kJ/mol)) + (1 mol × 0 kJ/mol)] = − **198.2 kJ.**

b)   This reaction has a negative enthalpy change, which corresponds with an **exothermic** reaction.

c)   The enthalpy change associated with the reaction of a specific amount of a reactant or product can be calculated by using the ratio of heat generated to amounts of material being reacted in the balanced chemical equation. In this specific case, the coefficients in the balanced equation not only represent the ratios of moles that react but also the actual amounts of matter, in moles, that are reacting to produce the heat change noted, that is, ΔH^{0}_{rxn}. In this case, 198.2 kJ of heat is liberated when 2 mol or 160.13 g of SO_3 are produced.

2.075 g SO_3 × (198.2 kJ / 160.13 g SO_3) = **2.568 kJ of heat will be released**

d)   This calculation is closely related to that in part c except this time you're given the heat change and asked to calculate the amount of product. The question asks for the answer in liters of gas, so the amount of material must be determined in moles and then the volume calculated using the ideal gas law, $PV = nRT$.

50.00 kJ × (2 mol SO_3/ 198.2 kJ) = 0.5045 mol of SO_3

For calculations involving gases, the Kelvin temperature scale must be used, so

Kelvin = 20.0 °C + 273.15 = 293.2 K

$V = nRT/P$ = (0.5045 mol × 0.08206 L·atm·mol^{-1}·K^{-1} × 293.2 K) / 1.00 atm = **12.1 L of SO_3**

2)   a)   This is a Hess' law calculation in which the sum of the enthalpy changes for a series of interconnected reactions equals the enthalpy change of a single reaction that represents the net chemical change of all the chemical reactions combined. In this particular case you must rearrange two of the reactions for which the enthalpy change is known. The first reaction given must be reversed and the second reaction, in which a mole of water is formed from its elements, must be reversed and the amounts doubled as follows:

CO_2 (g) + H_2 (g)→ CO (g) + H_2O (g) ΔH^0 = + 41.4 kJ

$2H_2O$ (g) → O_2 (g) + 2 H_2 (g) ΔH^0 = −2 × (−241.8 kJ) = 483.6 kJ

Note that the new enthalpy change for the first reaction is opposite in sign from that originally given because it has been reversed, and, the rearranged second reaction is the original value multiplied by –2 (the negative sign is for reversing the reaction and the 2 is because the quantities are doubled). All of the reactions can then be combined and their enthalpy changes added together to provide the overall enthalpy change for the process.

$$CO_2\ (g) + H_2\ (g) \rightarrow CO\ (g) + H_2O\ (g) \qquad\qquad \Delta H^0 = +\ 41.4\ kJ$$

$$C\ (s) + O_2\ (g) \rightarrow CO_2\ (g) \qquad\qquad\qquad\qquad \Delta H^0 = -\ 393.5\ kJ$$

$$2\ H_2O\ (g) \rightarrow O_2\ (g)\ +\ 2\ H_2\ (g) \qquad\qquad\qquad \Delta H^0 = +\ 483.6\ kJ$$

_____

$$\textbf{C\ (s) + H\textsubscript{2}O\ (g)} \rightarrow \textbf{CO\ (g) + H\textsubscript{2}\ (g)} \qquad\qquad \Delta H^0 = \ +\ \textbf{131.5\ kJ}$$

   b)   This reaction is endothermic, positive ΔH^0, which indicates that the products contain more heat energy than the reactants had, thus the positive flow of heat from the surroundings into the system during the reaction. The energy that flowed into the system, along with the energy initially present in all of the bonds, is now stored in the form of chemical potential energy.

3)   a)   The reaction must be **exothermic** because we see that the enthalpy change is a negative number.

   b)   The essence of finding a solution to this problem is to remember that in thermochemical equations the stated enthalpy change, –890.4 kJ in this case, correlates with the exact amount of material noted in the balanced equation; the coefficients in the balanced equation are identifying not only mole ratios that react but amounts of matter reacting as well. In this particular case the problem gives enough information to calculate, using the ideal gas equation, the number of moles of product desired. The heat change associated with this amount of matter can be calculated using the values from the thermochemical equation in a proportional manner.

$PV = nRT$ or, rearranging, $n = PV\ /\ RT$

T (in Kelvin) $= 21.0 + 273.15 = 294.2$ K

P (in atmospheres) $= 754$ mm Hg $\times$ (1 atm / 760 mm Hg) $= 0.992$ atm

$n = (0.992$ atm $\times\ 10.00$ L$)\ /\ (0.08206$ L·atm·mol^{-1}·K$^{-1} \times 294.2$ K$) = 0.411$ mol of CO_2

From the thermochemical equation:

$\Delta H = 0.411$ mol of $CO_2 \times (- 890.4$ kJ / 1 mol $CO_2) = $ **−366 kJ**

c)   This part of the problem can be solved in a similar fashion to part b. The one major change is that the problem asks for the enthalpy change that occurs during the formation of methane, but the thermochemical equation is written as a combustion reaction. Since this is the reverse process to that found in the equation, the sign of the enthalpy change must be changed.

310.0 g $CH_4 \times$ (1 mol CH_4 / 16.04 g CH_4) $\times$ (+890.4 kJ / 1 mol CH_4) = **1.721 $\times$ 10^4 kJ of heat would be required to form this quantity of methane according to this reaction.**

d)   Solution of this part of the question involves the determination of an amount of energy needed to heat 153 L of water from 22 °C to 55 °C, and then substituting that amount of energy into a calculation similar to parts b and c to calculate the amount of methane needed to provide this much energy. The heat absorbed by the water can be calculated by using the relationship $q = m \cdot s \cdot \Delta t$ (q is heat flow in joules, m is mass in grams, s is the specific heat of the substance in $J \cdot g^{-1} \cdot °C^{-1}$, and Δt is the temperature change of the substance in °C).

The mass of the water can be determined using its volume and density,

$m_{H_2O} = 153$ L $H_2O \times$ (1000 cm^3 / L) $\times$ (1.00 g / cm^3) = 1.53×10^5 g H_2O
The temperature change, Δt, is simply the initial temperature subtracted from the final.

$\Delta t = t_{final} - t_{initial} = 55$ °C $- 22$ °C $= 33$ °C

$q = 1.53 \times 10^5$ g $H_2O \times 4.184$ $J \cdot g^{-1} \cdot °C^{-1} \times 33$ °C $= 2.11 \times 10^7$ J or 2.11×10^4 kJ of heat is needed to raise the water from 22 °C to 55 °C

The amount of methane needed can then be calculated,

2.11×10^4 kJ $\times$ (1 mol CH_4 / 890.4 kJ) $\times$ (16.04 g CH_4 /1 mol CH_4) = **380. g CH_4**

In this calculation, a positive value for the enthalpy change was used because we are interested in the magnitude of the energy change only for this calculation. The sign convention for the thermochemical equations provides information about the relative amounts of energy in reactants and products. Including it here would lead to a negative mass of methane, which is clearly an impossibility.

4)   a) $NH_4^+ + OH^- \rightarrow NH_3 + H_2O$

b)   In this case heat will flow from the warm water to the cold water and the cold cup, heat spontaneously flows from warmer objects to colder ones. If the cup absorbed no heat, then the final temperature of the mixed water should be at the exact mid-point between their starting temperatures, since there were equal amounts of hot and cold water. If the mixture temperature is below the mid-point of the hot and cold water temperatures, then the cup must have absorbed some heat from the water, assuming no loss of heat to the lab surroundings. Based on the difference between the mixture's actual temperature and this theoretical mid-point temperature, you can calculate the amount of heat absorbed by the cup; it is equal to the amount of heat lost by the water. Heat flow, q, will be calculated as in 3d.

difference in water temperatures = 78.5 °C – 21.2 °C = 57.3 °C / 2 = 28.65 °C

mid-point temperature = 21.2 °C + 28.65 °C = 49.9 °C

actual measured mixture temperature = 48.5 °C

Since the mixed hot and cold water came up 1.4 °C short of the theoretical maximum, you can treat this situation as if the cup had absorbed the heat released when 100.0 g of water cooled from 49.9 °C to 48.5 °C.

q = 100.0 g H_2O × 4.184 J·g^{-1}·°C^{-1} × 1.4 °C = 586 J of heat were absorbed by the cup
This heat absorbed by the cup was associated with a temperature change for the cup of 48.5 °C – 21.2 °C = 27.3 °C. Its heat capacity is therefore the amount of heat absorbed divided by the temperature or

$C_{calorimeter}$ = 586 J / 27.3 °C = **21.5 J/°C**

c)   Since the final temperature of this reaction mixture is higher than the initial temperatures of the starting solutions, the reaction must be exothermic. The amount of heat given off by the reaction is equal to the heat absorbed by the solutions and by the calorimeter. The heat capacity of the solutions is equal to their combined mass times their specific heat. The total amount of heat absorbed by the solutions and container will be equal to the sum of their heat capacities multiplied by the temperature change.

mass of solutions = 100.0 mL × 1.04 g/mL = 104 g

heat capacity of the solutions = 104 g × 4.184 J·g^{-1}·°C^{-1} = 435 J/°C

temperature change of the solutions and cup = 23.5 °C – 22.3 °C = 1.2 °C

heat absorbed by the cup and solutions = (21.5 J/°C × 1.2 °C) + (435 J/°C × 1.2 °C) = 548 J.

548 J is the amount of heat gained by the cup and solutions, but it's also the amount of heat lost by the reaction. In this case 0.10 mol of acid and base were reacting (0.0500 L × 2.00 M = 0.10 mol). Dividing the amount of heat lost by the reaction by the number of moles reacting will provide the molar heat of neutralization or ΔH for this particular reaction.

548 J / 0.100 mol = 5480 J/mol or **5.48 kJ/mol**

d) In this case the measurement of the cup's heat capacity was significant. It corresponded to $(21.5/456) \times 100\%$ or 4.7% of the total heat capacity of the system under study. The values for molar heat of neutralization were measured to greater precision than ± 5%.

CHAPTER 7
QUANTUM THEORY AND
THE ELECTRONIC STRUCTURE OF ATOMS

This chapter reviews the development of modern atomic theory. The major topics relevant to the development of the atomic model include:

- The quantum theory of radiation
- The Bohr model of the atom
- The failure of the Bohr model: de Broglie and Heisenberg
- The Schrödinger quantum mechanical model of the atom
- The energy levels of the hydrogen atom
- Electron configurations, Hund's rule, and the Aufbau principle

The quantum theory of radiation

Light has a wave nature and a particle nature. The wave nature of light is expressed by the equation:

$$c = \upsilon\lambda \text{ where}$$

c is the speed of light, 3.0×10^{10} cm/sec, which is a constant
υ is the frequency of the radiation in Hertz or $\sec^{-1}$
λ is the wavelength of the radiation, usually expressed in nanometers

The particle nature of radiation, or the quantum theory, states that radiant energy is emitted by atoms and molecules in discrete amounts, **quanta**, rather than over a continuous range. The mathematical formulation of this concept is contained in the equation:

$$E = h\upsilon, \text{ where}$$

h is Planck's constant, 6.63×10^{-34} J·sec
υ is the frequency of the radiation in Hertz or $\sec^{-1}$
E is the energy associated with that specific frequency

This equation calculates the energy of one photon at a specified frequency.

Take Note: *The AP exam will occasionally ask you to apply one of these two equations for light. The formulas and constants for both the wave theory and the quantum theory are provided on the AP exam. However, it is important to be careful with the arithmetic and the units.*

The Bohr model of the atom

The evidence that light is quantized was employed by Niels Bohr to explain the flame test and the emission spectra of elements such as hydrogen. The emission spectra are the bright-line spectra produced by a gaseous element. Bohr developed a model of the hydrogen atom in which the energy of its single electron is quantized; that is, limited to certain energy values determined by an integer (n = 1, 2, 3,....), the **principal quantum number** (see diagram).

An *electron in its most stable state* (n = 1 for hydrogen) is said to be in the **ground state**. Bohr proposed that when energy is supplied to the hydrogen atom, the electron absorbs the energy and *moves to a higher quantum level* referred to as an **excited state**. When the electron drops from the higher energy state (excited state) to the lower, more stable energy state (ground state), the atom emits a specific amount of energy in the form of a photon that accounts for the specific lines of light in the emission spectrum.

Fig. 7.9

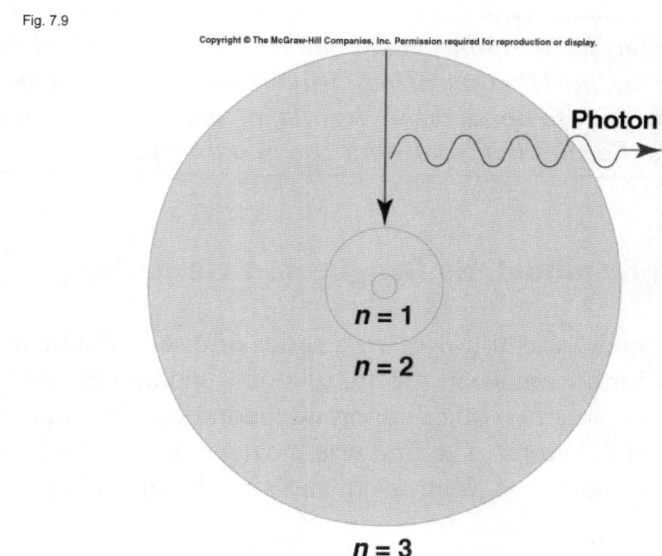

The Bohr model allows you to calculate the energy of the photon emitted by a particular electron falling from an excited to the ground state in hydrogen by the equation:

$$E = -R_H \left(\frac{1}{n^2} \right) \text{ where}$$

R_H is the constant 2.18×10^{-18} J and
n is the principal quantum level in which the electron is residing

Therefore, for the transition of the electron from the third level to the second level of hydrogen, the energy of the photon emitted can be calculated as follows:

$$\Delta E = E_{\text{final}} - E_{\text{initial}}$$

or

$$\Delta E = -R_H \left(\frac{1}{n_2^2} - \frac{1}{n_3^2} \right)$$

$$= -2.18 \times 10^{-18} \text{ J} \left(\frac{1}{2^2} - \frac{1}{3^2} \right)$$

$$= -3.03 \times 10^{-19} \text{ J}$$

The wavelength of this photon can also be calculated. Since $\Delta E = h\upsilon$ and $c = \upsilon\lambda$, then

$$\lambda = \frac{ch}{\Delta E} = \frac{(3.00 \times 10^8 \, m/sec)(6.63 \times 10^{-34} \, J \cdot sec)}{3.03 \times 10^{-19} \, J}$$

$$\lambda = 6.56 \times 10^{-7} \text{m} \times \frac{10^9 \, nm}{1 \, m} = 656 \text{ nm}$$

Take Note: *The equations for the Bohr model and the constants associated with these equations are provided on the AP exam. There is no reason for you to memorize them, but you must know how to apply them. It is important to be careful with the arithmetic and the units. Be sure that your units are in agreement.*

The failure of the Bohr model: de Broglie and Heisenberg

The success of the Bohr model was followed by a series of disappointments. The Bohr model could not account for the emission spectra of atoms containing more than one electron. In addition, there were two other important theoretical developments that affected the Bohr model of the atom. The first was the development by **Louis de Broglie** of an equation for the wave nature of electrons to explain why the energies of hydrogen electrons were quantized:

$$\lambda = \frac{h}{mv}$$

where λ is wavelength, h is Planck's constant, m is mass, and v is velocity of the moving mass. The implication of the de Broglie equation is that the electron in Bohr's atom could not be traveling at a fixed distance from the nucleus because it had a wave nature. The second theoretical relationship with implications for the Bohr atom was the **Heisenberg uncertainty principle**: *it is impossible to simultaneously know with certainty both the momentum p (sometimes defined as mass times velocity) and the position of a particle.* The Heisenberg uncertainty principle is expressed by the equation:

$$\Delta x \cdot \Delta p \geq \frac{h}{4\pi}$$

where Δx and Δp are the uncertainties in measuring position and momentum, respectively. Applying the Heisenberg uncertainty principle to the hydrogen atom results in the recognition that the electron does not orbit the nucleus in a well-defined path as the Bohr model of the atom predicted.

The Schrödinger quantum mechanical model of the atom

Bohr contributed significantly to the development of atomic theory, especially the idea that the energy of the electrons in the atom is quantized. However, his theory did not provide a complete description of the electronic behavior of atoms. Erwin Schrödinger used a complicated mathematical technique to formulate an equation that describes the behavior and energies of subatomic particles.

The solution of Schrödinger's complicated **quantum mechanical** equation for hydrogen produced a *picture of the atom with many possible energy states for the electron and the probability of its location in a particular region surrounding the nucleus*. These results could also be applied with reasonable accuracy to many-electron atoms.

Three numbers, derived from the mathematical solution of Schrödinger's equation for hydrogen, describe the distribution of electrons in hydrogen and other atoms. They are the

> **principal quantum number,** n
> the **angular momentum quantum number,** ℓ, and
> the **magnetic quantum number,** m_ℓ

A fourth quantum number, the **spin quantum number,** m_s, describes a specific electron and completes the picture. The significance and values of the quantum numbers are as follows:

- The **principal quantum number,** n, has integral values 1, 2, 3,…. *This number relates to the average distance of the electron from the nucleus.* The larger the value of n, the larger the average distance of the electron from the nucleus.

- The **angular momentum quantum number,** ℓ, *tells the shape of the orbitals.* For a given n, ℓ has integer values from 0 to $(n-1)$. The figure below relates the value of ℓ to the subshell.

ℓ	0	1	2	3	4	5
Name of orbital	s	p	d	f	g	h

- The **magnetic quantum number,** m_ℓ, *describes the orientation of the orbitals in space* and depends on the value of the angular momentum quantum number. For a certain value of ℓ, m_ℓ has all the integer values from $-\ell$ to $+\ell$, including 0. So if $\ell = 1$, m_ℓ has the values $-1, 0, 1$.

- The final quantum number, **the spin quantum number, m_s,** has only two possible values: $+\frac{1}{2}$ or $-\frac{1}{2}$.

(See Table 7.2 for a summary of the relationships among the quantum numbers.)

The four quantum numbers n, ℓ, m_ℓ, and m_s, provide the "address" of an electron in the atom. For example, for an electron in the 2s orbital, the quantum numbers are $n = 2$, $\ell = 0$, $m_\ell = 0$, and $m_s = +\frac{1}{2}$ or $-\frac{1}{2}$. Therefore, the two possible sets of quantum numbers to describe the 2s electrons are $(2, 0, 0, +\frac{1}{2})$ and $(2, 0, 0, -\frac{1}{2})$. Notice that the n, ℓ, and m_ℓ numbers are the same since the two electrons are in the same orbital. Since no two electrons can have the same set of four quantum numbers, the spin quantum number differentiates the two sets. The **Pauli exclusion principle** states that *no two electrons in the same atom can have the same four quantum numbers.* If the electrons are in the same orbital, they *must* have opposite spins.

Table 7.2 summarizes the quantum numbers and their interrelationships:

TABLE 7.2

Relation Between Quantum Numbers and Atomic Orbitals

n	ℓ	m_ℓ	Number of Orbitals	Atomic Orbital Designations
1	0	0	1	$1s$
2	0	0	1	$2s$
	1	$-1, 0, 1$	3	$2p_x, 2p_y, 2p_z$
3	0	0	1	$3s$
	1	$-1, 0, 1$	3	$3p_x, 3p_y, 3p_z$
	2	$-2, -1, 0, 1, 2$	5	$3d_{xy}, 3d_{yz}, 3d_{xz}, 3d_{x^2-y^2}, 3d_{z^2}$
⋮	⋮	⋮	⋮	⋮

The single *s orbital* for each energy level is *spherical* and centered on the nucleus. The three *p orbitals* are present at $n = 2$ and higher; *each orbital has two lobes*, and the pairs of lobes are arranged at right angles to each other. Starting with $n = 3$, there are five *d* orbitals with more complex shapes and orientations. The shapes of the *s*, *p*, and *d* orbitals are shown in the following figure:

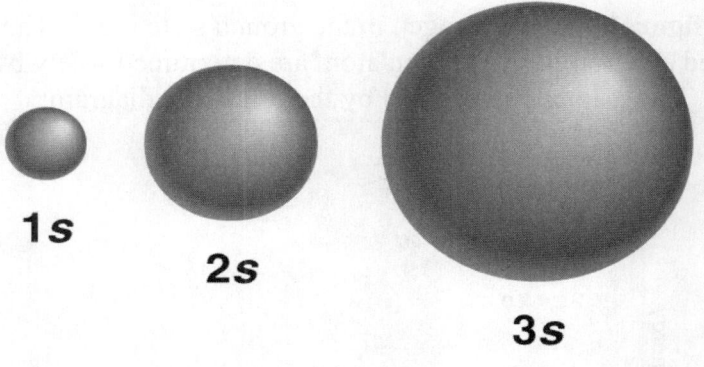

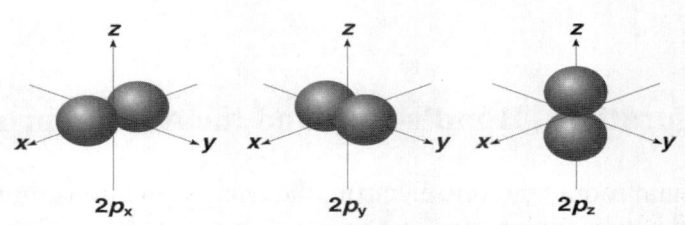

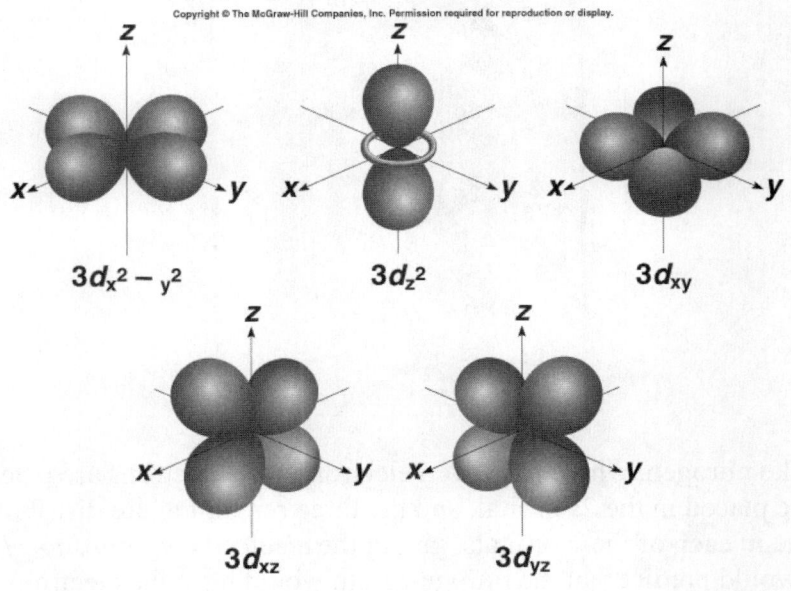

The energy levels of the hydrogen atom

The **electron configuration** for hydrogen in the ground state is $1s^1$. The available energy levels for an exited electron in a hydrogen atom are determined solely by the principal quantum number. They can be represented by the following diagram:

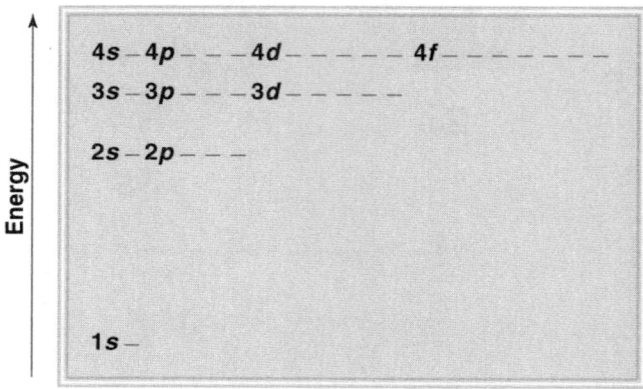

Electron configurations, Hund's rule, and the Aufbau principle

For atoms that contain more than one electron, the energy picture is more complex. The energy of repulsion between electrons in a many-electron system influences the overall energy of the atom, resulting in the following energy diagram:

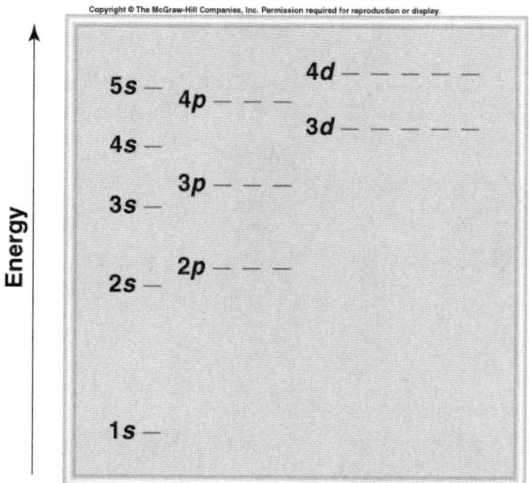

For an atom like nitrogen, which has seven electrons, two electrons are placed in the $1s$ orbital, two are placed in the $2s$ orbital, and the three remaining are distributed among the $2p$ orbitals, one in each of the p orbitals, giving the *electron configuration $1s^2 2s^2 2p^3$*. **Hund's rule** would predict that the nitrogen atom would have the electrons in the $2p$ subshell distributed with one electron in each of the three p orbitals (see diagram) because *when there are several orbitals of the same energy level, you place one electron in each of those orbitals before placing a second electron of opposite spin into any of those orbitals.*

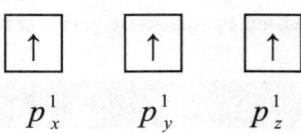

$$p_x^1 \qquad p_y^1 \qquad p_z^1$$

By comparison, the oxygen atom with eight electrons would have the *electron configuration $1s^2 2s^2 2p^4$*. The electron distribution would be as follows:

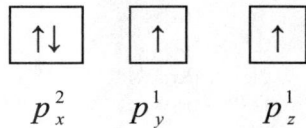

$$p_x^2 \qquad p_y^1 \qquad p_z^1$$

The **Aufbau principle** *dictates that as protons are added one by one to the nucleus to build up the elements, electrons are similarly added to atomic orbitals.* In the process, the ground state configurations for the elements emerge as shown in Table 7.3.

TABLE 7.3

The Ground-State Electron Configurations of the Elements*

Atomic Number	Symbol	Electron Configuration	Atomic Number	Symbol	Electron Configuration	Atomic Number	Symbol	Electron Configuration
1	H	$1s^1$	37	Rb	$[Kr]5s^1$	73	Ta	$[Xe]6s^24f^{14}5d^3$
2	He	$1s^2$	38	Sr	$[Kr]5s^2$	74	W	$[Xe]6s^24f^{14}5d^4$
3	Li	$[He]2s^1$	39	Y	$[Kr]5s^24d^1$	75	Re	$[Xe]6s^24f^{14}5d^5$
4	Be	$[He]2s^2$	40	Zr	$[Kr]5s^24d^2$	76	Os	$[Xe]6s^24f^{14}5d^6$
5	B	$[He]2s^22p^1$	41	Nb	$[Kr]5s^14d^4$	77	Ir	$[Xe]6s^24f^{14}5d^7$
6	C	$[He]2s^22p^2$	42	Mo	$[Kr]5s^14d^5$	78	Pt	$[Xe]6s^14f^{14}5d^9$
7	N	$[He]2s^22p^3$	43	Tc	$[Kr]5s^24d^5$	79	Au	$[Xe]6s^14f^{14}5d^{10}$
8	O	$[He]2s^22p^4$	44	Ru	$[Kr]5s^14d^7$	80	Hg	$[Xe]6s^24f^{14}5d^{10}$
9	F	$[He]2s^22p^5$	45	Rh	$[Kr]5s^14d^8$	81	Tl	$[Xe]6s^24f^{14}5d^{10}6p^1$
10	Ne	$[He]2s^22p^6$	46	Pd	$[Kr]4d^{10}$	82	Pb	$[Xe]6s^24f^{14}5d^{10}6p^2$
11	Na	$[Ne]3s^1$	47	Ag	$[Kr]5s^14d^{10}$	83	Bi	$[Xe]6s^24f^{14}5d^{10}6p^3$
12	Mg	$[Ne]3s^2$	48	Cd	$[Kr]5s^24d^{10}$	84	Po	$[Xe]6s^24f^{14}5d^{10}6p^4$
13	Al	$[Ne]3s^23p^1$	49	In	$[Kr]5s^24d^{10}5p^1$	85	At	$[Xe]6s^24f^{14}5d^{10}6p^5$
14	Si	$[Ne]3s^23p^2$	50	Sn	$[Kr]5s^24d^{10}5p^2$	86	Rn	$[Xe]6s^24f^{14}5d^{10}6p^6$
15	P	$[Ne]3s^23p^3$	51	Sb	$[Kr]5s^24d^{10}5p^3$	87	Fr	$[Rn]7s^1$
16	S	$[Ne]3s^23p^4$	52	Te	$[Kr]5s^24d^{10}5p^4$	88	Ra	$[Rn]7s^2$
17	Cl	$[Ne]3s^23p^5$	53	I	$[Kr]5s^24d^{10}5p^5$	89	Ac	$[Rn]7s^26d^1$
18	Ar	$[Ne]3s^23p^6$	54	Xe	$[Kr]5s^24d^{10}5p^6$	90	Th	$[Rn]7s^26d^2$
19	K	$[Ar]4s^1$	55	Cs	$[Xe]6s^1$	91	Pa	$[Rn]7s^25f^26d^1$
20	Ca	$[Ar]4s^2$	56	Ba	$[Xe]6s^2$	92	U	$[Rn]7s^25f^36d^1$
21	Sc	$[Ar]4s^23d^1$	57	La	$[Xe]6s^25d^1$	93	Np	$[Rn]7s^25f^46d^1$
22	Ti	$[Ar]4s^23d^2$	58	Ce	$[Xe]6s^24f^15d^1$	94	Pu	$[Rn]7s^25f^6$
23	V	$[Ar]4s^23d^3$	59	Pr	$[Xe]6s^24f^3$	95	Am	$[Rn]7s^25f^7$
24	Cr	$[Ar]4s^13d^5$	60	Nd	$[Xe]6s^24f^4$	96	Cm	$[Rn]7s^25f^76d^1$
25	Mn	$[Ar]4s^23d^5$	61	Pm	$[Xe]6s^24f^5$	97	Bk	$[Rn]7s^25f^9$
26	Fe	$[Ar]4s^23d^6$	62	Sm	$[Xe]6s^24f^6$	98	Cf	$[Rn]7s^25f^{10}$
27	Co	$[Ar]4s^23d^7$	63	Eu	$[Xe]6s^24f^7$	99	Es	$[Rn]7s^25f^{11}$
28	Ni	$[Ar]4s^23d^8$	64	Gd	$[Xe]6s^24f^75d^1$	100	Fm	$[Rn]7s^25f^{12}$
29	Cu	$[Ar]4s^13d^{10}$	65	Tb	$[Xe]6s^24f^9$	101	Md	$[Rn]7s^25f^{13}$
30	Zn	$[Ar]4s^23d^{10}$	66	Dy	$[Xe]6s^24f^{10}$	102	No	$[Rn]7s^25f^{14}$
31	Ga	$[Ar]4s^23d^{10}4p^1$	67	Ho	$[Xe]6s^24f^{11}$	103	Lr	$[Rn]7s^25f^{14}6d^1$
32	Ge	$[Ar]4s^23d^{10}4p^2$	68	Er	$[Xe]6s^24f^{12}$	104	Rf	$[Rn]7s^25f^{14}6d^2$
33	As	$[Ar]4s^23d^{10}4p^3$	69	Tm	$[Xe]6s^24f^{13}$	105	Db	$[Rn]7s^25f^{14}6d^3$
34	Se	$[Ar]4s^23d^{10}4p^4$	70	Yb	$[Xe]6s^24f^{14}$	106	Sg	$[Rn]7s^25f^{14}6d^4$
35	Br	$[Ar]4s^23d^{10}4p^5$	71	Lu	$[Xe]6s^24f^{14}5d^1$	107	Bh	$[Rn]7s^25f^{14}6d^5$
36	Kr	$[Ar]4s^23d^{10}4p^6$	72	Hf	$[Xe]6s^24f^{14}5d^2$	108	Hs	$[Rn]7s^25f^{14}6d^6$
						109	Mt	$[Rn]7s^25f^{14}6d^7$

*The symbol [He] is called the helium core and represents $1s^2$. [Ne] is called the neon core and represents $1s^22s^22p^6$. [Ar] is called the argon core and represents [Ne]$3s^23p^6$. [Kr] is called the krypton core and represents [Ar]$4s^23d^{10}4p^6$. [Xe] is called the xenon core and represents [Kr]$5s^24d^{10}5p^6$. [Rn] is called the radon core and represents [Xe]$6s^24f^{14}5d^{10}6p^6$.

Notice the **patterns** that emerge as the electrons fill the orbitals.

- The alkali metal family ends as s^1, the alkaline earth family as s^2.
- The groups 3A through 8A fill the p orbitals.
- The transition metals fill the d orbitals ($3d$, $4d$, $5d$, $6d$).

- The lanthanides and actinides fill the 4*f* and 5*f* orbitals, respectively.
- The value of *n* = the period number on the periodic table.

The order of filling of these orbitals can be read from the periodic table. Remember that **the period number** of the periodic table is the number of the *s* and *p* orbitals being filled, and the number of the ***d* orbitals** being filled is *one less than the period number, or (n–1)d.* For example, iodine is in the fifth period, hence its outer shell is $5s^2 4d^{10} 5p^5$. (See the ground-state configurations of the elements on the periodic table.)

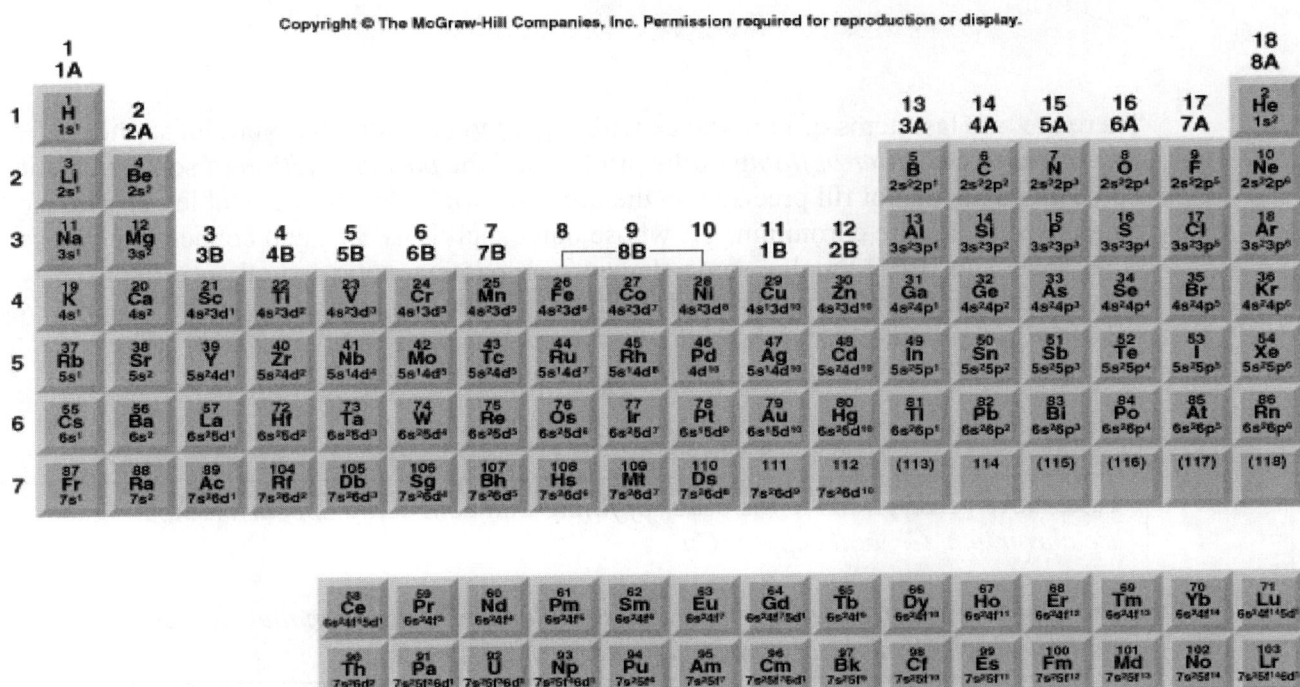

There are two last items of importance with regard to electron configurations: the *exceptions to the order of filling* of the orbitals and the *paramagnetism* of some elements. Some elements do not fill precisely in the order shown by the periodic table. The most notable examples are chromium, Cr, whose outer shell is $4s^13d^5$, and copper, Cu, whose outer shell is $4s^13d^{10}$. Finally, some elements are paramagnetic and others are diamagnetic. Elements or compounds such as Fe with *one or more unpaired electron spins* are **paramagnetic** and thus *attracted to a magnet*. Elements with all of their *electrons paired*, such as Zn, are **diamagnetic** and *not attracted to a magnet*.

SAMPLE MULTIPLE CHOICE QUESTIONS

1. Which set of quantum number is not correct?

    A. 3 0   0 –½
    B. 2 1   1 +½
    C. 3 2 –1 +½
    D. 2 1   0 –½
    E. 2 2   1 –½

2. Determine the ionization energy for a Bohr hydrogen atom ($R = 2.18 \times 10^{-18}$ J).

    A. 2.11×10^{-16} J
    B. 2.18×10^{-18} J
    C. 4.51×10^{-20} J
    D. 3.41×10^{-21} J
    E. 2.75×10^{-22} J

3. The correct electron configuration for the outer shell electrons of sulfur is:

    A.   [↑↓]   [↓↑] [↑↓] [↓]
          3s   3p   3p   3p

    B.   [↓↑]   [↑↓] [↑]   [↑]
          3s   3p   3p   3p

    C.   [↑]   [↑↓] [↓↑] [↑]
          3s   3p   3p   3p

    D.   [ ]   [↑↓] [↑↓] [↑↓]
          3s   3p   3p   3p

    E.   [↑]   [↑]   [↑]   [↑]
          3s   3p   3p   3p

4. Consider the following diagrams for the filling of $4p$ orbitals. Which one violates the Pauli exclusion principle?

A. [↑] [↑] [↑]

B. [↑] [↑] [↑↓]

C. [↑↓] [↑↓] [↑]

D. [↑↓] [↑↑] [↑↓]

E. [↑] [↑] [ ]

5. Which of the following elements would be paramagnetic?

A. Ne
B. Mg
C. Mn
D. Be
E. Zn

6. What is the maximum number of orbitals allowed when $\ell = 5$?

A. 2
B. 5
C. 9
D. 11
E. 13

7. Which of the following is not a member of the same family?

A. $1s^2 2s^2 2p^4$
B. $1s^2 2s^2 2p^6 3s^2 3p^4$
C. $1s^2 2s^2 2p^6 3s^2 3p^6 4s^2 3d^4$
D. $1s^2 2s^2 2p^6 3s^2 3p^6 4s^2 3d^{10} 4p^4$
E. $1s^2 2s^2 2p^6 3s^2 3p^6 4s^2 3d^{10} 4p^6 5s^2 4d^{10} 5p^4$

8.   The correct set of quantum numbers for the last electron in potassium could be:

      A. 4   0   0 $-\frac{1}{2}$
      B. 4   1   1 $-\frac{1}{2}$
      C. 4   0   1 $+\frac{1}{2}$
      D. 4   1 -1 $-\frac{1}{2}$
      E. 4   1   0 $+\frac{1}{2}$

9.   The complete ground state configuration for cadmium (48) is:

      A. $1s^2 2s^2 2p^6 3s^2 3p^6$
      B. $1s^2 2s^2 2p^6 3s^2 3p^6 4s^2 3d^9$
      C. $1s^2 2s^2 2p^6 3s^2 3p^6 4s^2 3d^{10} 4p^6 5s^1$
      D. $1s^2 2s^2 2p^6 3s^2 3p^6 4s^2 3d^{10} 4p^6 5s^2 4d^{10}$
      E. $1s^2 2s^2 2p^6 3s^2 3p^6 4s^2 3d^{10} 4p^6 5s^2 4d^{10} 5p^1$

10.   Blue light has a frequency of about 7.0×10^{14} Hz. What is the wavelength of this light in centimeters ($c = 3.0 \times 10^8$ m/sec)?

      A. 8.4×10^{-8} cm
      B. 3.1×10^{-12} cm
      C. 7.0×10^{-7} cm
      D. 2.0×10^{-24} cm
      E. 4.3×10^{-5} cm

Comprehension Questions

1)   In order to protect our eyes, modern sunglasses are designed to block transmission of ultraviolet light. They specifically protect our eyes from the UV radiation called UVA and UVB, two segments of the ultraviolet region of the electromagnetic spectrum that have average wavelengths of 360 and 305 nanometers, respectively. Calculate a) the average frequency of UVA radiation, b) the energy of this radiation in joules per photon, and c) its energy in joules per mole of photons.

2)   Explain why a single element, hydrogen, can have several different emission spectra (one of which, the Balmer series, is in the visible region of the electromagnetic spectrum).

3)   Detailed analysis of the various frequencies of light emitted by stars allows scientists to determine their chemical composition. a) Discuss/explain the processes that make this possible. b) What fundamental property of matter accounts for the ability to make these determinations?

4)   Provide sets of four quantum numbers (n, l, m_l, m_s) for all of the valence electrons in a silicon atom. Describe the ways in which these sets of quantum numbers obey the Pauli exclusion principle.

5)   Determine the ground-state electron configuration for the element selenium. Draw the orbital diagram for the ground state of selenium. Explain how Hund's rule, the Aufbau principle, and the Pauli exclusion principle helped you to create this electron configuration and orbital diagram.

6)   Compare the ground-state electron configuration of sulfur and oxygen to that of selenium determined in Question 5. What similarities or differences do you observe? How do you think this will affect the relative chemical and/or physical properties of these elements?

ANSWERS TO SAMPLE MULTIPLE CHOICE QUESTIONS

1.   E

        The allowed values for ℓ are the integers 0, 1, 2,….to (n–1). Therefore, if $n = 2$, the allowed values for ℓ are 0 and 1, and thus $n = 2$, $\ell = 2$ is not correct.

2.   B

        The Bohr equation allows for the calculation of the energy associated with the ionization of an electron from the ground state of a hydrogen atom:

$$\Delta E \quad = \quad -R_H \left(\frac{1}{n_f^2} - \frac{1}{n_i^2} \right)$$

$$= -2.18 \times 10^{-18} \text{ J} \left(\frac{1}{\infty^2} - \frac{1}{1^2} \right)$$

$$= 2.18 \times 10^{-18} \text{ J} \left(\frac{1}{1} \right)$$

$$= 2.18 \times 10^{-18} \text{ J}$$

The energy is positive because ionization energy must be added (+) to the atom to counteract the nuclear attraction.

3.   B

The electron configuration of the outer shell of sulfur is $3s^2 3p^4$. Obeying the Aufbau principle, the Pauli exclusion principle, and Hund's rule yields B.

4.   D

This choice violates the Pauli exclusion principle, which requires two electrons in the same orbital to have opposite spins. The middle p orbital contains two electrons with the same spin.

5.   C

The electron configurations of the elements in this question are:

Ne $1s^2 2s^2 2p^6$
Mg $1s^2 2s^2 2p^6 3s^2$
Mn $1s^2 2s^2 2p^6 3s^2 3p^6 4s^2 \mathbf{3d^5}$
Be $1s^2 2s^2$
Zn $1s^2 2s^2 2p^6 3s^2 3p^6 4s^2 3d^{10}$

Only Mn has unpaired electrons that result in the element being paramagnetic.

6.   D

The number of allowed orbitals are all the integer values between $-\ell$ and $+\ell$. Therefore $\ell = -5, -4, -3, -2, -1, 0, 1, 2, 3, 4, 5$, for a total of 11 orbitals.

7.   C

Members of the same family have the same number of valence electrons in exactly the same shape orbital. Only choice C has a d^4 outer orbital, while all the other elements end in p^4 and are members of the same family.

$$1s^22s^22p^4$$
$$1s^22s^22p^63s^23p^4$$
$$1s^22s^22p^63s^23p^64s^23\boldsymbol{d^4}$$
$$1s^22s^22p^63s^23p^64s^23d^{10}4p^4$$
$$1s^22s^22p^63s^23p^64s^23d^{10}4p^65s^24d^{10}5p^4$$

8.   A

The last electron in potassium is $4s^1$. If $n = 4$, $\ell = 0$. If $\ell = 0$, then m_ℓ must also equal 0. This restricts the choice to A.

9.   D

The last electron in cadmium, a transition metal in the fifth period of the periodic table, is a $4d^{10}$. All orbitals below $4d^{10}$ are filled.

10.   E

The wavelength and frequency are related by the speed of light according to the equation:

$$c = \lambda\upsilon$$

The speed of light is $3.0 \times 10^{10} \dfrac{cm}{\text{sec}}$ and the frequency is 7.0×10^{14} Hz or sec^{-1}.

Therefore,

$$3.0 \times 10^{10}\ \frac{cm}{\text{sec}} = \lambda\,(7.0 \times 10^{14}\ \text{sec}^{-1})$$

$$\lambda = 4.3 \times 10^{-5}\ \text{cm}$$

Answers to Comprehension Questions

1)   The relationship between frequency and wavelength of any wave is frequency (ν) $\times$ wavelength (λ) = speed at which wave is traveling. Since electromagnetic radiation, UV radiation in this case, travels at the speed of light, c, the relationship becomes $\nu \times \lambda$ = c, therefore

ν = c / λ = 3.00 $\times$ 10^8 m/sec / 360 $\times$ 10^{-9} m = **8.33 $\times$ 10^{14} sec^{-1}**

b) Energy and frequency/wavelength of a photon are related through Planck's constant, h, which has a value of 6.63 $\times$ 10^{-34} J·sec. The specific relationship is E_{photon} = hν, therefore

E_{photon} = 6.63 $\times$ 10^{-34} J·sec $\times$ 8.33 $\times$ 10^{14} sec^{-1} = **5.53 $\times$ 10^{-19} J/photon**

c) To calculate the energy per mole of photons you simply multiply the energy of a single photon by Avogadro's constant, N = 6.02 $\times$ 10^{23}.

5.53 $\times$ 10^{-19} J/photon $\times$ 6.02 $\times$ 10^{23} photons/mol = **3.33 $\times$ 10^5 J/mol of photons**

2)   Hydrogen has several emission spectra, the Lyman series, Balmer series, etc., corresponding to a variety of electron transitions. One specific set of spectral lines, the Lyman series, results from electrons emitting energy as they transition from a variety of excited states down to the ground state, $n = 1$, for the atom's electron. The specific frequency or wavelength of the light emitted correlates with differences in energy between the excited and ground states of each transition. The Balmer series, by contrast, results from electron transitions between excited states, $n = 3$ or higher, and the excited state $n = 2$. In a similar manner the other spectral series result from transitions between one excited state and another.

3)   a)   The electromagnetic radiation emitted by stars can be captured and examined through the use of radio telescopes. Since excited emissions from atoms of different elements provide a unique pattern of frequencies or wavelengths, almost a fingerprint of sorts, the specific elements that make up a given star can be determined from the combination of emission patterns observed.

        b)   The fundamental property of matter that makes this possible is the fact that the electrons of a given element have fixed and quantized ground and excited states.

And, the relative distance between these states, which determines the frequency of light emitted by excited atoms, is different for each element.

4)   Silicon has four valence electrons, two in a $3s$ orbital and two in two degenerate $3p$ orbitals. For the $3s$ electrons, the quantum numbers would be as follows: $n = 3$, $l = 0$, $m_l = 0$, and $m_s = +\frac{1}{2}$ and $n = 3$, $l = 0$, $m_l = 0$, and $m_s = -\frac{1}{2}$. Possible quantum numbers for the $3p$ electrons would be $n = 3$, $l = 1$, $m_l = -1$, and $m_s = +\frac{1}{2}$ and $n = 3$, $l = 1$, $m_l = 0$, and $m_s = +\frac{1}{2}$. For the $3p$ orbitals, you may optionally choose m_l values from -1, 0, or 1 as long as they are only used once until the electrons start pairing in these orbitals. And, the m_s values may either be $+\frac{1}{2}$ or $-\frac{1}{2}$, but they must be the same for Si as its $3p$ electrons will have parallel spins, that is, the same m_s value for both electrons.

The above sets of quantum numbers obey the Pauli exclusion principle in that the sets of numbers for each electron are unique. In the case of the $3s$ electrons, everything is the same but the spin, and in the $3p$ electrons, the m_l is different.

5)   The ground state electron configuration for Se is $1s^2\, 2s^2\, 2p^6\, 3s^2\, 3p^6\, 4s^2\, 3d^{10}\, 4p^4$. The orbital diagram is as follows:

$1s$		$2s$		$2p$	$2p$	$2p$		$3s$		$3p$	$3p$	$3p$		$4s$		$3d$	$3d$	$3d$	$3d$	$3d$		$4p$	$4p$	$4p$
↑↓		↑↓		↑↓	↑↓	↑↓		↑↓		↑↓	↑↓	↑↓		↑↓		↑↓	↑↓	↑↓	↑↓	↑↓		↑↓	↑	↑

The Aufbau principle caused the filling of lower energy orbitals prior to those of higher energy. The Pauli exclusion principle made sure that there were not two up arrows or two down arrows in a single box in the orbital diagram. This would have amounted to two electrons having the same four quantum numbers, and this is disallowed by the Pauli exclusion principle. Hund's rule can be accounted for in the fact that the $4p$ orbitals were all half-filled with electrons with paired spins, that is, all up arrows, before one of these three degenerate orbitals was filled with a second electron.

6)   Sulfur, selenium, and oxygen have different numbers of electrons, but they have similar valence electron configurations; that is, a filled s shell and a partially filled p shell. Since chemical reactivity is to a great extent controlled by the valence electrons of one atom interacting with those of another, you would expect that these elements would have similar patterns of chemical reactivity. This is in fact the case.

CHAPTER 8
THE PERIODIC CLASSIFICATION OF THE ELEMENTS

This chapter reviews the development of the periodic table and the periodic relationships among the elements that result from their electron configurations (see Chapter 7 of this review book). The topics reviewed include:

- The development of the periodic table
- The periodic classification of the elements
- Periodic variation of the physical properties of the elements
- Ionization energy
- Electron affinity
- The periodic variation of the chemical properties of the representative elements

Many of the physical and chemical properties of the elements can be related to the electron configuration of their atoms. Elements in the same group usually display similar behavior because they have the same number of **valence shell electrons**, that is, *the same number of electrons in the outer shell of the atom that are involved in chemical change.*

The development of the periodic table

John Newlands, one of the first people to attempt to order the elements, did so by arranging the elements by atomic mass and found that every eighth element had similar properties. This relationship was referred to as the "law of octaves." The relationship broke down after the element calcium.

Dmitri Mendeleev proposed a more extensive tabulation of the elements based upon both atomic mass and the regular, periodic recurrence of properties. He studied the properties of the elements so thoroughly that he correctly predicted the existence of unknown elements when he perceived a gap in the table.

However, there were some inconsistencies in Mendeleev's table. For example, the mass of argon (39.95 amu) is greater than that of potassium (39.10 amu). If the table is arranged based solely on mass, then the noble gas argon would appear in the position occupied by the active alkali metal potassium. Mendeleev arranged these two elements based upon their group behavior rather than mass. This and other discrepancies were eliminated by **Henry Moseley's** subsequent discovery of the **atomic number** (the *number of protons*) of an element. Arranging the table on the basis of atomic number places argon ($Z = 18$) in the noble gas family and potassium ($Z = 19$) in the alkali metal family, as they should be. *The modern periodic table is ordered on the basis of atomic number.*

The periodic classification of the elements

If you examine the periodic table and the electron configurations (see Chapter 7 of this review book), you will see many patterns emerging from the ground-state configurations of the elements. Based on the type of subshell being filled, the elements can be assigned to one of the following categories:

- The **representative elements** (Groups 1A to 7A) *fill the s or p subshells of the highest principal quantum number.*

- **Noble gases** (Group 8A), with the exception of helium, have *completely filled p subshells or ns^2np^6, where n is the highest quantum number.*

- **Transition metals** (Groups 1B and 3B to 8B) *fill the d subshells.*

- The **lanthanide** and **actinide elements** *fill the f subshells.* These are also referred to as the **inner transition elements.**

- The **metalloids** are those *elements whose properties are intermediate between those of metals and nonmentals (B, Si, Ge, As, Sb, Te, Po, At).*

Copyright © The McGraw-Hill Companies, Inc. Permission required for reproduction or display.

If you examine the periodic table, clear *patterns* emerge when you look at the *electron configurations for a particular group.*

- All of the **Group 1A** elements are similar. They have a noble gas core and an ns^1 outer shell electron configuration (one valence electron).
- The **Group 2A** elements have a noble gas core and an ns^2 outer electron configuration (two valence electrons).
- **Groups 3A to 7A** elements are filling *p* subshells.
- The **Group 8A**, or the noble gas family, are all ns^2np^6 configurations (eight valence electrons). Helium, with a configuration of $1s^2$, is an exception. The noble gas configuration is a very stable electronic state.

Elements that have the same number of outer electrons, the valence electrons, have similar chemical and physical properties (periodicity) and a similar ability to bond. (Newer periodic tables assign the number 1 to the alkali metals and number the families consecutively across the table, ending with the number 18 for the noble gas group.)

The configuration of an element's valence electrons directly affects the properties of its atoms and the manner in which the element reacts. Examine the formation of *metallic cations* and *nonmetallic anions* in the following tables:

Metallic Element	Configuration of the atom	Cation	Configuration of the cation
Na	$[Ne]3s^1$	Na^+	$[Ne]$
Ca	$[Ar]4s^2$	Ca^{2+}	$[Ar]$
Al	$[Ne]3s^23p^1$	Al^{3+}	$[Ne]$

Nonmetallic Element	Configuration of the atom	Anion	Configuration of the anion
F	$[He]2s^22p^5$	F^-	$[He]2s^22p^6$ or $[Ne]$
O	$[He]2s^22p^4$	O^{2-}	$[He]2s^22p^6$ or $[Ne]$
N	$[He]2s^22p^3$	N^{3-}	$[He]2s^22p^6$ or $[Ne]$

Metals have only a *few electrons* beyond the noble gas core. They tend to *lose electrons*, assume the configuration of a noble gas (ns^2np^6) and *acquire a positive charge* equal to the number of electrons lost. Nonmetals, on the other hand, have an electron configuration that is usually one or two electrons less than a noble gas. *Nonmetals tend to gain electrons* to obtain an ns^2np^6 noble gas configuration and *acquire a negative charge* equal to the number of electrons gained. *Ions that have exactly the same ground-state electron configuration* are said to be **isoelectronic**.

Example 1. Writing electron configurations of ions.

Write the electron configurations for the potassium ion, argon atom, and magnesium ion. Which of these species are isoelectronic?

General Strategy	Solution to Example 1
Write the electron configuration for the <u>atoms</u> of each element.	K $\quad 1s^22s^22p^63s^23p^64s^1$ Ar $1s^22s^22p^63s^23p^6$ Mg $1s^22s^22p^63s^2$
Write the configuration for the <u>ions</u> that each element would form.	K^+ $\quad 1s^22s^22p^63s^23p^6$ Ar $\quad 1s^22s^22p^63s^23p^6$ Mg^{2+} $1s^22s^22p^6$
Those ions and atoms that have the exact same electron configuration are isoelectronic.	K^+ and Ar have the exact same electron configuration. They are isoelectronic.

Transition metals such as zinc have an outer electron configuration of $[Ar]4s^23d^{10}$. Notice that the principal quantum number for the *d* subshell is *(n − 1)d*. Thus, *transition metals form ions by removing the 4s electrons before losing 3d electrons*. The electron configuration for the Zn^{2+} ion is $[Ar]3d^{10}$. Other transition metals behave in a similar fashion.

Periodic variation of the physical properties of the elements

Effective nuclear charge is *the amount of nuclear charge actually felt by the valence electrons*. Two factors affect the amount of nuclear charge felt by the valence electrons, the distance the electrons are from the nucleus and the amount of shielding by the core electrons. *The attraction to the nucleus felt by the valence shell electrons declines as a function of 1/distance^2*. Therefore, the larger the atom, the greater the distance of the valence electrons from the nucleus and the smaller the attraction of the nucleus for those valence electrons. In addition, in a large atom like rubidium, the valence $5s^1$ electron does not feel the attraction of all 37 protons due to the **shielding** provided by the many electrons between the nucleus and the one valence electron.

The impact of these two factors can be seen in the trend in **atomic radius,** which is *half the distance between two nuclei in two adjacent atoms.* Across a period, one proton is added to the nucleus and one electron is added to the outermost orbital. Electrons in the same energy level do not shield each other from the nucleus. This means that as you go across the period, a greater force of attraction is exerted by the nucleus on the outer electrons due to an increasing number of protons. The result is that *the atomic radius decreases from left to right in a period.* However, in a group each successive element down the group has its valence electrons in a principal energy level with a larger n value, meaning that there is a greater distance between the nucleus and the valence electrons. The combined effect of a greater distance from the nucleus and increased shielding results in a larger radius. Consequently, *the atomic radius increases from top to bottom in a group.*

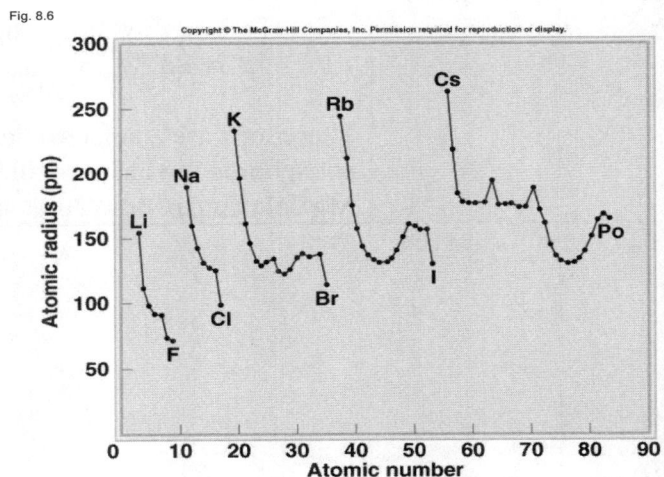

Fig. 8.6

The **ionic radius** is *the radius of a cation or anion.* When the atom loses or acquires electrons, the resulting ion changes in size from the original atom. If you compare the ionic radius of the isoelectronic cations Na^+, Mg^{2+}, and Al^{3+}, you find that as the positive charge on the isoelectronic cations increases, indicating a greater loss of electrons, the ionic radius becomes smaller. The proton/electron ratio increases from Na^+ to Al^{3+} and the extra protons exert a stronger pull on the electron cloud, resulting in a smaller ionic radius. Conversely, for the isoelectronic anions N^{3-}, O^{2-}, and F^-, the greater the negative charge, the smaller the proton/electron ratio. The N^{3-} ion has the smallest proton/electron ratio and the largest radius.

Example 2. Determination of ion size.

Which ion in each pair has the larger radius?
a) Na^+ or K^+
b) S^{2-} or P^{3-}

General Strategy	Solution to Example 2
Examine the electron configuration of each ion. Check the principal quantum number and determine the proton/electron ratio.	a. $\quad Na^+\ 1s^2 2s^2 \underline{2p^6} \qquad$ 11 p/10e$^-$ $\quad K^+\ 1s^2 2s^2 2p^6 3s^2 \underline{3p^6} \quad$ 12p/18e$^-$
Then determine which ion is larger. The lower the proton/electron (p/e–) ratio, the larger the ion.	K^+ is filling more electrons into a higher energy level, n. Therefore, it has the larger radius.
The higher the principal quantum number, the larger the ion.	b. $\quad S^{2-}\ 1s^2 2s^2 2p^6 3s^2 3p^6 \quad$ 16p/18e$^-$ $\quad P^{3-}\ 1s^2 2s^2 2p^6 3s^2 3p^6 \quad$ 15p/18e$^-$ These ions are isoelectronic. The phosphorus ion has a smaller proton/electron ratio and a larger radius.

Ionization energy

Ionization energy is *the minimum energy ($\frac{kJ}{mol}$) required to remove 1 mol of electrons from 1 mol of gaseous atoms of an element in the ground state*: $X(g) \rightarrow X(g)^+ + 1e^-$.
The magnitude of the ionization energy is a measure of "how tightly" the electrons are held in the atom. The higher the ionization energy, the more difficult it is to remove the electron. For a many-electron system, the **first ionization energy** is *the energy needed to remove the first electron* from the atom. Metals with few valence electrons and large radii tend to have low first ionization energies. *First ionization energies generally increase from left to right in a period.* The nuclear charge is increasing by one with each successive element, but the distance between the nucleus and the valence electrons is relatively unchanged. The result is an increasing attraction between the nucleus and the valence electrons, which requires more energy to remove an electron. The *first ionization energy decreases from top to bottom in a group* because each successive element has its valence electrons in a principal energy level with a larger n value. There are also more shielding electrons between the nucleus and the valence electrons, reducing the attraction between the nucleus and the valence electrons.

Fig. 8.11

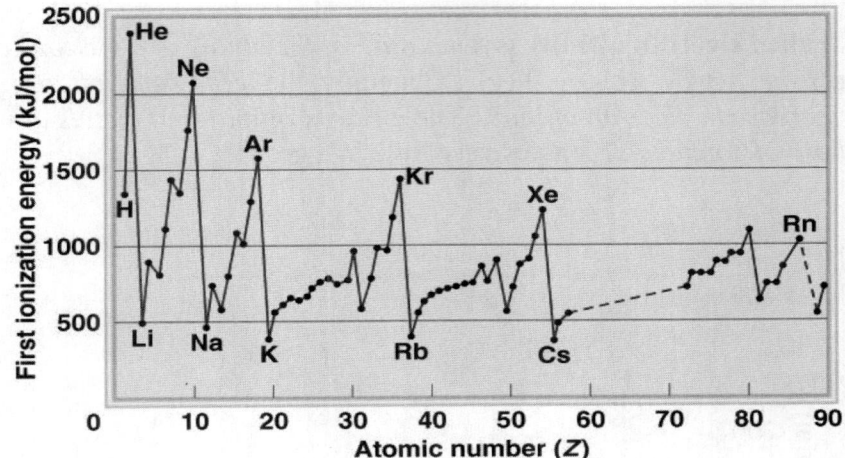

There are some anomalies in the trends for first ionization energies. For example, there is a drop in the ionization energy from Group 2A to Group 3A because the p^1 electron of Group 3A is slightly further from the nucleus than the s^2 electrons in Group 2A. The slight increase in distance combined with a small increase in shielding due to the s^2 electrons helps to diminish the nuclear attraction. The result is a lower ionization energy for Group 3A. There is also a drop in ionization energy from Group 5A to Group 6A. Group 6A contains the first electron *pair* in the *p* subshell. This spin pair increases the electrostatic repulsion, which is a destabilizing factor, making it easier to remove an electron from the valence shell of Group 6A elements.

Take Note: *The AP exam frequently asks about trends in ionization energy, ion size, and atomic radius. If the question asks you to determine if the ionization energy is increasing or decreasing, be sure to begin you answer by stating whether the trend is increasing or decreasing. If asked to explain the trend, be sure that you explain why the trend is increasing or decreasing and do not simply restate that the trend is increasing or decreasing.*

It is possible to remove more than one electron from an atom. *Because the nuclear charge remains constant, more energy is needed to remove another electron from the positively charged ion.* Thus, ionization energies (I) always increase in the following order:

$$I_1 < I_2 < I_3 \ldots$$

Electron affinity

The property called **electron affinity** is *a measure of the ability of a gaseous atom to accept an electron*: $X(g) + 1e^- \rightarrow X(g)^-$. Generally, the *trend from left to right in a period is an increase in electron affinity*. The *electron affinities of metals are generally lower than those of nonmetals*. Values vary little in a group.

Fig. 8.12

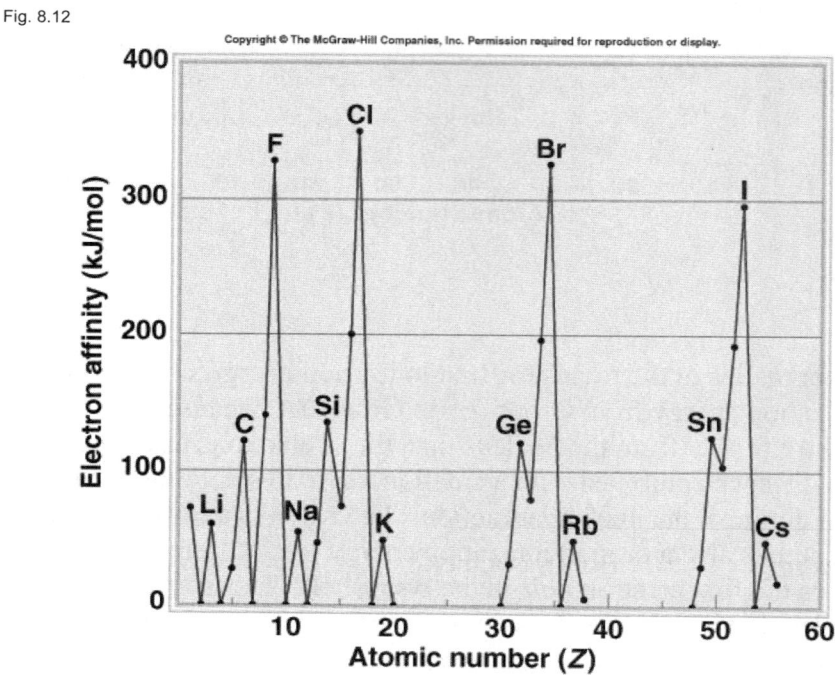

Copyright © The McGraw-Hill Companies, Inc. Permission required for reproduction or display.

The periodic variation of the chemical properties of the representative elements

Group	Valence Shell Configuration	Common Reactions and Properties
Group 1A M = Alkali metals	ns^1	These elements have low ionization energies and low electron affinities. $2M\,(s) + 2H_2O \rightarrow 2MOH\,(aq) + H_2\,(g)$ $4M\,(s) + O_2\,(g) \rightarrow 2M_2O\,(s)$ Alkali metals become more reactive as you go from top to bottom in the family. Lithium forms predominantly oxides, but potassium, rubidium, and cesium can also form peroxides, O_2^{2-}, and superoxides, O_2^-.

Group 2A M = Alkaline earth metals	ns^2	These elements have low ionization energies and low electron affinities. The alkaline earth metals vary quite markedly in their reactions with water. Beryllium does not react; magnesium reacts slowly with steam; calcium, strontium, and barium are reactive with cold water. $M\ (s)\ +\ 2H_2O\ (l)\ \rightarrow\ M(OH)_2\ (aq)\ +\ H_2(g)$ The reactivities of the alkaline earth metals toward oxygen increase from top to bottom in the group. $2M\ (s)\ +\ O_2\ (g)\ \rightarrow\ 2MO\ (s)$ The reactivities of the alkaline earth metals toward acid increase from top to bottom in a group. $M\ (s)\ +\ 2H^+\ (aq) \rightarrow\ M^{2+}\ (aq)\ +\ H_2\ (g)$ Calcium, strontium, and barium also react similarly with water. $M\ (s)\ +\ 2H_2O\ (l) \rightarrow\ M(OH)_2\ (aq)\ +\ H_2\ (g)$
Group 3A Boron family	ns^2np^1	Boron is a metalloid and unreactive toward oxygen and water. Aluminum forms only tripositive ions, but many elements further down in the group form more stable unipositive ions. $4Al\ (s)\ +\ 3O_2\ (g)\ \rightarrow\ 2Al_2O_3\ (s)$ $2Al\ (s)\ +\ 6H^+\ (aq)\ \rightarrow\ 2Al^{3+}\ (aq)\ +\ 3H_2\ (g)$
Group 4A	ns^2np^2	Carbon is a nonmetal. Silicon and germanium are metalloids. The metallic elements tin and lead do not react with water but do react with acids to liberate hydrogen. $Sn\ (s)\ +\ 2H^+\ (aq)\ \rightarrow\ Sn^{2+}\ (aq)\ +\ H_2\ (g)$ $Pb\ (s)\ +\ 2H^+\ (aq)\ \rightarrow\ Pb^{2+}\ (aq)\ +\ H_2\ (g)$ For carbon and silicon, the +4 oxidation state is more stable. For tin and lead, the +2 oxidation state is more stable.

Group 5A	ns^2np^3	Nitrogen and phosphorus are nonmetals. Diatomic nitrogen, N_2, forms five oxides and phosphorus forms two oxides. The important oxoacids HNO_3 and H_3PO_4 are formed when the following oxides are placed in water: $$N_2O_5\ (s)\ +\ H_2O\ (l) \rightarrow\ 2HNO_3\ (aq)$$ $$P_4O_{10}\ (s)\ +\ H_2O\ (l) \rightarrow 4H_3PO_4\ (aq)$$ Arsenic and antimony are metalloids. Bismuth is a metal.
Group 6A	ns^2np^4	Oxygen, sulfur, and selenium are nonmetals. Tellurium and polonium (radioactive) are metalloids. Oxygen, sulfur, selenium, and tellurium often form di-negative ions (O^{2-}, S^{2-}, Se^{2-}, Te^{2-}). The elements in this group, especially oxygen, form a large number of molecular compounds with nonmetals. Nonmetal oxides form acids in water. For example, sulfuric acid is formed when sulfur trioxide reacts with water: $$SO_3\ (g)\ +\ H_2O\ (l) \rightarrow\ H_2SO_4\ (aq)$$
Group 7A (Halogens)	ns^2np^5	All of the halogens are nonmetals that form diatomic molecules at room temperature, X_2, where X denotes a halogen. All the halogens are highly colored and poisonous. They have high ionization energies and high electron affinities. Halogens react with alkali and alkaline earth elements to form halide salts. They all exhibit an oxidation state of -1 in binary compounds. They react with hydrogen to form binary acids. $$H_2\ (g)\ +\ X_2\ (g)\ \rightarrow\ 2HX\ (aq)$$

| Group 8A

Noble gases | ns^2np^6 | All the noble gases exist as monatomic atoms. Note that xenon, radon, and krypton have been observed to form a few unstable compounds.

The ionization energies of the noble gases are among the highest.

These gases have no tendency to accept extra electrons. |
|---|---|---|

Notice in the chart that the period 2 elements frequently have different chemical properties than the rest of their group. This is the result of the unusually small size of the atoms in the second period.

Finally, an examination of the chemistry of the representative elements in the chart shows a *trend in their oxides* that is summarized in Table 8.4. **Metal oxides** are **basic anhydrides**, *bases without water*. **Nonmetal oxides** are **acid anhydrides**, *acids without water*.

Table 8.4

TABLE 8.4 — Some Properties of Oxides of the Third-Period Elements

	Na_2O	MgO	Al_2O_3	SiO_2	P_4O_{10}	SO_3	Cl_2O_7
Type of compound	← Ionic →			← Molecular →			
Structure	← Extensive three-dimensional →			← Discrete molecular units →			
Melting point (°C)	1275	2800	2045	1610	580	16.8	−91.5
Boiling point (°C)	?	3600	2980	2230	?	44.8	82
Acid-base nature	Basic	Basic	Amphoteric	← Acidic →			

Take Note: *You need to have a strong working knowledge of the periodic table to be successful on the AP exam. If you are intimately familiar with the table, you can predict important trends in chemical behavior, including the size of atoms and ions, the potential products of reactions, acid-base behavior, trends in ionization and electronegativity, bond type, and chemical formulas.*

SAMPLE MULTIPLE CHOICE QUESTIONS

For questions 1–4.

      A. Cr
      B. O
      C. Cl
      D. Xe
      E. Ba

1.   Which element is usually unreactive but has been observed to form unstable halides?

2.   Which element forms an oxide that produces a soluble base when placed in water?

3.   Which element forms an acid with the general formula HXO_4?

4.   Which element can exist in more than one oxidation state in a binary compound?

5.   Element X has the following electron configuration: $1s^2 2s^2 2p^6 3s^2 3p^4$. The formula of the compound it would form with aluminum is:

      A. AlX_3
      B. Al_3X_2
      C. AlX
      D. Al_2X_3
      E. Al_2X

6.   Which element has the highest second ionization energy?

      A. F
      B. Be
      C. Li
      D. Al
      E. Ca

7.   Which element is most easily oxidized?

      A. Al
      B. Rb
      C. Co
      D. Li
      E. Ag

8.   Phosphorus reacts with oxygen to form a compound. Which of the following
         statements is not true of the product?

         A.   It is an acidic anhydride
         B.   It is molecular
         C.   It has covalent bonds
         D.   A likely formula is P_2O_5
         E.   It is gaseous

9.   Place the following elements in order of increasing first ionization energy: F,
         K, P, Ca, and Ne.

         A.   Ne < Ca < P < F < K
         B.   K< Ca < P < F < Ne
         C.   Ca < P < F < Ne < K
         D.   P < F < K < Ne < Ca
         E.   P < Ca < K < Ne < F

10.   List the following ions in order of increasing ionic radius: P^{3-}, K^+, Cl^-, Ca^{2+},
         and S^{2-}.

         A. $P^{3-} < S^{2-} < Cl^- < K^+ < Ca^{2+}$
         B. $K^+ < Ca^{2+} < Cl^- < S^{2-} < P^{3-}$
         C. $Ca^{2+} < Cl^- < K^+ < S^{2-} < P^{3-}$
         D. $Ca^{2+} < K^+ < Cl^- < S^{2-} < P^{3-}$
         E. They are all the same radius because they are isoelectronic.

Comprehension Questions

1)   Explain what is meant by the term "periodic" property?

2)   Chlorine, argon, and potassium have atomic numbers 17, 18, and 19, respectively.
Write the complete electron configurations of a chloride ion, an argon atom, and a
potassium ion without consulting a periodic table. What relationship do you see between
the electron configurations of these three species? What is this relationship called?

3)   What similarities in chemical properties would you expect to see among the
elements of a single group and why?

4)   Arrange the following atoms and ions in order of increasing radius: S^{2-}, Ar, and Ne. Explain the predicted trend in terms of nuclear charge and electronic structure of each species.

5)   How does first ionization energy vary across a period? Within a group?

ANSWERS TO SAMPLE MULTIPLE CHOICE QUESTIONS

1. D

Noble gases are unreactive. Xenon (Xe) with its large electron cloud has been observed to form compounds with oxygen and the halogens.

2. E

Soluble metal oxides in water produce basic solutions. Barium oxide will undergo a reaction to produce the soluble base barium hydroxide. Chromium oxide is insoluble.

3. C

Chlorine is the only element in the list that would form a monoprotic oxoacid, $HClO_4$.

4. A

Transition metals tend to have multiple oxidation states. Chromium is the only transition metal in the list.

5. D

X ends in a p^4 configuration, putting it in Group 6A. It is a nonmetal and its most likely oxidation number is –2. Since aluminum is +3, the formula is Al_2X_3.

6. C

Lithium is a very small atom that has only one valence electron, resulting in a fairly high first ionization energy. The second ionization energy is very high because the second electron is being removed from the $1s^2$ noble gas structure. The $1s$ shell is much closer to the nucleus and, therefore, the electrons are very tightly held. In addition, the positive charge on the lithium ion, Li^+, makes the removal of the second electron require more energy.

7. B

Rubidium, with a larger n value for its valence shell, places the one valence electron far from the nucleus ($5s^1$), where it experiences a lower effective nuclear charge and a lower ionization energy. It is easily oxidized.

8. E

The most likely formulas of the oxide are P_2O_5 and P_4O_{10}. The molar masses are substantial in either case. This is most likely a solid product.

9. B

Overall, the order is K<Ca<P<F<Ne.

Potassium, K, and calcium, Ca, are in the fourth period. The ionization energy generally increases as you go from left to right in a period due to the increasing nuclear charge. Therefore, the order of increasing ionization energy for these two elements is: K< Ca. Phosphorus, P, is in the third period. It has a lower n value and thus a higher ionization energy. The electrons are closer to the nucleus and more tightly held. Fluorine and neon are in the second period and their valence electrons are closer to the nucleus (lower n value) and have a higher ionization energy than P. Since ionization energy increases as you go from left to right in a period due to increasing nuclear charge, F < Ne.

10. D

$$Ca^{2+} < K^+ < Cl^- < S^{2-} < P^{3-}$$

20	19	17	16	15	protons
18	18	18	18	18	electrons

All of the ions here are isoelectronic: $1s^2 2s^2 2p^6 3s^2 3p^6$. What differs is their proton/electron ratio. The more protons, the tighter the electron cloud is pulled in, resulting in a smaller ion. The fewer the number of protons (i.e., more electrons), the larger the electron cloud.

Answers to Comprehension Questions

1)   A periodic property is one that varies or repeats at regular intervals.

2)   The electron configurations of all three species are the same, $1s^2 2s^2 2p^6 3s^2 3p^6$. This relationship is called isoelectronic.

3)   You would expect elements in a group to form similar ions with similar charges. They should bond to other elements in similar ways, that is, if one member of the group forms an oxide with the formula X_2O_3, then other members of the group would be likely to form oxides with similar formulas. The reason for these similarities is that the properties noted are related to outer, or valence, electron properties, and elements of a group have similar valence electron configurations. The only difference in valence electron configuration between members of a group is that the principal quantum number of the outermost electrons increases by one with each successive period.

4)   When comparing neon and argon, argon should be the larger atom, as atomic radius increases when moving down a period. The sulfide ion is isoelectronic with argon, but has two fewer protons in the nucleus when compared to argon. This causes sulfide to be quite a bit larger than argon. So the relative order, from smallest to largest, is Ne, Ar, S^{2-}. Argon is larger than neon because it has an additional eight electrons in orbits that are at a greater average distance from the nucleus than the outermost electrons of neon. Argon and sulfide are isoelectronic, but the nuclear charge of sulfide is smaller, so all of the electrons are at a greater average distance from the nucleus, leading to a larger radius.

5)   First ionization energy generally increases across a period, and decreases down a group.

CHAPTER 9
CHEMICAL BONDING I: BASIC CONCEPTS

This chapter is an introduction to chemical bonding. Chapter 10 provides a more detailed treatment of the different theories of chemical bonding. The topics dealing with basic chemical bonding reviewed in this chapter include:

- Lewis dot symbols
- Electronegativity
- The ionic bond
- The lattice energy of ionic compounds
- The covalent bond
- Writing Lewis dot structures
- Formal charge
- Multiple bonds and resonance
- Exceptions to the octet rule
- Bond energy calculations

Lewis dot symbols

Valence electrons, the *electrons in the outermost energy shell*, are involved in chemical bonding. The valence electrons of the elements are depicted in the Lewis dot diagrams illustrated in Figure 9.1.

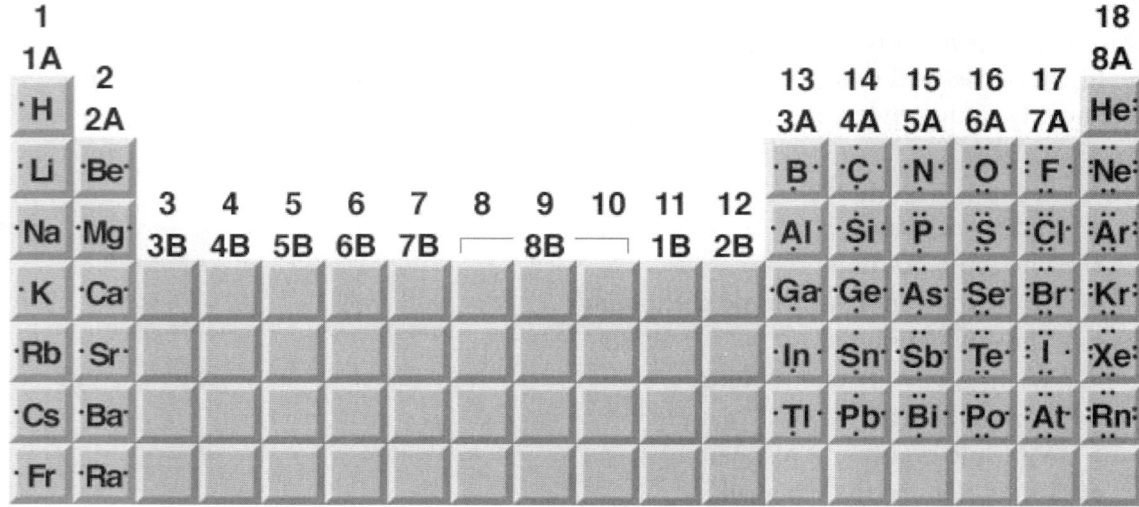

Figure 9.1

Electronegativity

The type of bond that forms between atoms depends on the differences in electronegativity of the atoms involved. **Electronegativity** is the *ability of an atom to attract toward itself the electrons of another atom in a chemical bond.* The most electronegative elements are the halogens and the elements O, N, and S. The least electronegative elements are the alkali metals.

Electronegativity generally increases as you go across a period of the periodic table. The nuclear charge increases but the distance between the nucleus and valence electrons stays relatively unchanged. Therefore, there is increased attraction between the nucleus and the valence electrons. *Electronegativity decreases as you go down a family.* The increased distance (higher energy shells) and increased shielding (due to the filled inner electron shells) result in a decrease in the attraction between the nucleus and the valence electrons.

Two other properties that follow trends similar to electronegativity are electron affinity and ionization energy. **Electron affinity** refers to the *energy change that accompanies a gaseous atom or ion gaining an electron* [X (g) + e⁻ → X⁻ (g)]. **Ionization energy** refers to the *energy required to remove an electron from a gaseous atom in the ground state* [X (g) → X⁺(g) + e⁻] . (See the periodic trend discussion in Chapter 8 of this review book.)

The ionic bond

An ionic bond forms when a very active metal reacts with a very electronegative nonmetal. Ionic bonds usually form when Family 1A or Family 2A metals combine with a halogen, S, N, or O. The metal essentially transfers an electron(s) to the nonmetal. The *electrostatic force that exists between the metal cation and the nonmetal anion is* called the **ionic bond**. For example:

$$Na \cdot \ + \quad :\overset{.}{\underset{..}{F}}: \ \rightarrow \ Na^+ \quad :\overset{..}{\underset{..}{F}}:^{-}$$

Note that when the Na atom transfers its valence electron to the fluorine atom, both ions achieve a noble gas configuration, s^2p^6. In this case both ions are isoelectronic with the noble gas neon. The tendency for atoms to lose or gain electrons to achieve a stable noble gas configuration (eight electrons, s^2p^6) is referred to as the **octet rule**.

The lattice energy of ionic compounds

Ionic bonds are strong bonds. It requires a lot of energy to break an ionic bond. Therefore, ionic solids tend to have high melting and boiling points. The **lattice energy** is the *energy required to separate 1 mol of solid ionic compounds into gaseous ions*. When comparing ionic compounds and predicting the relative sizes of their lattice energies, there are two major principles to consider:

    1.   the greater the ion charge, the greater the force of attraction between ions; and
    2.   the greater the size of the ion, the less the force of attraction between the ions.
You can observe these trends in Table 9.1.

TABLE 9.1

Lattice Energies and Melting Points of Some Alkali Metal and Alkaline Earth Metal Halides and Oxides

Compound	Lattice Energy (kJ/mol)	Melting Point (°C)
LiF	1017	845
LiCl	828	610
LiBr	787	550
LiI	732	450
NaCl	788	801
NaBr	736	750
NaI	686	662
KCl	699	772
KBr	689	735
KI	632	680
$MgCl_2$	2527	714
Na_2O	2570	Sub*
MgO	3890	2800

* Na_2O sublimes at 1275 °C.

Example 1. Predicting relative lattice energies.

Arrange the following in order of increasing lattice energy and justify your arrangement: NaF, KCl, and BaF_2.

Solution to Example 1.

KCl < NaF < BaF$_2$

The lattice energy for NaF is greater than that for KCl because NaF is the smaller ion (i.e., the ions are packed closer together in NaF). When the ions are close together, they experience a greater force of attraction and it takes more energy to separate them. The lattice energy of BaF_2 is greater than that for NaF because the 2^+ to 1^- ion attraction is a greater force than the 1^+ to 1^- ion attraction. The greater the force of attraction, the more energy is required to separate the ions.

The covalent bond

Instead of electrons being considered as <u>transferred</u> from one atom to another as they are in an ionic bond, *electrons may be <u>shared equally</u> between two atoms,* forming a

- 146 -

nonpolar covalent bond, or *electrons may be <u>shared unequally</u>,* forming a **polar covalent bond.**

Which type of bond is formed depends on the differences in electronegativity of the atoms making up the bond. If the difference in electronegativity values is greater than or equal to 2.0, the bond is considered ionic. In general, *Family 1A and Family 2A* metals (except Be, which generally forms covalent bonds) *form ionic bonds* when reacting with the very electronegative elements *N, O, S, and the halogens.* If two elements are close to one another on the periodic table, they tend to form predominately nonpolar covalent bonds. Diatomic molecules such as N_2 have a 100% covalent character. If the elements are further apart on the periodic table, such as H_2S or CO_2, they tend to form polar covalent bonds. Usually hydrogen forms covalent bonds where H has a +1 oxidation state, such as in HF and HCl. Hydrogen forms an ionic bond only when it has a -1 oxidation state, which occurs in hydrides such as sodium hydride, NaH.

In general, ionic compounds are solids with high boiling and melting points, are soluble in polar solvents like water, conduct in the molten and dissolved states, and are nonconductors as solids. In contrast, covalent compounds are often liquids or gases, have low boiling and melting points, do not dissolve in polar solvents, and are poor electrical conductors. You can observe these differences by contrasting NaCl and CCl_4 in Table 9.3.

TABLE 9.3

Comparison of Some General Properties of an Ionic Compound and a Covalent Compound

Property	NaCl	CCl_4
Appearance	White solid	Colorless liquid
Melting point (°C)	801	-23
Molar heat of fusion* (kJ/mol)	30.2	2.5
Boiling point (°C)	1413	76.5
Molar heat of vaporization* (kJ/mol)	600	30
Density (g/cm³)	2.17	1.59
Solubility in water	High	Very low
Electrical conductivity		
Solid	Poor	Poor
Liquid	Good	Poor

* Molar heat of fusion and molar heat of vaporization are the amounts of heat needed to melt 1 mole of the solid and to vaporize 1 mole of the liquid, respectively.

Take Note: *Both ionic and covalent bonds and the general properties associated with each bond type are frequently tested in the Multiple Choice and Free Response sections of the exam. If you have an ionic compound, its properties are likely to be similar to those of NaCl; if you have a covalent compound, its properties are likely to be similar to those of CCl₄.*

Writing Lewis dot structures

Writing Lewis dot structures for compounds allows you to predict their correct molecular geometries (see Chapter 10 in this review book). These geometries have implications for the physical and chemical properties of the compound.

To write a Lewis dot structure:
1. Determine the total number of valence electrons in the molecule or polyatomic ion. For negative polyatomic ions, add an electron for each unit charge.
2. Draw a plausible skeletal structure with single bonds only. Add the appropriate number of electrons around each atom (most atoms will have eight). Hydrogen has a maximum of two electrons and thus is always at a terminal part of the structure.
3. Count all the electrons in the skeletal structure and place any 'leftover' electrons as pairs on the central atom.*
4. Check for compliance with the octet rule. Exceptions to the octet rule are discussed later in this chapter.

*The *electrons in a structure that are not involved in bonding* are called **lone pairs** or **nonbonding electrons**.

Example 2. Writing a Lewis dot structure.

Write the Lewis dot diagram for H_2CO_3.

Solution to Example 2.

Applying the four guidelines above:

1. Determine the total number of valence electrons: $2 H + C + 3 O = 2(1) + 4 + 3(6) = 24$ valence electrons.

2. Draw the skeletal structure and then add the electrons to the skeletal structure:

$$H - O - C - O - H$$
$$|$$
$$O$$

$$H - \overset{..}{\underset{..}{O}} - C - \overset{..}{\underset{..}{O}} - H$$
$$|$$
$$\overset{}{\underset{..}{:O:}}$$

3. There are 24 valence electrons in the structure so there are no 'leftover' electrons to be placed on the central atom. H has two valence electrons and each O has eight valence electrons. However, C only has six valence electrons and thus has not met the octet rule. If you makes a double bond for C, then C is also consistent with the octet rule.

4. Structure with electrons consistent with octet rule.

$$H — \overset{..}{\underset{..}{O}} — \overset{}{\underset{\overset{||}{:O:}}{C}} — \overset{..}{\underset{..}{O}} — H$$

Formal charge

There is often more than one way to write a Lewis dot structure. One technique to determine which is the most likely structure involves determining the formal charges associated with each of the atoms in the structure.

Formal charge = (*number of valence electrons in the free atom*) – (*number of nonbonding electrons*) – (*½ the number of bonding electrons*). The sum of the formal charges is zero for a neutral molecule or equal to the charge of the polyatomic ion. The most likely structures have atoms with a formal charge of 0 or a charge consistent with the relative electronegativity of the atom.

Example 3. Determining formal charge.

Determine the formal charge on each atom in the following two possible Lewis structures for $COCl_2$. State which is the most likely structure. Justify your choice.

Possible structure (a):   Possible structure (b):

$$: \overset{..}{\underset{..}{Cl}} — \overset{}{\underset{\overset{||}{:O:}}{C}} — \overset{..}{\underset{..}{Cl}} :$$ $$: \overset{..}{\underset{..}{Cl}} — \overset{}{\underset{\overset{||}{:C:}}{O}} — \overset{..}{\underset{..}{Cl}} :$$

Solution to Example 3

Formal charge bookkeeping for structure (a):

	Cl	C	Cl	O
1. The number (#) of valence electrons in the free atom.	7	4	7	6
2. # of nonbonding electrons	6	0	6	4
3. ½ the # of bonding electrons (is equivalent to the # of bonds)	1	4	1	2
Formal charge = 1. – 2. – 3.	0	0	0	0

Formal charge bookkeeping for structure (b):

	Cl	O	Cl	C
1. The number (#) of valence electrons in the free atom.	7	6	7	4
2. # of nonbonding electrons	6	0	6	4
3. ½ the # of bonding electrons (is equivalent to the # of bonds)	1	4	1	2
Formal charge = 1. – 2. – 3.	*0*	*2*	*0*	*–2*

Structure (a) is more likely than structure (b) since all atoms have a formal charge of zero in structure (a). In structure (b), the electronegative element, oxygen, has a positive formal charge, while the less electronegative element, carbon, has a negative formal charge. Structure (b) is not a likely structure.

Multiple bonds and resonance

Some molecules satisfy the octet rule by the formation of multiple bonds. Common elements that form multiple bonds are *S, N, O, and C*. A **double bond** is formed when *two pairs of electrons are shared.* A double bond is made of *one sigma and one pi bond* and has a *bond order of 2*. When *three pairs of electrons are shared,* a **triple bond** is formed. A triple bond is made of *one sigma and two pi bonds* and has a *bond order of 3.*

Two common examples of molecules with double bonds are CO_2 and C_2H_4:

$$:\ddot{O}=C=\ddot{O}: \quad \text{and} \quad \begin{matrix} H & H \\ | & | \\ H-C & =C-H \end{matrix}$$

Three common examples of molecules with triple bonds are N_2, CO, and C_2H_2:
$:N \equiv N:$ and $:C \equiv O:$ and $H - C \equiv C - H$

There are times when no *one* Lewis dot structure is applicable. For example, NO_2^- has 18 valence electrons [$5 + 2(6) + 1 = 18$] and you can write two likely structures:

$$\left[:\ddot{O}-N=\ddot{O}: \right]^{-} \quad \Leftrightarrow \quad \left[:\ddot{O}=N-\ddot{O}: \right]^{-}$$

These structures imply that one single bond and one double bond are present. This is not actually the case. In fact, two identical N - O bonds are formed that are shorter than a single bond and longer than a double bond. The bond order for the NO_2^- polyatomic ion is 1.5. A molecule or polyatomic ion that *cannot be represented by only one Lewis structure* is said to display **resonance** and the species' true structure is a *composite of the structures drawn*. Other common examples of structures that *exhibit resonance are C_6H_6 (benzene), CO_3^{2-}, and NO_3^-.*

Exceptions to the octet rule

There are three exceptions to the octet rule:

1. molecules in which the central atom does not have eight valence electrons;
2. molecules with an odd number of electrons; and
3. molecules in which the central atom has more than eight valence electrons.

These exceptions are summarized in the table:

Type of exception	Common central element(s) that exhibit the exception	Molecules that exhibit the exception	Comments
1. Incomplete octet	Be, B, Al	BeH_2, BF_3, AlI_3	Be forms linear molecules. B and Al form trigonal planar molecules. Trigonal planar molecules act as Lewis acids (electron pair acceptors).
2 Odd number of electrons	N	NO, NO_2	These tend to be very reactive molecules.
3. Expanded octet	S, Cl, Br, I, Xe	SF_4, SF_6, XeF_4	The central atom must have a *d* subshell to expand into. For a discussion of how these molecules form, see Chapter 10 of this review book.

Bond energy calculations

Bond energy is *a measure of the strength of a covalent bond* and bond energy values can be used to estimate the enthalpy of gaseous reactions. Multiple bonds contain significantly more energy than single bonds. It takes energy to break a bond. The breaking of a bond is an endothermic process and therefore the *reactant molecule bond energies are assigned positive values.* Energy is released when a bond is formed. Bond formation is an exothermic process and therefore the *product molecule bond energies are assigned negative values.* Some sample bond energies are given below:

Bond	Bond energy (kJ/mol)
H — H	436
O — O	142
O = O	499
H — O	460
H — F	568
H — Cl	432

Example 4. Estimating the enthalpy of reaction using bond energies.

Given the reaction $H_2 (g) + \frac{1}{2}O_2 (g) \rightarrow H_2O (g)$, predict the approximate enthalpy of reaction for the combustion of hydrogen gas using bond energies.

Solution to Example 4.

The general strategy for using bond energies to estimate the ΔH_{rxn} involves three steps:
1. Determine the types of bonds broken and formed.
2. Assign positive values to the bonds that are broken (energy is put into system) and assign negative values to the bonds that are formed (energy is released). Multiply the bond energy values to reflect the molar relationships (coefficients) of the balanced chemical equation.
3. Algebraically sum the bond energies to determine the ΔH_{rxn}.

General Strategy	Solution to Example 4
1. Identify the bonds in the reactant and product molecules.	$H_2 (g) + \frac{1}{2}O_2 (g) \rightarrow H_2O (g)$ H –H ½ O=O 2 H – O
2. Assign + values to the bond energies of the reactant molecules. Assign – values to the bond energies of the product molecules. Multiply the bond energy values to reflect the molar relationships (coefficients) of the balanced chemical equation.	$H_2 (g) + \frac{1}{2}O_2 (g) \rightarrow H_2O (g)$ H –H ½ O=O 2 H – O **+436** **+ ½ (499)** **– 2(460)**
3. Sum the bond energies to determine the ΔH_{rxn}	$+436 + \frac{1}{2} (499) - 2(460) = -235 \ kJ/mol = \Delta H_{rxn}$

SAMPLE MULTIPLE CHOICE QUESTIONS

Questions 1 – 4 refer to the following compounds.

> A. O_3
> B. HF
> C. CsF
> D. SrF_2
> E. NH_3

1. Which compound has the highest lattice energy?
2. Which compound exhibits hydrogen bonding and can act as a Lewis base?
3. Which compound exhibits the largest electronegativity difference in its bonding?
4. Which compound exhibits resonance?

5. Which molecule has the shortest bonds?

> A. NO
> B. CO
> C. CO_2
> D. C_2H_4
> E. O_3

6. Which compound contains both ionic and covalent bonds?

> A. NaCl
> B. $NaNO_3$
> C. NaH
> D. H_3PO_4
> E. None of these

7. Which is the correct expression for calculating the ΔH_f for NH_3 (g) in kJ/mol based on bond energy data? A partial table of bond energies is provided.

Bond	Bond Energy (kJ/mol)
H $-$ H	436
H $-$ N	393
N $-$ N	193
N $=$ N	418
N $\equiv$ N	941

> A. 3 (393) – 1 ½ (436) – ½ (941)
> B. 393 – 436 – 941
> C. 3 (436) + 941 – 6 (393)
> D. 1 ½ (436) + ½ (941) – 3 (393)
> E. None of these is correct

8. All of the following are true about RbF *except:*

    A. It conducts electricity in the solid state.
    B. It has a high boiling point.
    C. It dissolves in NH_3 *(aq)*.
    D. It is an electrolyte.
    E. It conducts electricity in the molten state.

9. Arrange the following according to *increasing* bond polarity:

    HF   CsF Br_2 SO_2 SrO

    A. $CsF < SrO < SO_2 < HF < Br_2$
    B. $Br_2 < HF < SO_2 < SrO < CsF$
    C. $Br_2 < SO_2 < HF < SrO < CsF$
    D. $Br_2 < SO_2 < HF < CsF < SrO$
    E. $Br_2 < HF < SO_2 < CsF < SrO$

10. Which of the following statements are true about I_2?

    I. I_2 is a solid at standard conditions.
    II. I_2 dissolves in CCl_4.
    III. I_2 exhibits nonpolar covalent bonding.
    IV. I_2 has a very high boiling point.
    V. I_2 is a conductor in the molten state.

    A. Statements I, II, and III are true.
    B. Statements I and III are true.
    C. Statements I, II, III, and IV are true.
    D. Statements I, II, and V are true.
    E. All statements are true.

Comprehension Questions

1) a) What is the octet rule, and what are its limitations?

    b) What is formal charge, and how can it be used to differentiate between various Lewis structures for a compound?

    c) What are the advantages and limitations of using resonance structures to represent compounds?

2) The following questions refer to the formation of phosphorus pentachloride from phosphorus trichloride and elemental chlorine as depicted below.

$$PCl_3 \ (g) + Cl_2 \ (g) \rightarrow PCl_5 \ (g)$$

Bond Type	Bond Dissociation Energy (kJ / mol)
P — Cl	326
Cl — Cl	243

a) Draw Lewis structures for elemental chlorine and phosphorus pentachloride.

b) Estimate the enthalpy change associated with the above reaction using the bond dissociation energies provided. (Hint: this involves a calculation.)

3) a) Draw Lewis structures for the following two compounds: NaCl and BrCl.

b) NaCl has a boiling point of 1413 °C, whereas BrCl has a boiling point of 5 °C. Account for this difference using your understanding of chemical bonding and molecular structure.

ANSWERS TO SAMPLE MULTIPLE CHOICE QUESTIONS

1. D

The lattice energy of the ionic solid SrF_2 is greater than that of the ionic solid CsF because the 2^+ to 1^- ion attraction in SrF_2 is a stronger force than the 1^+ to 1^- ion attraction in CsF.

2. E

NH_3 exhibits hydrogen bonding. Hydrogen bonding occurs among molecules in which the hydrogen atom is bonded to F, O, or N atoms. NH_3 has a lone pair of electrons and *can act as an electron pair donor, that is, a Lewis base.*

3. C

When Family 1A metals combine with the most electronegative element on the periodic table, F, an ionic bond results due to the large difference in electronegativity between the atoms.

4. A

O_3 exhibits resonance. Its true structure is a composite of the following two structures:

$$: \ddot{O} = \ddot{O} - \ddot{O} : \quad \Leftrightarrow \quad : \ddot{O} - \ddot{O} = \ddot{O} :$$

5. B

CO has a triple bond. A triple bond is shorter than a double bond (as in NO, CO_2, and C_2H_4), which is shorter than the bonds in O_3 (which has a bond order of 1.5 as shown in question #4).

6. B

The bond between Na^+ and NO_3^- is ionic. The bond between N and O in NO_3^- is covalent.

7. D

	$1\frac{1}{2}H_2\ (g)\ +\ \frac{1}{2}N_2\ (g) \rightarrow NH_3\ (g)$		
Bonds:	$1\frac{1}{2}$ H—H	$\frac{1}{2}$ N≡N	3 N – H
Bond energy	$+1\frac{1}{2}\ (436)$	$+\ \frac{1}{2}\ (941)$	$-\ 3(393)$
Sum	$+1\frac{1}{2}\ (436)\ +\ \frac{1}{2}\ (941) - 3(393)\ =\Delta H_{rxn}$		

Be sure not to confuse these bond energy calculations with ΔH_f where $\Delta H_{rxn} = \sum \Delta H_{f,\ products}\ -\ \sum \Delta H_{f,\ reactants}$.

8. A

RbF is an ionic solid and <u>cannot</u> conduct in the solid state. The ions are fixed in the crystalline structure and are not free to move, thus an ionic solid cannot conduct electricity. The characteristics listed in B through E are true for ionic solids.

9. C

$Br_2 < SO_2 < HF < SrO < CsF$ The relative increase in polarity of the molecules is based on the increasingly large differences in electronegativity between the atoms. Br atoms have no electronegativity difference and thus the Br- Br bond is the least polar (i.e., 100% covalent character). CsF has the largest electronegativity difference between the atoms (Family 1A metal combining with the very electronegative element, F) and thus forms the most polar bond. CsF is an ionic solid. SrO has the next largest electronegativity difference after CsF. Family 2A metals tend to form ionic solids when combined with very electronegative elements such as F, O, or N. HF is more polar than SO_2.

10. A

Iodine is a nonpolar compound. It, like all nonpolar compounds, dissolves in nonpolar solvents such as CCl_4. A relatively large molecule, I_2 is a solid at room temperature and pressure (in contrast to the halogens F_2 and Cl_2, which are gases and Br_2 which is a liquid at these conditions).

Answers to Comprehension Questions

1) a)   The octet rule, formulated by Lewis, states that atoms other than hydrogen tend to form bonds until they are surrounded by eight valence electrons. The problem with the octet rule is that many compounds do not follow it. Elements in Group 2 form compounds with four valence electrons, and elements in Group 3 tend to form compounds with six valence electrons. Also, as you move to the third period and higher, elements often form compounds with 10 or more valence electrons.

   b)   Formal charge is an electron accounting system used when drawing Lewis structures. Electrons associated with an atom in a structure are counted (all of the lone pair electrons and half of the shared electrons) and compared to the number of valence electrons the atom had in its elemental form. The difference between these two values is the formal charge (negative for excess electrons assigned to the atom in a structure and positive for less than the original number).
         If more than one valid Lewis structure can be drawn for a molecular formula, the structure with the smallest absolute value sum of formal charge is considered to be the most accurate representation of bonding for the structure.

   c)   Resonance structures are used when two or more equivalent Lewis structures can be drawn for a compound (the only difference between the structures is the distribution of several multiple bonds). It is understood that the true structure for the compound is a hybrid of the several resonance structures drawn. The disadvantage of these structures is that they insinuate that there are several possible isomeric forms of a molecule rather than a single structure. This disadvantage can sometimes be overcome by using dashed lines to represent bonds with fractional bond orders.

2) a)

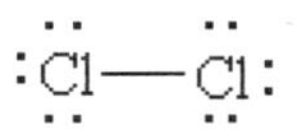

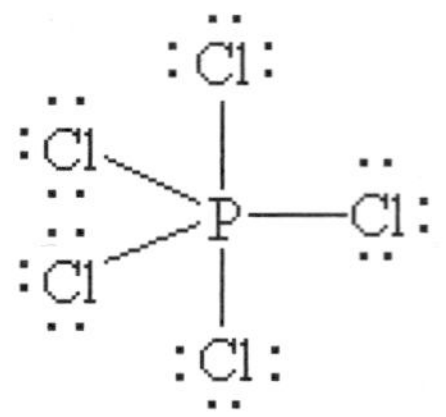

:Cl—Cl:

```
        :Cl:
         |
  :Cl\   |
     >P—Cl:
  :Cl/   |
         |
        :Cl:
```

b) When estimating the enthalpy change for a reaction using bond energies, the idea is to subtract the total energy released when all new bonds are formed from the total energy required to break all of the bonds that must be broken to accomplish the transformation. The formula used for the calculation is shown:

$$\Delta H^0_{rxn} = \Sigma\,BE(\text{reactants}) - \Sigma\,BE(\text{products}) \qquad BE = \text{bond energy}$$

In this case 1 mol of Cl – Cl bonds must be broken and 2 mol of P – Cl bonds must be formed.

$$\mathbf{\Delta H^0_{rxn} = (1\,mol \times 243\,kJ/mol) - (2\,mol \times 326\,kJ/mol) = -409\,kJ}$$

3) a)

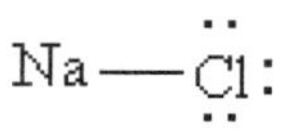

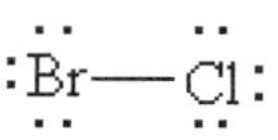

Na—Cl: :Br—Cl:

b) The Lewis structure for NaCl represents an ionic compound. Ionic compounds exist in a three-dimensional form known as a crystal lattice in which alternating positive and negative ions form strong attractions to a number of oppositely charged particles. BrCl, on the other hand, is a molecular compound. The bond between the two atoms within a molecule is quite strong; however, the attractions that exist between these diatomic molecules are relatively weak (i.e., it is easy to separate one BrCl molecule from the next). The reason NaCl is ionic and BrCl is molecular has mostly to do with the fact that there is a very large difference in electronegativity between Na and Cl, and there is a very small electronegativity difference between Br and Cl.

Since it is very hard to separate particle from particle in NaCl, it requires the input of significant amounts of energy to cause it to change from the liquid to the vapor phase (i.e., boiling) when compared with BrCl, thus the significantly higher boiling point of NaCl.

CHAPTER 10
CHEMICAL BONDING II:
MOLECULAR GEOMETRY AND HYBRIDIZATION
OF ATOMIC ORBITALS

This chapter reviews molecular geometry, which is important in predicting the physical and chemical properties such as boiling point, melting point, and the reactivity of molecules. Important topics include:

- Molecular geometry and VSEPR
- Dipole moments
- Hybridization of atomic orbitals
- Hybridization in molecules containing double and triple bonds

Molecular geometry and VSEPR

Molecular geometry is the *three-dimensional arrangement of atoms* in a molecule. The general principle governing the geometry of a molecule is the *minimization of the repulsion between electron domains*. An **electron domain** is *composed of two electrons* (either in a chemical bond or an unbonded electron pair). This approach to determining molecular geometry is called the **valence-shell electron-pair repulsion (VSEPR)** model.

The _geometry of the molecule_ *is based on the* _geometry of the atoms_ *to each other*. The three-dimensional arrangement of the atoms is determined by two main factors:
1. the presence of neighboring atoms
2. the presence of lone pairs of electrons on the central atom.

Electron domains repel each other. Like charges repel. The most stable configuration is the one that maximizes the distances between the domains. Depending on the number of domains, there are five basic geometries that minimize repulsions:

- *two domains*   *linear (180° angle)*
- *three domains*   *trigonal planar (120° angles)*
- *four domains*   *tetrahederal (109.5° angles)*
- . *five domains*   *trigonal bipyramidal (120° and 90° angles)*
- *six domains*   *octahedral (90° angles)*

Molecular geometry when the central atom has no lone pairs

When the central atom in a molecule has no lone pairs of electrons, there are five basic *molecular geometries*, as seen in Table 10.1.

TABLE 10.1

	Arrangement of Electron Pairs About a Central Atom (A) in a Molecule and Geometry of Some Simple Molecules and Ions in Which the Central Atom Has No Lone Pairs			
Number of Electron Pairs	**Arrangement of Electron Pairs***	**Molecular Geometry***	**Examples**	
2	180° :—A—: Linear	B—A—B Linear	$BeCl_2$, $HgCl_2$	
3	120° Trigonal planar	Trigonal planar	BF_3	
4	109.5° Tetrahedral	Tetrahedral	CH_4, NH_4^+	
5	90° 120° Trigonal bipyramidal	Trigonal bipyramidal	PCl_5	
6	90° 90° Octahedral	Octahedral	SF_6	

* The colored lines are used only to show the overall shapes; they do not represent bonds.

Molecular geometry when the central atom has one or more lone pairs

However, the situation changes when the central atom has one or more lone electron pairs. Lone-pair electron to lone-pair electron repulsions are slightly greater than bonding pair to bonding pair repulsions and this affects the molecular geometry. For instance, H_2O has two bonded atoms and two lone pairs of electrons on the central atom, O. Since the four electron domains are not identical, the electron domain geometry is nonideal tetrahedral and the *geometry of the molecule is bent*. Because the unbonded electron pairs have a greater repulsion to each other than the H — O bonds have to each other, the bond angle in H — O — H is 104.5°, not 109.5° as in an ideal tetrahedral structure.

Table 10.2 provides a visual review of the geometries that result when the central atom has one or more lone pairs.

TABLE 10.2

Geometry of Simple Molecules and Ions in Which the Central Atom Has One or More Lone Pairs

Class of molecule	Total number of electron pairs	Number of bonding pairs	Number of lone pairs	Arrangement of electron pairs*	Geometry	Examples
AB_2E	3	2	1	Trigonal planar	Bent	SO_2
AB_3E	4	3	1	Tetrahedral	Trigonal pyramidal	NH_3
AB_2E_2	4	2	2	Tetrahedral	Bent	H_2O
AB_4E	5	4	1	Trigonal bipyramidal	Distorted tetrahedron (or seesaw)	SF_4
AB_3E_2	5	3	2	Trigonal bipyramidal	T-shaped	ClF_3
AB_2E_3	5	2	3	Trigonal bipyramidal	Linear	I_3^-
AB_5E	6	5	1	Octahedral	Square pyramidal	BrF_5
AB_4E_2	6	4	2	Octahedral	Square planar	XeF_4

* The colored lines are used to show the overall shape, not bonds.

Another way to represent the information in Table 10.2 is:

- **three electron domains**   (nonideal trigonal planar is the geometry of the domains)
  two bonding pairs and one lone pair   Molecular geometry: <u>bent</u>

- **four electron domains**   (nonideal tetrahederal is the geometry of the domains)
  three bonding pairs and one lone pair   Molecular geometry : <u>trigonal pyramidal</u>
  two bonding pairs and two lone pairs   Molecular geometry: <u>bent</u>

- **five electron domains**   (nonideal trigonal bipyramidal is the geometry of the domains. Note that lone pairs assume equatorial positions.)
  four bonding pairs and one lone pair   Molecular geometry: <u>seesaw (distorted tetrahedron)</u>

- 161 -

three bonding pairs and two lone pairs   *Molecular geometry: <u>T-shaped</u>*
two bonding pairs and three lone pairs   *Molecular geometry: <u>linear</u>*

- **six domains** (nonideal octahedral is the geometry for the domains)
  five bonding pairs and one lone pair   *Molecular geometry: <u>Square pyramidal</u>*

  four bonding pairs and two lone pairs   *Molecular geometry: <u>Square planar</u>*

Now that you are familiar with Table 10.1 and Table 10.2, you can proceed to predict molecular geometries for molecules that you have never seen. The strategy for predicting molecular geometries using the VSEPR model is:

1. Draw the *Lewis structure* (see Chapter 9 in this review book).
2. Count the number of *electron domains around the central atom*. Double and triple bonds count as one domain. Use Table 10.1 to determine the overall geometry of the electron domains.
3. If there are no lone pairs on the central atom, the electron domain geometry is the molecular geometry (the geometric arrangement of the atoms to each other). If lone pairs are present on the central atom, use Table 10.2 to predict the molecular geometry.

Example 1. Predicting molecular geometries using the VSEPR model.

Predict the geometry of NO_3^-.

Solution to Example 1.
1. The Lewis structure is a resonance structure.

$$\left[:\overset{..}{O} - \overset{..}{N} = \overset{..}{O}: \right]^{-} \quad \Leftrightarrow \quad \left[:\overset{..}{O} = \overset{..}{N} - \overset{..}{O}: \right]^{-}$$

2. There are a total of three domains around N, the central atom. The double bond, the single bond, and the lone electron pair each count as one domain. Since the three domains are not identical, the electron domain geometry is nonideal trigonal planar. It is not an ideal trigonal planar geometry because the unbonded electron pair has greater repulsion with the $N - O$ bonds than the chemical bonds have to each other.

3. The *molecular geometry* (geometry of the atoms to each other) is *bent*. The bond angle is less than $120°$.

Note that *all* the electron domains (bonded and unbonded electrons) are vital in determining the bond angle, but the *name of the shape* (molecular geometry) only takes into account *the bonded domains*.

Dipole moments

Dipole moments *are the quantitative measure of the polarity of a molecule.* Dipole moments depend on the differences in electronegativity of the atoms that form bonds in the molecule and the overall geometry of the molecule. Diatomic molecules containing atoms of different elements with significantly different electronegativities *have a dipole moment* and are called **polar molecules** (such as HCl). Diatomic molecules that contain atoms of the same element *do not have a dipole* and are called **nonpolar molecules** (such as H_2).

Molecules can contain polar bonds but still be nonpolar as a whole. Bond moments are vector quantities and if they are symmetrically arranged in the molecule, they cancel. For instance, the bonds between C and O in CO_2 are polar, but the linear symmetry of the two bonds [O=C=O] means the bond moments cancel. CO_2 is a nonpolar molecule and has no dipole moment. In general, if the central atom is surrounded by atoms of the same element and has one of several geometries (linear, trigonal planar, tetrahedral, trigonal bipyramidal, or octahedral), the bond moments cancel and the molecule is nonpolar. Conversely, if the central atom has a lone pair of electrons the molecule tends to be polar and does have a dipole moment.

Example 2. Predicting dipole moments.

Predict which of the following molecules have a dipole moment: CHF_3 , H_2S, BeF_2, SF_4 , PCl_5.

Solution to Example 2.
Strategy: Look at the electronegativity differences of the elements that form each bond in the molecule. Determine the relative polarity of each bond. Determine the geometry of the molecule. Remember that you can have polar bonds form between atoms, but if the dipole moments cancel out due to the symmetry of the structure, then the molecule is nonpolar.

For *CHF_3*: C is the central atom. F is very electronegative, H is not. There are four bonded electron domains around C. The bonds are not identical and the bond angles will not all be 109.5°. The bond moments do not cancel. Therefore, the molecule *has a dipole moment.*

For *H_2S*: S is the central atom. The H — S bond is polar. The S atom has two bonded electron domains and two unbonded domains (two lone electron pairs). The electron domains form a nonideal tetrahedral structure. The molecule has bent geometry. The bond moments do not cancel and the molecule *has a dipole moment.*

For *BeF₂*: Be is the central atom. The Be — F bond is polar. The two Be –F bonds are in a linear arrangement. Be has no lone electron pairs. The bond moments cancel out. The molecule has *no dipole moment.*

For *SF₄*: S is the central atom. The S — F bond is polar. There are four bonded electron domains and one unbonded domain (one lone pair of electrons on S). These five electron domains have a nonideal trigonal bipyramidal structure. The molecular geometry is seesaw (also called a distorted tetrahedron). The bond moments do not cancel. The molecule *has a dipole moment.*

For *PCl₅*: P is the central atom. The P — Cl bond is polar. There are five identical bonded electron domains around the central atom. They have a trigonal bipyramidal structure. The bond moments cancel out in this highly symmetrical structure. The molecule *has no dipole moment.*

Hybridization of atomic orbitals

The **valence bond theory** assumes that the *electrons in a molecule occupy atomic orbitals of the individual atoms.* For example, in H_2 the covalent bond is formed by the overlap of the two $1s$ orbitals; in F_2, the covalent bond is formed by the overlap of the $2p$ orbitals. A stable molecule forms from the reacting atoms when the potential energy of the system reaches a minimum.

Hybridization is a theoretical model to further explain bonding. **Hybridization** is the *mixing of at least two nonequivalent atomic orbitals to produce hypothetical* **hybrid orbitals** such as sp, sp^2, sp^3, sp^3d, sp^3d^2. Table 10.4 summarizes important hybrid orbitals and their shapes (see Chapter 7 of this review book for quantum theory).

TABLE 10.4

Important Hybrid Orbitals and Their Shapes

Pure Atomic Orbitals of the Central Atom	Hybridization of the Central Atom	Number of Hybrid Orbitals	Shape of Hybrid Orbitals	Examples
s, p	sp	2	180° Linear	$BeCl_2$
s, p, p	sp^2	3	120° Trigonal planar	BF_3
s, p, p, p	sp^3	4	109.5° Tetrahedral	CH_4, NH_4^+
s, p, p, p, d	sp^3d	5	90° 120° Trigonal bipyramidal	PCl_5
s, p, p, p, d, d	sp^3d^2	6	90° 90° Octahedral	SF_6

The general strategy for determining the hybridization of the central atom and the geometry of the molecule is:

1. Write the ground-state electron configuration of the central atom. You can use orbital notation as well.
2. Predict the overall geometry based on the VSEPR method. The number of electron domains determines the overall geometry as shown in Table 10.1.
3. Hybridize the central atom to obtain the appropriate geometry with all the bonded atoms.
4. Use Table 10.4 to determine the type of hybridization and resulting geometry.

Example 3. Determining hybridization and molecular geometry.

Determine the hybridization of the central atom and the geometry of the molecule for the following molecules: CH_4, BCl_3, SeF_6.

Solution to Example 3.

Strategy: Use steps 1–4 of the general strategy.

For *CH_4*:
The ground-state electron configuration for the central atom, carbon, is $1s^2 2s^2 2p^2$. C has four valence electrons. The two half-filled $2p$ orbitals could form two bonds with H, producing CH_2. However, the simplest molecule of C and H found in nature is CH_4, methane. There are four equivalent bonds around C. The *C atom is hybridized*; the $2s$ orbital is mixed with the three $2p$ orbitals, creating *four sp^3 hybrid orbitals* around C. Four identical electron domains result in *a tetrahedral structure with 109.5° angles* (see Table 10.1). Remember that sp^3 *hybridization yields tetrahedral structures* (see Table 10.4).

You can also use orbital notation (see Chapter 7 of this review book) to visualize the hybridized central atom.

The orbital notation of the ground state of the central atom C (*not hybridized*) is:

↑↓		↑		↑		
$2s$		$2p$		$2p$		$2p$

The orbital notation of the central atom C *(four sp^3 hybrid orbitals) is*:

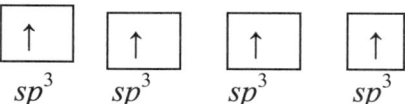

↑	↑	↑	↑
sp^3	sp^3	sp^3	sp^3

Notice how there are four equivalent half-filled orbitals available to bond with the four H atoms and form CH_4.

For *BCl_3*:
The ground-state electron configuration for the central atom, boron, is: $1s^2 2s^2 2p^1$. B has three valence electrons. The one half-filled $2p$ orbital could form one bond with Cl, producing BCl. However, the molecule found in nature is BCl_3. There are three equivalent bonds around B. The promotion of one of the $2s$ electrons into the $2p$ orbital yields *three sp^2 hybrid orbitals*. With three identical electron domains around the central atom, a *trigonal planar structure* with 120° angles is formed (see Table 10.1). Remember that sp^2 *hybridization yields trigonal planar structures* (see Table 10.4).

The orbital notation of the ground state of the central atom B (*not hybridized*) is:

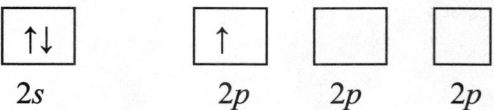

2s   2p   2p   2p

The orbital notation of the central atom B (*three sp^2 hybrid orbitals*) is:

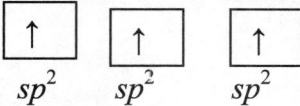

sp^2 sp^2 sp^2

For SeF_6 :

The ground-state electron configuration for the central atom, selenium, is: $[Ar]4s^2 3d^{10} 4p^4$. Se has six valence electrons. This ground-state configuration has the two half-filled p orbitals available for bonding, resulting in SeF_2. To form SeF_6, hybridization into the d sublevel occurs. One of the $4s$ electrons and one of the $4p$ electrons are promoted into the $4d$ sublevel, resulting in six $sp^3 d^2$ *hybrid orbitals*. With six identical electron domains around the central atom, an *octahedral structure with 90° angles* forms (see Table 10.1). Remember that $sp^3 d^2$ *hybridization yields octahedral structures* (see Table 10.4).

Hybridization in molecules containing double and triple bonds

Sigma bonds *lie in the plane between two nuclei*. **Pi bonds** are formed by the *overlap of the p orbitals above and below the plane of the molecule* and are present in multiple bonds. Single bonds between atoms form sigma bonds. For example, CH_4 has four sigma bonds. Double bonds between atoms form one sigma and one pi bond. For example, the C atoms in ethylene, C_2H_4, have a double bond (C is sp^2 hybridized and has a trigonal planar geometry). A triple bond is made from one sigma and two pi bonds. For example, the C atoms in acetylene, C_2H_2, have a triple bond (C is sp hybridized and has a linear geometry).

Take Note: *Molecular orbital theory is no longer part of the prescribed AP curriculum and, therefore, is not reviewed in this review book.*

SAMPLE MULTIPLE CHOICE QUESTIONS

Questions 1 – 4 refer to the following geometries.

A.   Linear
B.   Trigonal planar

C.   Tetrahdedral
D.   Angular (bent)
E.   Trigonal pyramid

1. The CO_2 molecule has this geometry.
2. The SF_2 molecule has this geometry.
3. The SO_3 molecule has this geometry.
4. The PH_3 molecule has this geometry.

5. Which of the following molecules has polar bonds but is a nonpolar molecule?

   A. NH_3
   B. $BeCl_2$
   C. CH_3F
   D. SF_4
   E.   None of these

6. Arrange the following molecules in order of increasing bond angle:

   H_2O, BF_3, SiF_4, NH_3, BeF_2

   A. $H_2O < BF_3 < SiF_4 < NH_3 < BeF_2$
   B. $BeF_2 < H_2O < BF_3 < SiF_4 < NH_3$
   C. $NH_3 < H_2O < BF_3 < SiF_4 < BeF_2$
   D. $H_2O < NH_3 < BF_3 < SiF_4 < BeF_2$
   E. $H_2O < NH_3 < SiF_4 < BF_3 < BeF_2$

7. Which of the following molecules has two pi bonds between the same two atoms?

   A. CO_2
   B. C_2H_4
   C. SF_2
   D. O_3
   E. CO

8. Which of the following molecules contains a central atom that is sp^3d hybridized?

   A. SF_6
   B. PF_5
   C. BrF_3
   D. PF_5 and BrF_3
   E.   All of these

Questions 9 and 10 refer to the following molecular geometries.

     A.   Trigonal bipyramid
     B.   Octahedron
     C.   Square pyramid
     D.   T-shaped
     E.   Seesaw

9. What is the molecular geometry of SF_4?

10. What is the molecular geometry of IF_5?

Comprehension Questions

1) a) Define VSEPR and explain how it helps to predict molecular geometry.

    b) What factors influence molecular polarity, and how does molecular polarity affect the physical properties of a compound?

2) a) Draw Lewis structures for BH_3 and NH_3.

    b) What is the hybridization and bond angle about the central atom for each of these molecules?

    c) What differences in physical and chemical properties would you predict for these two compounds based on differences in structure?

3) The following questions refer to the chlorination of ethylene depicted below.

$$C_2H_4 \ (g) + Cl_2 \ (g) \rightarrow C_2H_4Cl_2 \ (g)$$

Bond Type	Bond Dissociation Energy (kJ/mol)
C — C	346
C = C	602
C — H	414
Cl — Cl	243
C — Cl	327

a) Draw Lewis structures for ethylene, C_2H_4 (g), and dichloroethane, $C_2H_4Cl_2$ (g).

b) i)   What is the hybridization of the carbon atoms in the structure of ethylene?

    ii)   Predict the bond angle about the carbon atoms in ethylene.

    iii)   What name is given to the molecular geometry around the carbon atoms in ethylene?

c) i)   What is the hybridization of the carbon atoms in the structure of dichloroethane?

    ii)   Predict the bond angle about the carbon atoms in dichloroethane.

    iii)   What name is given to the molecular geometry around the carbon atoms in dichloroethane?

d) Estimate the molar enthalpy of chlorination using bond dissociation energies.

ANSWERS TO SAMPLE MULTIPLE CHOICE QUESTIONS

1. A

    CO_2 is a *linear molecule* with two domains around the central atom. :O=C=O:

2. D

    SF_2 is an *angular (bent) molecule* with four electron domains. Two of the domains are unbonded electron pairs and two are F atoms. The $S-F-S$ angle is less than than $109.5°$ since the two lone electron pairs repel each other more than the bonded electrons do. This structure should remind you of H_2O, which has a bond angle of $104.5°$.

3. B

    SO_3 is a *trigonal planar molecule* with three electron domains around the central atom. The structure exhibits resonance (i.e., there are three equivalent bonds around the central atom S).

    :O $-$ S$-$O:
    $\quad\quad\ \ ||$
    $\quad\quad$:O:

4. E

    PH_3 is a *trigonal pyramid* with four domains around the central atom, P, consisting of one unbonded electron pair and three bonded electron pairs. The bond angle between the atoms is less than $109.5°$ due to the electrostatic repulsion of the lone electron pair, which is greater than the repulsion between the bonded electron pairs. This structure is similar to NH_3, which has an approximate bond angle of $107°$.

5. B

    $BeCl_2$ has polar bonds due to the electronegativity difference between Be and Cl. The molecule is *sp* hybridized, which results in a linear geometry. The dipole moments cancel and the *molecule is nonpolar*. Don't be mislead by SF_4 (see Question 9).

6. E

	H_2O <	NH_3	< SiF_4	< BF_3	< BeF_2
geometry	Angular	trigonal pyramid	tetrahedral	trigonal planar	linear
Bond angles	*104°*	*107°*	*109°*	*120°*	*180°*

7. E

    CO has a triple bond and therefore one sigma bond and two pi bonds.

8. D

PF₅ and BrF₃ both have sp³d hybridization.
P has a ground-state electron configuration of $[Ne]3s^23p^3$, which is hybridized to $3s3p^33d$ and thus has *five* half-filled sp^3d orbitals.
Br has a ground-state electron configuration of $[Ar]4s^23d^{10}4p^5$, which is also hybridized to five sp^3d orbitals. In the case of Br, there are *three* half-filled orbitals and two lone electron pairs.

9. E

SF₄ has a seesaw shape. S has an ground-state electron configuration of $[Ne]3s^23p^4$, which is hybridized to $3s3p^33d$. One of the five sp^3d orbitals contains a lone pair of electrons. The five electron domains have a trigonal bipyramidal structure. The *molecular* geometry is a seesaw shape.

10. C

IF₅ is a square pyramid. Iodine has an ground-state electron configuration of $[Kr]5s^24d^{10}5p^5$, which is hybridized to sp^3d^2. One of the six sp^3d^2 orbitals contains a lone pair of electrons. The six electron domains have an octahedral structure. The *molecular* geometry is a square pyramid.

Answers to Comprehension Questions

1) a) VSEPR is an acronym that stands for *valence-shell electron-pair repulsion*, and is the principal theory guiding our understanding of factors that influence molecular shape. The outer regions of all atoms are essentially electron clouds. When atoms are bonded together to form molecules, the outer, negatively charged electrons of one atom become attracted to the positive charge of a neighboring nucleus. These electrons, however, are repelled by the valence electrons of the new adjacent atom. This tug-of-war between attraction to another nucleus versus repulsion from other electrons ultimately ends in an energy-minimized system in which multiple atoms bonded to a single central atom arrange themselves as far away from each other as possible without losing contact with the central atom. The resulting shapes of molecules are predictable and reproducible.

b) The factors affecting molecular polarity are polarity of individual bonds coupled with symmetry of the molecule. Sometimes a single polarized bond between two atoms with significantly differing electronegativity, can bring about overall molecule polarization. Alternatively, if polarized bonds are symmetrically distributed around a central atom, a molecule may have no net polarity.

A second factor influencing molecular polarity is asymmetric distribution of lone pairs. Whether a molecule contains polar or nonpolar bonds it can be quite polar based on the presence of lone pairs. The "side" of the molecule is usually going to be the negative end, that is, the side with the greatest electron density. Again, if the lone pairs are symmetrically arranged about the central atom, their polarizing effect can be cancelled, resulting in a nonpolar molecule.

2) a)

b) Borane, BH_3, has sp^2 hybridization and a bond angle of 120°. Ammonia is sp^3 hybridized with a bond angle of about 107°. Ammonia's bond angle deviates from the usual 109.5° bond angle for sp^3 hybridization due to the presence of the lone pair (which has a slightly greater repulsive effect than a hydrogen atom; see VSEPR).

c) Because of its symmetry, borane is a nonpolar molecule. The nonpolar character along with its low molecular weight should cause it to be a gas at room temperature, and it would likely be soluble in nonpolar (i.e., organic) solvents and poorly soluble in polar solvents such as water. Also, because it is two electrons short of an octet, it might be expected to behave chemically as a Lewis acid.

Ammonia is a polar molecule, based on the fact that it has polarized bonds and a single lone pair (both of these tending to leave excess electron density on the side of the molecule where the lone pair is located). This would lead to the prediction that ammonia would have an anomalously high boiling point compared to other molecules with a molar mass of 17, and that it would be soluble in polar solvents such as water. Chemically, ammonia should react by donating a pair of electrons, that is, act as a Lewis base.

3) a)

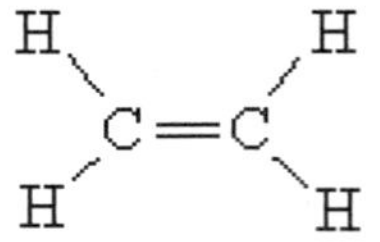

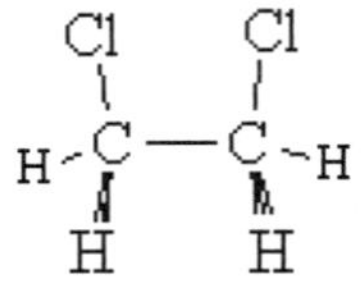

b) i)   Ethylene's carbon atoms are sp^2 hybridized.

   ii)   The bond angle is approximately 120°.

   iii)   The shape at the carbon centers is called trigonal or trigonal planar.

c) i)   Dichloroethane's carbon atoms are sp^3 hybridized.

   ii)   The bond angles are approximately 109.5°.

   iii)   The shape at the carbon centers is called tetrahedral.

d) When estimating the enthalpy change for a reaction using bond energies, the idea is to subtract the total energy released when all new bonds are formed from the total energy required to break all of the bonds that must be broken to accomplish the transformation (see answers to Chapter 9 problems for an additional example). The formula used for the calculation is seen below:

$$\Delta H^0_{rxn} = \Sigma\, BE(\text{reactants}) - \Sigma\, BE(\text{products}) \qquad BE = \text{bond energy}$$

In this case 1 mol of Cl — Cl bonds and 1 mol of C = C double bonds must be broken, and 2 mol of C — Cl bonds and 1 mol of C — C single bonds must be formed.

$\Delta H^0_{rxn} = [(1 \text{ mol} \times 243\,\text{kJ/mol}) + (1 \text{ mol} \times 602\,\text{kJ/mol})] - [(2 \text{ mol} \times 327\,\text{kJ/mol}) + (1 \text{ mol} \times 346\,\text{kJ/mol})] = \mathbf{-155\,kJ}$

CHAPTER 11
INTERMOLECULAR FORCES, LIQUIDS, AND SOLIDS

This chapter reviews the structure and properties of liquids and solids. The focus is on intermolecular forces to explain the observed properties. Important topics include:

- The three states of matter
- Intermolecular forces
- Properties of liquids and solids
- Types of crystals
- Phase changes and phase diagrams

The three states of matter

The three states of matter are gas, liquid, and solid. Table 11.1 provides an overview of their properties.

TABLE 11.1

Characteristic Properties of Gases, Liquids, and Solids

State of Matter	Volume/Shape	Density	Compressibility	Motion of Molecules
Gas	Assumes the volume and shape of its container	Low	Very compressible	Very free motion
Liquid	Has a definite volume but assumes the shape of its container	High	Only slightly compressible	Slide past one another freely
Solid	Has a definite volume and shape	High	Virtually incompressible	Vibrate about fixed positions

Intermolecular forces

Intermolecular forces are *attractive forces between molecules* and determine macroscopic properties such as melting and boiling points. **Intramolecular forces** *hold atoms together in a molecule* (i.e., chemical bonds) and stabilize individual molecules. Molecules in the gaseous state are far apart, in constant random motion, and have few, if any, attractive intermolecular forces. Liquids and solids, on the other hand, are condensed states of matter and have strong intermolecular forces.

There are several types of intermolecular forces. The chart gives an overview of this topic:

Intramolecular force	Intermolecular force	Examples
Ionic bond	Electrostatic force	$NaCl$, BaF_2
Covalent bond	Hydrogen bonding	H_2O, NH_3

| | Dipole-dipole | H_2S, PH_3 |
| | Dispersion forces | He, SF_6 |

| Covalent bond forming network solid | Covalent bond | C(diamond), SiO_2 |

| Metallic bond | "Sea of electrons" | Ag, Na |

Take Note: *You need to be clear on the difference between intramolecular forces (i.e., chemical bond types) and intermolecular forces. Questions dealing with these forces are frequently asked in both sections of the AP exam.*
For example, when you are asked about boiling point and melting point trends, you need to focus on the type of intermolecular force present. If you are asked about lattice energies and relative bond energies, you need to focus on the intramolecular forces.

Hydrogen bonding *is a relatively strong intermolecular force that exists between molecules* where the very electronegative elements of *N, O, or F are covalently bonded to H.* Hydrogen bonding occurs between hydrogen and the unbonded electron pairs of nearby N, O, or F molecules. Examples of molecules exhibiting hydrogen bonding are H_2O, NH_3, and CH_3OH.

Dipole-dipole forces are *attractive forces between polar molecules.* The negative dipole of one molecule is attracted to the positive dipole of a nearby neighbor. The larger the dipole moment, the stronger the force of attraction is. Examples of molecules exhibiting dipole-dipole interactions are H_2S, CO, and HBr.

Note that hydrogen bonding is really a type of very strong dipole-dipole interaction. O and S are in the same chemical family, so you would expect similar properties. H_2O has a normal boiling point of 100 °C, and H_2S has a boiling point of –60 °C. This variation in boiling point is due to the difference in the strength of their intermolecular forces.

Ion-dipole forces *are attractive forces between an ion and a polar molecule.* Hydration where water molecules orient themselves around an ion is an example.

Dispersion forces (also called London dispersion forces) are attractive forces among nonpolar molecules and noble gas atoms. The dispersion forces arise as a result of temporary dipoles induced in atoms and molecules. The strength of the dispersion force depends on the polarizability of the molecule. In general, the polarizability increases as the total number of electrons increases. Examples of atoms and molecules exhibiting dispersion forces are He, Kr, CO_2, and SF_6. As the polarizability of the molecule increases (because of the increase in the number of electrons), so does the melting point, due to the increasing strength of the dispersion forces.

Example 1. Boiling and melting point trends.

    a. Arrange the following according to increasing boiling point: H_2O, H_2S, HF, He, CO_2. Justify your ranking.

    b. Arrange the following according to increasing melting point: CH_4, CI_4, CF_4. Justify your ranking.

Solution to Example 1.

You are *assessing the relative strength of the intermolecular forces when predicting relative melting and boiling points*. The stronger the forces between molecules, the more energy you have to put into the system to separate the molecules. In general, dispersion forces are the weakest and hydrogen bonding is the strongest intermolecular force.

    a. *$He < CO_2 < H_2S < HF < H_2O$.* He and CO_2 only have dispersion forces, with CO_2 being the more polarizable due to its larger number of electrons. H_2S has dipole-dipole interactions. HF and H_2O both have hydrogen bonding. H_2O has the stronger intermolecular force because it has two H atoms that can interact with nearby neighbors while HF can only form one H bond per molecule.

    b. *$CH_4 < CF_4 < CI_4$.* All three molecules are nonpolar and thus only have dispersion forces between them. The bigger the molecule, the more electrons and thus the more polarizable the molecule is. The larger the temporary dipole, the stronger the intermolecular force and thus the higher the melting point.

Properties of liquids and solids

Surface tension is the measure of the *elastic force in the surface of a liquid*. Liquids with strong intermolecular forces have high surface tension. It is what makes water bead into drops. Another property of liquids is **viscosity,** which is *a measure of a fluid's resistance to flow*. Fluids with strong intermolecular forces tend to have higher viscosities.

Water is an excellent solvent and has several unique properties. It has unusually high surface tension, a relatively large specific heat capacity (4.184 J/g·°C), an unusually high boiling point, and liquid water is actually denser than its solid form. The three-dimensional structure of ice separates the molecules further in the solid state than in the liquid state and thus the solid is less dense.

Solids are *crystalline* where the particles (atoms, molecules, or ions) *occupy specific positions in an ordered arrangement,* or **amorphous** where there is *no well defined arrangement of particles* such as in glass. **X-ray diffraction** is the *technique used to deduce the structure of a crystalline solid by observation* of the *scattering patterns of x-rays by the units of a crystalline solid.*

Types of crystals

There are four types of crystals: ionic, covalent, molecular, and metallic. Melting point, density, and hardness are among the properties determined by the kind of forces that hold the particles together. Table 11.4 is a summary of crystal types and associated properties.

TABLE 11.4

Types of Crystals and General Properties

Type of Crystal	Force(s) Holding the Units Together	General Properties	Examples
Ionic	Electrostatic attraction	Hard, brittle, high melting point, poor conductor of heat and electricity	$NaCl$, LiF, MgO, $CaCO_3$
Covalent	Covalent bond	Hard, high melting point, poor conductor of heat and electricity	C (diamond),[†] SiO_2 (quartz)
Molecular*	Dispersion forces, dipole-dipole forces, hydrogen bonds	Soft, low melting point, poor conductor of heat and electricity	Ar, CO_2, I_2, H_2O, $C_{12}H_{22}O_{11}$ (sucrose)
Metallic	Metallic bond	Soft to hard, low to high melting point, good conductor of heat and electricity	All metallic elements; for example, Na, Mg, Fe, Cu

* Included in this category are crystals made up of individual atoms.

† Diamond is a good thermal conductor.

> **Take Note**: *You need to be very familiar with the contents of Table 11.4 for the AP exam. In both sections of the AP exam you are likely to be asked questions in which you must identify the intramolecular forces and the intermolecular forces and predict general properties for a particular molecule.*

Phase changes and phase diagrams

Phase changes *are transformations from one phase to another.* They require that energy be added or removed. They are physical changes and the molecules remain chemically unchanged. An example is when ice is heated: heat + H_2O (*s*) → H_2O (*l*).

Endothermic transformations (where energy is absorbed) include:

melting	*(solid → liquid)*	ΔH_{fusion}
boiling	*(liquid → gas)*	$\Delta H_{vaporization}$
sublimation	*(solid → gas)*	

Exothermic transformations (where energy is released) include:
solidification *(liquid → solid)* $-\Delta H_{fusion}$
condensation *(gas → liquid)* $-\Delta H_{vaporization}$
deposition *(gas → solid)*

Liquid-vapor equilibrium. In a liquid at any given temperature, there are faster and slower moving molecules. The faster moving molecules may escape the liquid phase and become a gas (evaporation). If the liquid is in a closed container, some of the gas molecules return to the liquid phase. The **vapor pressure** has been attained when the *rate of evaporation is equal to the rate of condensation*. Equilibrium has been established at this point. The vapor pressure of a particular liquid depends only on the temperature of that liquid. When the *vapor pressure of the liquid is equal to the external pressure*, **boiling** occurs. Water has a vapor pressure of 0.5 atm at 82 °C. As the water temperature increases, so does the vapor pressure. At 100 °C, the vapor pressure of water is 1 atm. When the atmospheric pressure (i.e., the external pressure) is 1 atm, water boils at 100 °C, which is called the **normal boiling point** (*the boiling point of a liquid when the external pressure is 1 atm*). If the water is heated high on a mountain, where the atmospheric pressure is less, it will boil at a lower temperature. The molar heat of vaporization, ΔH_{vap}, is the energy required to vaporize 1 mol of a liquid. If the intermolecular attractions are strong, a liquid has a high ΔH_{vap}.

Liquid-solid equilibrium. The **melting point** of the solid and the **freezing point** of the liquid is the *temperature at which solid and liquid phases coexist in a dynamic equilibrium: solid ⇔ liquid. The* **molar heat of fusion,** ΔH_{fus}**,** is the *energy required to melt 1 mol of a solid.* Strong intermolecular forces mean relatively high ΔH_{fus} values.

Solid-vapor equilibrium. **Sublimation** is the process in which molecules go directly from the solid into the vapor phase. Deposition is when the molecules go directly from the vapor to the solid state. The **molar heat of sublimation,** ΔH_{sub}**,** is the *energy required to sublime 1 mol of a solid.*

Heating curves. When heated, most solids change to liquids and eventually to gases. A heating curve shows all three states of matter and their phase changes as a function of temperature. Figure 11.38 is a typical heating curve for a pure substance.

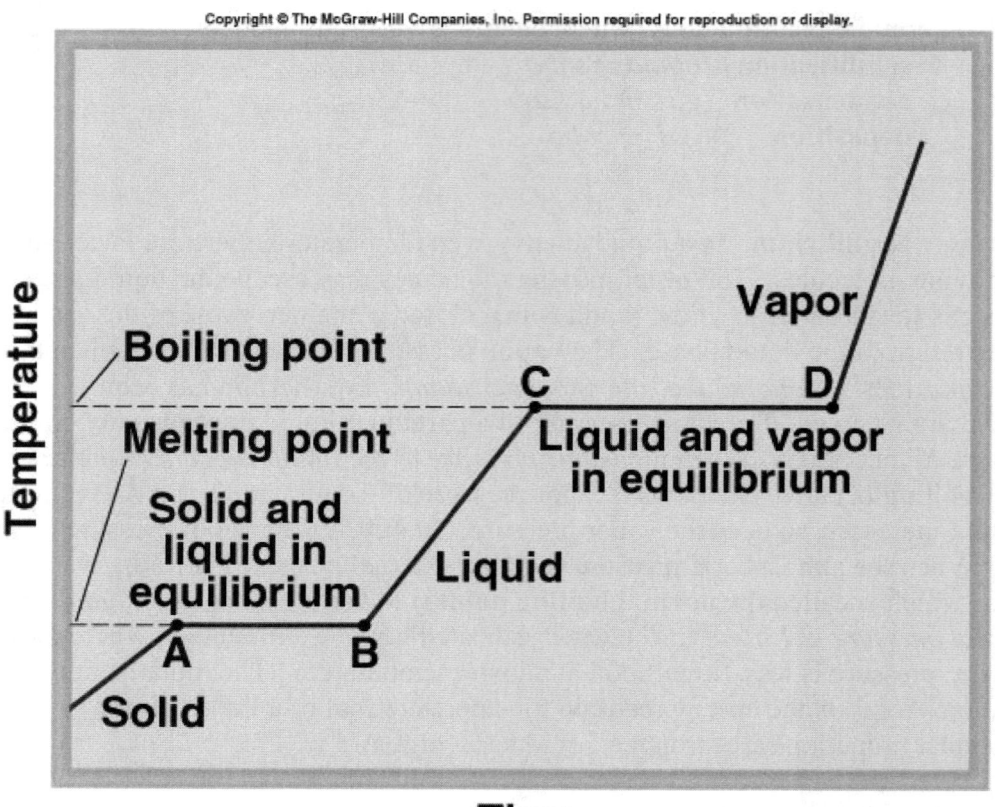

Figure 11.38

The areas on the curve are:

Point on the curve	State or change of state	Calculations relating to a particular part of the curve
Below point A	Solid is being heated	$q = m\,\Delta T\,C_{solid}$
From point A to B	Melting	$q = m\,\Delta H_{fus}$
From point B to C	Liquid is being heated	$q = m\,\Delta T\,C_{liquid}$
From point C to D	Boiling	$q = m\,\Delta H_{vap}$
Above point D	Gas is being heated	$q = m\,\Delta T\,C_{gas}$

(q = heat, m = mass, T = temperature, C = specific heat capacity, ΔH_{fus} = heat of fusion , ΔH_{vap} = heat of vaporization)

Example 2. Heating curve.

a. Calculate the amount of heat required to heat 20.0 g of ice from $-10.0\ °C$ to $15.0\ °C$. (The specific heat capacity for ice is 2.03 J/g·°C and water is 4.18 J/g·°C. The $\Delta H_{fusion} = 6.01$ kJ/mol and the $\Delta H_{vap} = 40.8$ kJ/mol for water.)

b. On the heating curve for water, do you expect the length of the plateau from point A to point B to be: less than, equal to, more than, or can't determine without further data, when compared to the plateau from point C to D? Explain your choice.

General Strategy	Solution to Example 2a

You have to divide your calculations into three stages:
    heating the solid ($q_1 = m \, \Delta T \, C_{solid}$)
    melting the solid ($q_2 = m \, \Delta H_{fusion}$)
    heating the liquid ($q_3 = m \, \Delta T \, C_{liquid}$)
 	
Note that you need to keep the units consistent: grams vs. moles and kilojoules vs. joules.
	Heating the solid from $-10\ °C$ to $0\ °C$
	$q_1 = m \, \Delta T \, C_{solid}$
	   $= (20.0\ g)\ (10\ °C)\ (2.03\ J/g·°C)$
	   $= 406\ J$
	
	Melting the solid (ice) at $0\ °C$
	$20.0\ g \times \dfrac{1\ mol\ H_2O}{18.0\ g\ H_2O} = 1.11\ mol\ H_2O$
	$q_2 = m \, \Delta H_{fusion} = (1.11\ mol)(\ 6.01\ kJ/mol)$
	      $= 6.67\ kJ = 6670\ J$
	
	Heating the liquid from $0\ °C$ to $15\ °C$
	$q_3 = m \, \Delta T \, C_{liquid}$
	    $= (20.0\ g)\ (15\ °C)\ (4.18\ J/g·°C)$
	    $= 1250\ J$

Add up the heat required for each stage to get the overall heat required: $q_{total} = q_1 + q_2 + q_3$	$q_{total} = q_1 + q_2 + q_3$ $q_{total} = 406\ J + 6670\ J + 1250\ J = 8330\ J$

Solution to Example 2b. The heating curve length from point A to B represents melting and will *be less* than the heating curve length from point C to D, which represents boiling. It takes 6.01 kJ to melt 1 mol of ice whereas it takes 40.8 kJ to vaporize (boil) 1 mol of liquid water at 100 °C. It takes much less energy to turn a solid into a liquid than it takes to turn a liquid into a gas. Thus, the plateau length that represents time will be shorter for the solid-to-liquid conversion.

Phase diagrams show the *conditions of temperature and pressure under which a substance exists as a solid, liquid, and gas.* The point at which all three phases coexist is the **triple point.** The line separating the solid and liquid phases represents melting at various pressures, the line between the liquid and vapor phases represents boiling, and the line between solid and vapor phases represents sublimation. The **critical temperature** is the temperature above which the vapor cannot be liquefied. A phase diagram for CO_2 is given in Figure 11.41.

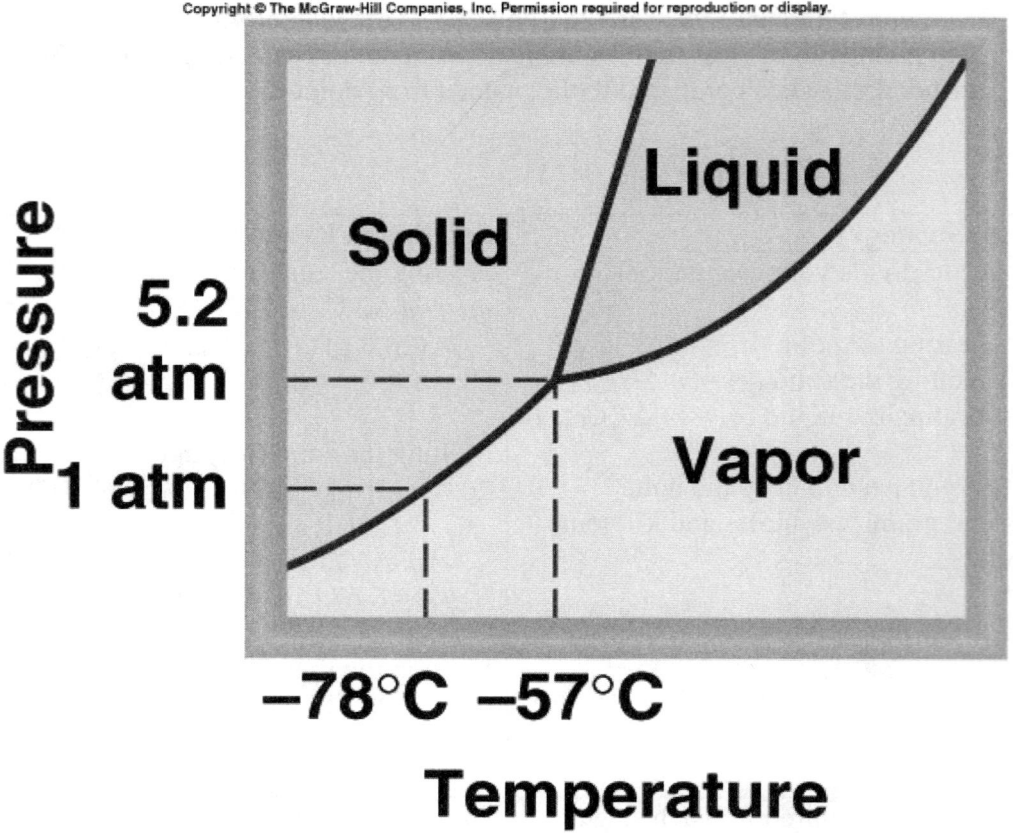

Figure 11.41

Example 3. Interpreting a phase diagram.

Using Figure 11.41, the phase diagram for CO_2, answer the following questions:

    a. At 5.4 atm and –65 °C, what is the phase of CO_2?
    b. What are the pressure and temperature at the triple point?
    c. In what phase does CO_2 exist at STP?
    d. Does solid CO_2 float in liquid CO_2? Explain your answer.
    e. Can dry ice, CO_2 (*s*), melt on your desk? Explain.

Solution to Example 3.
    As a solution strategy, you need to look at the T and P axes.
    a. Solid
    b. 5.2 atm and –57 °C
    c. STP is 0 °C and 1 atm of pressure. CO_2 is a gas at STP.
    d. No, it does not float. The solid-liquid interface has a positive slope.
    e. No, it cannot melt. The liquid state does not exist below 5.2 atm. Dry ice sublimes at room conditions.

SAMPLE MULTIPLE CHOICE QUESTIONS

1. Which of the following molecules displays hydrogen bonding?

   A. HF and CH_3CH_2CHO
   B. HF
   C. CH_3CH_2CHO
   D. H_2S
   E. All of the above

2. Rank the following according to *increasing* boiling point:
   He   Kr H_2O NH_3 CO   NaCl

   A. He < Kr < H_2O < NH_3 < CO < NaCl
   B. He < Kr < NH_3 < H_2O < CO < NaCl
   C. He < Kr < CO < NH_3 < H_2O < NaCl
   D. He < Kr < CO < H_2O < NH_3 < NaCl
   E. Kr < He < CO < H_2O < NH_3 < NaCl

3. Which of the following is an *incorrect* association?

   A. $I_2 (s) \Leftrightarrow I_2 (g)$ Sublimation
   B. $H_2O (l) \Leftrightarrow H_2O (s)$ ΔH_{fus}
   C. $H_2O (g) \Leftrightarrow H_2O (l)$ Condensation
   D. $CH_3OH (l) \Leftrightarrow CH_3OH (g)$ ΔH_{vap}
   E. $I_2 (g) \Leftrightarrow I_2 (s)$ Deposition

4. Arrange the following according to *increasing* melting point.
   CBr_4 Ne CH_4 CI_4 CF_4

   A. CBr_4 < CF_4 < CI_4 < CH_4 < Ne
   B. CI_4 < CBr_4 < CF_4 < CH_4 < Ne
   C. Ne < CI_4 < CBr_4 < CF_4 < CH_4
   D. Ne < CI_4 < CBr_4 < CF_4 < CH_4
   E. Ne < CH_4 < CF_4 < CBr_4 < CI_4

5. Which of the following statements is *incorrect?*

   A. Materials with a high specific heat can absorb a substantial amount of heat while undergoing only a small change in temperature.
   B. Liquids with weak intermolecular forces tend to have high vapor pressures.
   C. Liquids with weak intermolecular forces tend to have high boiling points.
   D. Liquids with strong intermolecular forces tend to have high surface tension.
   E. Cohesion forces refer to intermolecular attractions between like molecules.

6. Which of the following statements does not apply to $CCl_4 (l)$?

A. Has a definite volume and assumes the shape of its container
B. Is a nonpolar molecule with polar bonds
C. Has dipole-dipole interactions
D. Dissolves solid iodine
E. Has a lower normal boiling point than water

7. Which of the following molecules has a permanent dipole moment?

A. CO_2
B. CH_4
C. H_2S
D. BeH_2
E. N_2

8. Which of the following statements about phase diagrams is *false?*

A. The line separating liquid and vapor represents boiling.
B. The line separating solid and vapor represents sublimation.
C. A negative slope line between the solid and liquid indicates that the solid will float in the liquid.
D. The point at which all three phases coexist is called the critical point.
E. All of the statements are true.

9. Which of the following pairs have the same intramolecular forces and intermolecular forces?

A. NH_3 and PH_3
B. CO_2 and CO
C. H_2S and CO_2
D. C (diamond) and C (graphite)
E. CO_2 and CBr_4

10. Which of the following statements are true about diamond and graphite?

I. They both have covalent bonds.
II. They are allotropes of each other.
III. They are isotopes.
IV. In graphite, the C atoms are six-membered rings.

A. Statements I, II, and IV are true.
B. Statements I, II, and III are true.
C. Statements I and III are true.
D. Statements I and II are true.
E. All the statements are true.

Comprehension Questions

1)   Which of the following substances would have the highest melting point:
$C_{18}H_{38}$, H_2O, $MgCl_2$, N_2O_5? Explain your reasoning.

2)   Droplets of water form on the inside of a 2 L bottle of soda as it sits on a shelf
inside a supermarket. Explain why these drops have formed on the inside of the
container. Be sure to include a discussion of intermolecular forces, phase transitions, and
energy requirements for this process.

3)   The following questions refer to the phase diagram.

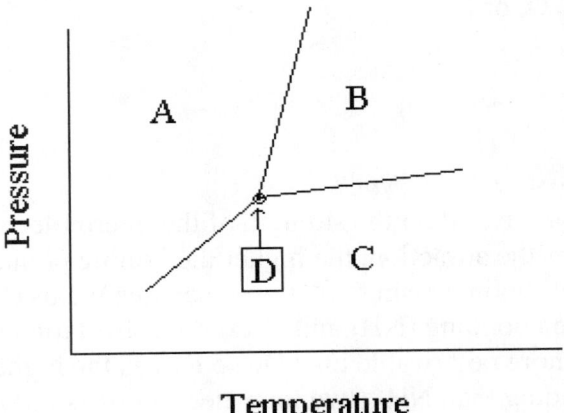

a)   Provide names and descriptions of the labeled regions (or points), A
through D, on the diagram.

b)   Describe, at a molecular level, what happens as a substance transitions
from region A to B and from region C to B. Include types of intermolecular forces, a
name for the process, and the sign of ΔH^0 for the transition.

c)   What is the name of the process that occurs when a substance moves from
region A to region C without passing through region B or point D? What is the sign of
ΔH^0 for this process?

4)   a)   Rank the following inter molecular and intramolecular forces in terms of increasing average strength: dipole–induced dipole, dispersion, dipole-dipole, covalent, and ion-dipole.

      b)   Give examples of substances (i.e., chemical formula or description) that experience each of the above forces.

ANSWERS TO SAMPLE MULTIPLE CHOICE QUESTIONS

1.   B

HF meets the criterion of H being bonded to F, O, or N. CH_3CH_2CHO has H and O present but the H is not attached to the O and thus there is no hydrogen bonding. H_2S lacks the presence of F, O, or N.

2.   C

He < Kr < CO < NH₃ < H₂O < NaCl
Relative boiling points are based on the strength of the intermolecular attractions between molecules. The stronger the attraction, the higher the boiling point. The strengths of the intermolecular attractions in increasing order are: van der Waals (He and Kr) < dipole-dipole (CO) < Hydrogen bonding (NH_3 and H_2O) < electrostatic interactions in ionic bonding (NaCl). Kr is more polarizable than He so Kr has the higher boiling point. H_2O has more hydrogen bonding than NH_3 (two lone electron pairs on oxygen versus one on nitrogen) so water has the higher boiling point.

3.   B

ΔH_{fus}, the heat of fusion, refers to the change from solid to liquid. The equation in reference to water is: H_2O (*s*) $\Leftrightarrow$ H_2O (*l*). When the equation is written in reverse, the heat change is equal to $-\Delta H_{fus}$.

4.   E

Ne < CH₄ < CF₄ < CBr₄ < CI₄ These are all nonpolar molecules and one noble gas (Ne). The intermolecular forces are dispersion forces. The magnitude of the dispersion force depends on the polarizability of the species. The more electrons there are, the more polarizable the species, the stronger the intermolecular attraction and thus the higher the melting point.

5.   C

Liquids with weak intermolecular forces tend to have *low* boiling points. If the force between the liquid molecules is weak, relatively little energy is needed to separate the liquid molecules. Thus they tend to have relatively low boiling points.

6.   C

CCl_4 (*l*), carbon tetrachloride, is a liquid at room conditions. The C—Cl bond is polar but the molecule is nonpolar due to the symmetry of the tetrahedral geometry where the dipoles cancel out. Since CCl_4 (*l*) is a nonpolar molecule it exhibits intermolecular *dispersion forces*.

7.   C

H_2S has a geometry similar to water. There are two lone electron pairs on the central atom and two bonded pairs. The shape is bent and there is a permanent dipole moment.

8.   D

The point at which all three phases coexist is called the *triple point*.

9.   E

The intramolecular bonds are covalent for all the molecules but the intermolecular forces differ in all except E.

		Intermolecular forces in the two molecules:
A.	NH_3 and PH_3	Hydrogen bonding and dipole-dipole
B.	CO_2 and CO	dispersion forces and dipole-dipole
C.	H_2S and CO_2	dipole-dipole and dispersion forces
D.	C (diamond) and C (graphite)	network solid and covalent with dispersion forces
E.	*CO_2 and CBr_4*	*dispersion forces* and *dispersion forces*

10.   A

Diamond and graphite are allotropes, not isotopes of one another. Allotropes are different forms of the same element in the same physical state. Isotopes are atoms of the same element that have different numbers of neutrons.

Answers to Comprehension Questions

1)   Magnesium chloride, $MgCl_2$, would have the highest melting point as it is an ionic compound, and almost all ionic compounds are solids at room temperature with high melting points. All of the forces holding each atom of the substance together are very strong ionic bonds.

        This choice makes sense versus the others by a process of elimination. Water has a lower melting point than an ionic compound, as it is a liquid at room temperature. Dinitrogen pentoxide is a small (low molecular weight) molecule and experiences only weak intermolecular forces, which causes it to be a gas at room temperature. The hydrocarbon, $C_{18}H_{38}$, is a molecular substance (it is made from only nonmetallic elements), and therefore only experiences weak dispersion forces to hold one molecule to the next. Because of its moderate molecular weight, it is probably a solid at room temperature, but is likely a relatively low-melting, waxy solid.

2)   These water droplets are condensation from the water vapor above the soft drink trapped within the bottle. Even though the soda is well below its boiling point (on a shelf in a supermarket), some of the water molecules at the surface have sufficient energy to make the liquid-to-gas phase change. Molecules in a sample such as this have a variety of kinetic energies, the average of which is below that necessary for the substance to be at its boiling point. The energy needed for vaporization is equal to that amount required to overcome the hydrogen bonds holding one molecule to the next in the soda. In the vapor phase, the intermolecular forces are negligible. Evaporation from the liquid to the vapor phase is an endothermic process, and condensation on the interior surface of the bottle is an exothermic process. As the water condenses on the interior surface of the bottle, new intermolecular forces form between water molecules and the plastic of the bottle. This is probably a dipole–induced dipole type of interaction.

3)   a)   In a phase diagram, region A represents various temperature and pressure combinations under which the substance will be in the solid phase. Region B represents the liquid phase, and region C represents the gas phase. D is called the triple point, and represents a pressure-temperature combination at which all three phases, solid, liquid, and gas, are in equilibrium with each other. See Table 11.1 for a description of the basic properties of each phase of matter.

        b)   The transition from A to B is called melting or fusion, and is an endothermic process. In both phases, there are significant intermolecular forces, but they are disrupted as melting occurs. Atoms are still in contact with one another, but in the liquid phase they can move relative to each other whereas this is not possible, to a great extent, in the solid phase.

The transition from B to C is called vaporization or boiling, and is also an endothermic process. In the liquid phase, there are significant intermolecular forces, but they are disrupted as boiling occurs. Atoms in the gas phase are separated by relatively large distances and there are negligible intermolecular forces. The sign of ΔH^0 for both of these processes would be positive since they are endothermic.

    c)   The process is called sublimation, and it is endothermic.

4)   a)   From weakest to strongest they are: dispersion, dipole–induced dipole, dipole-dipole, ion-dipole, and covalent.

    b)   Dispersion–methane, CH_4 (intermolecular force)

Dipole–induced dipole, a solution of oxygen gas in water (intermolecular force)

Dipole-dipole, pure water (intermolecular force)

Ion-dipole, a solution of sodium chloride in water (intermolecular force)

Covalent, carbon dioxide (intramolecular force)

CHAPTER 12
PHYSICAL PROPERTIES OF SOLUTIONS

This chapter reviews the properties of solutions. The main topics include:

- Vocabulary associated with solutions
- The solution process
- Concentration units
- Temperature and pressure effects on solubility
- Colligative properties

Vocabulary associated with solutions

There are some basic terms relating to solutions that you need to know. The terms are:

- **Solution** is a *homogenous mixture*. Examples include air, saltwater, and red gold (an alloy of copper and gold). Notice that solutions can be gaseous, liquid, or solid.

- **Solute** is the *component of lesser amount* in a solution. Salt is the solute in a saltwater solution.

- **Solvent** is the *component of greater amount* in a solution. Water is the solvent in a saltwater solution. Water is the solvent in all aqueous solutions.

- **Saturated solution** means that a dynamic equilibrium exists where the *maximum amount of solute is dissolved* in solution at a given temperature.

- **Unsaturated solution** means that *less than the maximum amount of solute is dissolved* in solution at a given temperature.

- **Supersaturated solution** means that *more than the maximum amount of solute is dissolved* in solution at a given temperature. This is not a stable solution and when a seed crystal is introduced, all the excess solute immediately precipitates out of solution.

- **Miscible** refers to *solutions in which a solute and solvent dissolve completely in one another*. An example is vinegar and water.

- **Immiscible** refers to *mixtures in which a solute and solvent do not mix*, for example, oil and vinegar.

- **Solvation** is the process in which *an ion or molecule is surrounded by solvent molecules* in a specific manner.

- **Hydration** is the process in which *an ion or molecule is surrounded by <u>water molecules</u>* in a specific manner.

The solution process

A general guideline for determining whether a particular solute will dissolve in a particular solvent is that *'like dissolves like'*. This means that solvents dissolve solutes with like or similar bond types. *Polar solvents dissolve polar or ionic solutes.* For example, methanol (CH_3OH, a polar molecule) and salt (NaCl, an ionic crystal) both dissolve in the polar solvent water. The positive end of the solvent molecule is oriented so that is surrounds the negative solute ion or is near the negative end of the polar solute molecule. *Nonpolar solvents dissolve nonpolar solutes.* For example, I_2 will dissolve in CCl_4. Oil will not dissolve in water since nonpolar substances such as oil are not miscible with polar substances such as water.

When the solute dissolves in the solvent, the solute particles disperse throughout the solvent. The ease with which this happens depends on the strength of the intermolecular forces between the solvent-solvent, solute-solute, and solvent-solute species. *Changes in energy and disorder are the two factors driving the solution process.* As depicted in Figure 12.2, for steps 1 and 2 energy is required to overcome the attractive forces and separate the particles of the solute and solvent. These processes are always endothermic. Step 3 of the solution process where solute and solvent particles mix, can be exothermic or endothermic. The heat of solution is the sum of steps 1, 2, and 3 and is given by the formula: $\Delta H_{solution} = \Delta H_1 + \Delta H_2 + \Delta H_3$. The solution process and the energy associated with the process are diagrammed in Figure 12.2.

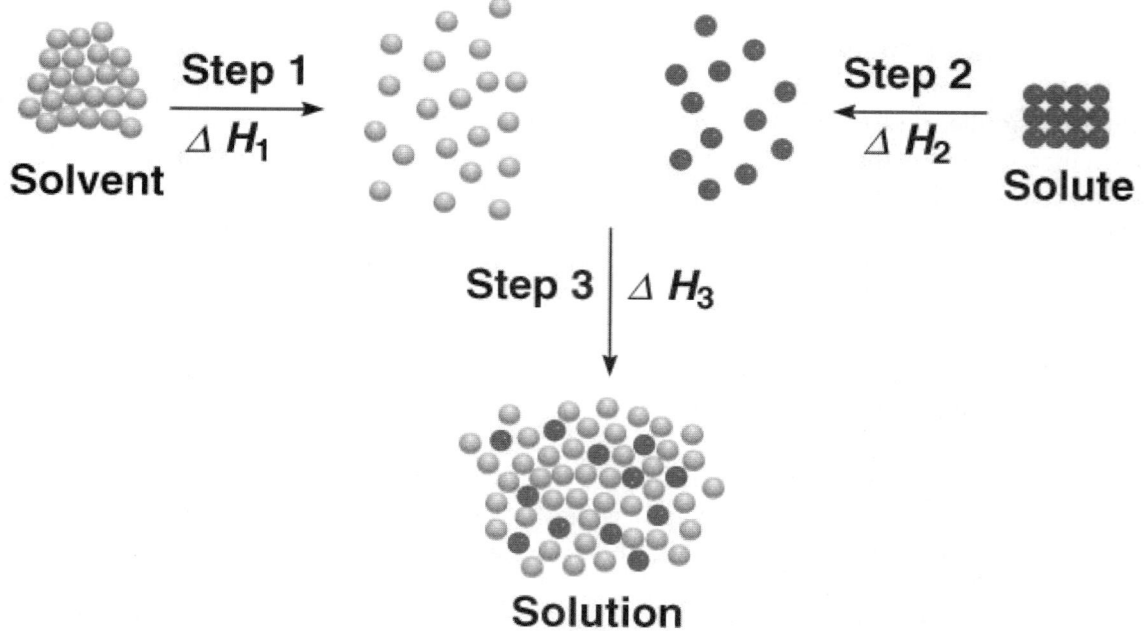

Figure 12.2

Example 1. Predicting solubilities.

Which of the following solutes will dissolve in C_6H_{14} (*l*): KF, H_2O, I_2, CH_3CH_2OH?

Solution to Example 1. *Only I_2 will dissolve in C_6H_{14} (l),* since both molecules are
nonpolar. Ionic compounds (such as KF) and polar covalent compounds (H_2O and
CH_3CH_2OH) will not dissolve in nonpolar solvents.

Concentration units

The major concentration units that are used to describe solutions are:

> **Molarity, M** $=$ $\dfrac{\text{moles of solute}}{\text{liter of solution}}$

> **Molality, *m*** $=$ $\dfrac{\text{moles of solute}}{\text{kg of solvent}}$

> **Percent by mass** $=$ $\dfrac{\text{mass of solute}}{\text{mass of solution}}$

> **Mole fraction** of component *y* in solution $=$ $\dfrac{\text{moles of component } y}{\text{total moles of all components}}$

Example 2. Concentration of solution calculations.

A concentrated solution of hydrochloric acid is 36% HCl by weight and has a density of 1.2 g/mL. Determine the following:

- a. molarity (M)
- b. molality (*m*)
- c. mole fraction of HCl in this concentrated acid solution

General Strategy	Solution to Example 2a
Define the concentration unit asked for.	Molarity = M = $\dfrac{\text{moles of solute}}{\text{liters of solution}}$
Rearrange the data to get the *desired* units. In this case, convert the grams of solute (HCl) to moles of solute by dividing by molecular weight.	To determine moles of solute: 36% HCl means there are 36 g HCl in 100. g of solution. 36 g HCl × $\dfrac{\text{1 mol HCl}}{\text{36 g HCl}}$ = 1.0 mol HCl
Use density to convert grams of solution to mL of solution.	To determine the liters of solution: The solution has a density of 1.2 g mL of solution. 100. g solution × $\dfrac{\text{1 mL}}{\text{1.2 g}}$ = 83 mL solution 83 mL solution × $\dfrac{\text{1.0 L}}{\text{1000 mL}}$ = 0.083 L
Calculate molarity.	M = $\dfrac{\text{1.0 mol HCl}}{\text{0.083 L}}$ = *12 M HCl*

General Strategy	Solution to Example 2b
Define the concentration unit asked for.	Molality = *m* = $\dfrac{\text{moles of solute}}{\text{kg of solvent}}$
Rearrange the data to get *desired* units.	Moles of solute are calculated above and are equal to 12 mol HCl in a liter of solution. To determine kg of solvent: The solution is 36% HCl and thus it is 64% water. 1.2 g/mL = 1200 g/L = density of solution 1200 g/L × 0.64 = 770 g H_2O = 0.77 kg H_2O in a liter of solution
Calculate molality.	*m* = $\dfrac{\text{12 mol HCl}}{\text{0.77 kg } H_2O}$ = *16 m*

General Strategy	Solution to Example 2c
Define the concentration unit asked for.	Mole fraction of HCl = $\dfrac{\text{moles of HCl}}{\text{total mol}}$
Rearrange the data to get *desired* units.	Moles of solute (HCl) = 12 mol Moles of solvent = $\dfrac{770 \text{ g H}_2\text{O}}{18 \text{ g/mol}}$ = 43 mol
Calculate mole fraction of HCl = moles HCl divided by total moles present.	*Mole fraction of HCl* = $\dfrac{12 \text{ mol}}{(12 + 43)\text{mol}}$ = *0.22*

Temperature and pressure effects on solubility

Solubility and temperature. Gases become less soluble with an increase in temperature. Most but not all solids become more soluble with an increase in temperature, as seen in Figure 12.3.

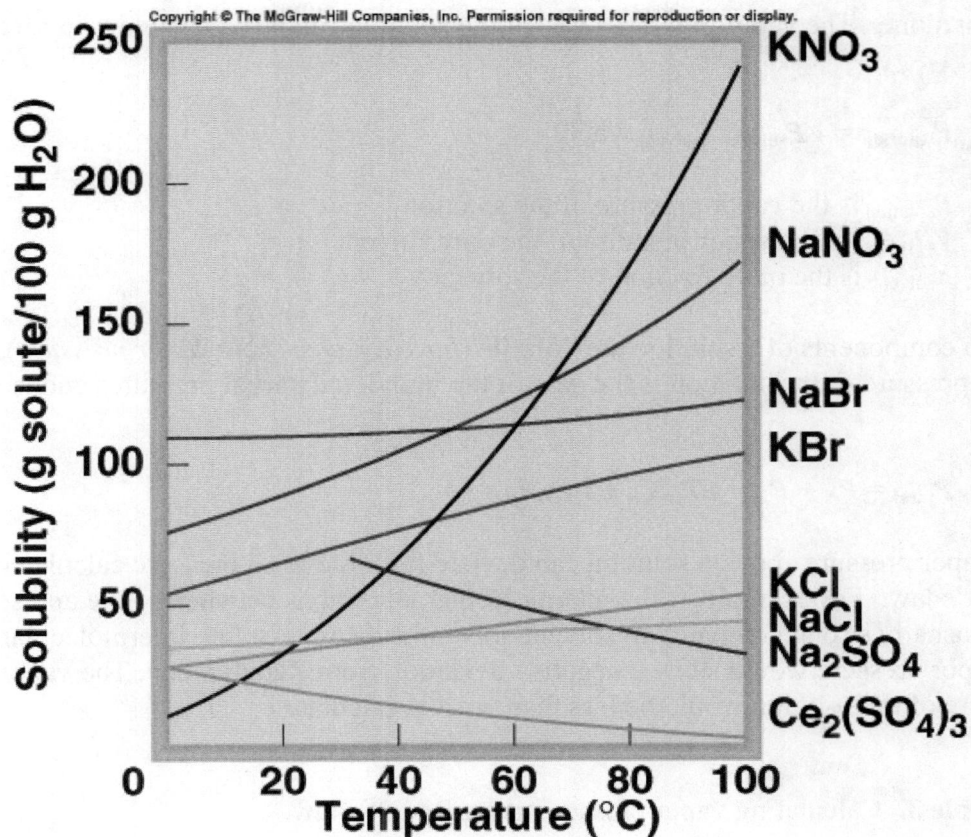

Figure 12.3

Solubility and pressure. The solubility of liquids and solids is not appreciably affected by pressure changes. The solubility of gases is greatly affected by pressure changes. The gas solubility in a liquid can be calculated using **Henry's law**:

$c = kP$ where

c is the molar concentration of the dissolved gas (in mol/L)
k is the Henry's law constant for the gas (in mol / L·atm)
P is the pressure of the gas above the solution (in atm)

Henry's law is not applicable if the gas reacts with water, such as is the case with NH_3.

Colligative properties

Colligative properties are properties that *depend only on the number of particles.* Examples of colligative properties are *vapor pressure lowering, freezing point depression, boiling point elevation, and osmotic pressure.*

Vapor pressure lowering. When a **nonvolatile** (*does not have a measurable vapor pressure*) solute is added to a solvent, the solution has a lower vapor pressure than the solvent alone. The vapor pressure (VP) of the solution can be calculated using **Raoult's law:**

$$P_{solution} = VP_{solvent}X_{solvent} \text{ where}$$

$P_{solution}$ is the vapor pressure of the solution
$VP_{solvent}$ is the vapor pressure of the pure solvent
$X_{solvent}$ is the mole fraction of the solvent

If both components of a solution are **volatile** (*have a measurable vapor pressure*), the vapor pressure of the solution is the sum of the individual partial pressures and is given by:

$$P_{total} = P_A + P_B = VP_A X_A + VP_B X_B$$

The vapor pressure above a solution can deviate from the ideal pressure calculated from Raoult's law. For example, if the intermolecular attractions between solute and solvent are unusually strong compared to solvent-solvent and solute-solute intermolecular forces, the vapor pressure would show a negative deviation from Raoult's law. The vapor pressure of the solution would be less than what is calculated.

Example 3. Calculating vapor pressure using Raoult's law.

Determine the vapor pressure of a solution at 40 °C that is 5.0% NaCl. (The vapor pressure of water at 40 °C is 56 torr.)

General Strategy	Solution to Example 3
Write the formula.	$P_{solution} = VP_{solvent}X_{solvent}$
Determine the individual variables in the formula.	The vapor pressure of the pure solvent is given.

To determine the mole fraction of the solvent calculate:
   the moles of solvent,
   the moles of solute,
   the moles of solute particles.
 	
$X_{solvent} = \dfrac{\text{moles of solvent}}{\text{total number of moles in solution}}$
	the moles of solvent:
	$95 \text{ g H}_2\text{O} \times \dfrac{1 \text{ mol H}_2\text{O}}{18 \text{ g H}_2\text{O}} = 5.3 \text{ mol H}_2\text{O}$
	
	the moles of solute:
	$5.0 \text{ g NaCl} \times \dfrac{1 \text{ mol NaCl}}{58 \text{ g NaCl}} = 0.086 \text{ mol NaCl}$
	
	total number of moles of solute particles:
	NaCl ionizes into *two* ions as shown:
	$\text{NaCl} \rightarrow \text{Na}^+ + \text{Cl}^-$
	$0.086 \text{ mol NaCl} \times 2 = 0.17 \text{ mol of solute particles}$

	$X_{solvent} = \dfrac{(5.3 \text{ mol } H_2O)}{(5.3 + 0.17)} = 0.97$
Solve for *desired* quantity.	$P_{solution} = (56 \text{ torr}) (0.97) = 54 \text{ torr}$

Freezing point depression. When a *solute is added to a solvent* the *freezing point is lowered* according to the equation:

$$\Delta T_f = K_f m^* \text{ where}$$

ΔT_f is the freezing point depression
K_f is the molal freezing point constant
m^* (molality here refers to the <u>total moles of *solute particles* per kg of solvent</u>,
which is not necessarily the molality of the solution. For example, a *1 m solution* of NaCl solution is <u>2 *m* in terms of solute particles</u> since NaCl dissociates into two ions, Na$^+$ and Cl$^-$ in water.)

The **freezing temperature of the solution** = $T_{f, solution} = T_{f, pure\ solvent} - \Delta T_f$

Boiling point elevation. When a *solute is added to a solvent* the *boiling point is raised* according to the equation:

$$\Delta T_b = K_b m^* \text{ where}$$

ΔT_b is the boiling point elevation
K_b is the molal boiling point constant
m^* (molality here refers to the <u>total moles of *solute particles* per kg of solvent</u>,
which is not necessarily the molality of the solution.)

The **boiling temperature of the solution** = $T_{b, solution} = \Delta T_b + T_{b, pure\ solvent}$

Example 4. Freezing point depression calculation.

Calculate the freezing point of a solution containing 7.5 g of KCl in 100.0 g of water.
(The K_f of water is 1.86 °C/m.)

General Strategy	**Solution to Example 4**
Write the formula.	$\Delta T_f = K_f m$
Determine the variables in the formula.	K_f is a constant and given.

To determine the molality calculate:
    the moles of solute
    the *total* moles of solute particles
	The moles of solute:
	    $7.5 \text{ g KCl} \times \dfrac{1 \text{ mol KCl}}{75 \text{ g KCl}} = 0.10 \text{ mol KCl}$
	The *total* moles of solute particles:
	KCl dissociates into *two* ions and thus 0.10 mol of KCl produces *0.20 mol of solute*

	particles: $KCl \rightarrow K^+ + Cl^-$
	0.10 mol + 0.10 mol
To determine kg of solvent: $g \times \dfrac{1 \text{ kg}}{1000 g}$	kg of solvent = 100.0 g = 0.1000 kg
$m = \dfrac{\text{moles of solute particles}}{\text{kg}_{\text{ solvent}}}$	$m = \dfrac{0.20 \text{ mol of solute particles}}{0.10 \text{ kg solvent}} = 2.0 \ m$
Calculate the *desired* quantity. $\Delta T_f = K_f m$ $T_f = 0 \ °C - \Delta T_f$	$\Delta T_f = (1.86 \ °C/m)(2.0 \ m) = 3.7 \ °C$ $T_f = 0 \ °C - 3.7 \ °C = -3.7 \ °C$

Both the freezing point depression and boiling point elevation data can be used to determine the molar mass of a nondissociating solute. This is done by substituting for the molality variable and solving for molar mass:

$$m = \frac{(\text{moles of solute})}{\text{kg}_{\text{ solvent}}} = \frac{(\text{grams of solute})}{(\text{molar mass of solute})\text{kg}_{\text{ solvent}}}$$

$$\Delta T_f = K_f m = \frac{(K_f)(g_{\text{ solute}})}{(M)(\text{kg}_{\text{ solvent}})}$$

Rewriting the above equation in terms of molar mass yields:

$$\text{Molar mass} = M = \frac{(K_f)(g_{\text{ solute}})}{(\Delta T_f)(\text{kg}_{\text{ solvent}})}$$

You can do a similar kind of substitution with boiling point elevation data:

$$\text{Molar mass} = M = \frac{(K_b)(g_{\text{ solute}})}{(\Delta T_b)(\text{kg}_{\text{ solvent}})}$$

The general strategy when solving problems like this is to substitute into the equation and solve for the desired variable as demonstrated in Example 4.

SAMPLE MULTIPLE CHOICE QUESTIONS

1.   Which of the following substances is soluble in CCl_4?

        A. CH_2O
        B.   NaCl
        C. NH_3
        D. NH_3 and NaCl
        E.   None of the above

2.   Concentrated sulfuric acid is 98% H_2SO_4 by mass. The density of the solution is 1.8 g/mL. What is the molarity of this solution?

A.   12 M
B.   15 M
C.   17 M
D.   18 M
E.   None of the above

3.   Statements I – IV are based on the solubility chart, Figure 12.3. Refer to the chart to assess the validity of the statements.

I.   The salt whose solubility is most unaffected by temperature is NaBr.
II.   When KNO_3 is cooled from 80 °C to 30 °C, approximately 120 g settle out of solution per 100 g of water.
III. Na_2SO_4 and $Ce_2(SO_4)_3$ do not follow the usual pattern of solid solubility.
IV.   The dissolving process for most salts is an endothermic process.

A.   I and II are true statements.
B.   II and III are true statements.
C.   I, II, and III are true statements.
D.   III and IV are true statements.
E.   All the statements are true.

4.   2.50 g of a compound are dissolved in 25.0 g of benzene. The solution freezes at 4.3 °C. Which of the following is the correct expression for calculating the molar mass of the solute? (K_f for benzene is 5.12 °C /m and the normal freezing point is 5.5 °C.)

A. $\dfrac{(5.12)\,(2.50)}{(5.5 - 4.3)\,(25)}$

B. $\dfrac{(5.12)\,(2.50)}{(5.5 - 4.3)\,(0.025)}$

C. $\dfrac{(5.12)}{(5.5 - 4.3)\,(0.025)}$

D. $\dfrac{(5.5 - 4.3)\,(0.025)}{(5.12)\,(2.50)}$

E. $\dfrac{(5.5 - 4.3)\,(25)}{(5.12)\,(2.50)}$

5.   Arrange the following aqueous solutions in order of increasing boiling point:

0.20 M CH₃OH   O.20 M NaOH   O.20 M MgCl₂   O.20 M CH₃COOH
A.   0.20 M CH₃COOH < 0.20 M CH₃OH < 0.20 M NaOH < 0.20 M MgCl₂
B.   0.20 M CH₃OH < 0.20 M NaOH < 0.20 M CH₃COOH < 0.20 M MgCl₂
C.   0.20 M CH₃OH < 0.20 M CH₃COOH < 0.20 M NaOH < 0.20 M MgCl₂
D.   0.20 M MgCl₂ < 0.20 M NaOH < 0.20 M CH₃COOH < 0.20 M CH₃OH
E.   None of the above are correct.

6.   What is the approximate freezing point of a solution prepared by adding 14 g of Na_2SO_4 to 100.0 g of H_2O? (The molar mass of Na_2SO_4 is 142 g/mol. The molal freezing point constant for water is 1.86 °C /m.)

A.   −0.19 °C
B.   −0.56 °C
C.   −1.9 °C
D.   −3.8 °C
E.   −5.6 °C

7.   58 g of NaCl are added to 180 g of water at 38 °C. Which of the following is the correct expression for calculating the vapor pressure of the solution at 38 °C? (The molar mass of NaCl is 58 g/mol. The vapor pressure of water at 38 °C is 50. torr.)

A. $\dfrac{50\ (10)}{10}$

B. $\dfrac{50\ (10)}{11}$

C. $\dfrac{50\ (10)}{12}$

D. $\dfrac{50\ (1)}{10}$

E. $\dfrac{50\ (2)}{12}$

8.   Which of the following statements is incorrect?

A. Henry's law can be used to calculate the solubility of SO_2 gas in water.
B. A gas will decrease in solubility as the temperature of the solution is increased.

C. A gas will increase in solubility as the partial pressure of the gas above the solution increases.

D. When the vapor pressure of a liquid is equal to the external pressure boiling occurs.

E. It takes longer to hard boil an egg at high altitude than at sea level because water boils at a lower temperature at high altitude.

9. A sample of 18 g of $C_6H_{12}O_6$ is dissolved in 180 g of water. Which of the following is the correct expression for calculating the mass percent of sugar in this solution? (The molecular mass of $C_6H_{12}O_6$ is 180 g/mol.)

A. $\dfrac{18 \times 100\%}{180}$

B. $\dfrac{18 \times 100\%}{198}$

C. $\dfrac{18 \times 100\%}{0.180}$

D. $\dfrac{18 \times 100\%}{0.198}$

E. None of the above are correct.

10. A solution is made by mixing acetone (C_3H_6O) and methanol (CH_3OH) in a closed container at 25 °C. The acetone mole fraction in the liquid solution is 0.50. Assume ideal behavior for both the liquid and gas phase. What is the mole fraction of acetone in the gas phase above the solution at 25 °C? (The vapor pressures at 25 °C for pure acetone and pure methanol are 270 torr and 140 torr, respectively.)

A. 0.5

B. $\dfrac{135}{205}$

C. $\dfrac{70}{205}$

D. 0.8

E. None of these

Comprehension Questions

1) Concentrated sulfuric acid solution, H_2SO_4 (*aq*), has a concentration of approximately 18.2 M. Its density is 1.83 g/mL.

    a) What is the concentration of sulfuric acid expressed as a mass percent?

    b) What is the concentration of sulfuric acid expressed in terms of molality?

    c)   i) Comment on the difference between the molarity and molality values.

        ii) Is the relationship seen between these two values common for solutions used in the lab? Why or why not?

2) Saturated aqueous solutions of the gas hydrogen chloride (also known as hydrochloric acid or muriatic acid) have a concentration of 37% by mass at 20 °C, and a density of 1.18 g/mL.

    a) Express the concentration of hydrochloric acid solution in terms of molarity.

    b) Care must be taken when opening. New bottles of this strong acid for many reasons. Not the least of which is the fact that as they are opened, especially on a hot day, they can be under pressure and release a significant quantity of gaseous HCl, which is dangerous to breathe and can even burn your skin. Explain the buildup of pressure in the unopened hydrochloric acid containers.

3) The data for the solvent benzene given in the chart may be useful in answering the questions that follow.

Solvent	Normal freezing point (°C)	K_f (°C/m)	Normal boiling point (°C)	K_b (°C/m)	Density (g/mL)
benzene	5.5	5.12	80.1	2.53	0.879

a) What is the freezing point of a solution made up by dissolving 16.4 g of triethylamine, $(C_2H_5)_3N$, in 455 mL of benzene?

b) 7.45 g of an unknown molecular solute was dissolved into 87.9 g of benzene. The boiling point of this mixture was measured to be 83.8 °C. What is the molar mass of the unknown solute?

4) Solubility data for sodium nitrate in water are given.

Temperature (°C)	Solubility of $NaNO_3$ (g of solute / 100 g H_2O)
0	74
20	87
40	105
60	124
80	145

a) Based on the above data, would the following process have a positive or negative value for ΔH^0? Explain your reasoning.

$$NaNO_3(s) + H_2O(l) \rightarrow Na^+(aq) + NO_3^-(aq)$$

b) What is the general trend for solubility of solids in water as temperature increases? What is the trend for solubility of gases?

c) When 10.74 g of sodium nitrate was dissolved in 246.1 mL of pure water, the resulting freezing point of the solution was – 1.91 °C. Calculate the value of the freezing point depression constant, K_f, for water.

d) What would be the freezing point of a solution created by dissolving 5.00 g of magnesium nitrate, $Mg(NO_3)_2$, in 100.0 mL of pure water (the density of water is 1.00 g/mL)?

ANSWERS TO SAMPLE MULTIPLE CHOICE QUESTIONS

1.   E

CCl$_4$ is a nonpolar solvent. Nonpolar solutes dissolve in nonpolar solvents. The solutes listed are polar (CH$_2$O, NH$_3$) or ionic (NaCl) molecules and thus will not dissolve in CCl$_4$.

2.   D

$$\frac{1.8\,g}{mL} \times \frac{1000\,mL}{1\,L} \times \frac{98}{100} \times \frac{1\,mol\,H_2SO_4}{98\,g\,H_2SO_4} = 18\,M\,H_2SO_4$$

3.   E

All the statements are true.

I. NaBr slope is a flat line as temperature increases so there is very little change in solubility.

II. At 80 °C, approximately 170 g of solute are dissolved in 100 g of water. At 30 °C, approximately 50 g of solute are dissolved. 120 g of solute settle out as the temperature goes from 80 °C to 30 °C.

III. Na$_2$SO$_4$ and Ce$_2$(SO$_4$)$_3$ become less soluble as temperature increases. Most solids become more soluble as temperature increases.

IV. The forward reaction (dissolving the solid) is usually favored as temperature increases. According to Le Chatelier's principle, when a stress is placed on a system at equilibrium, the system reacts to relieve the stress. In this case, the stress is heat. When heat is added to the system, the system reacts to absorb it, favoring the endothermic direction (the dissolving process).

4.   B

$$\Delta T_f = K_f m = \frac{(K_f)(g_{solute})}{(M)(kg_{solvent})}$$

Rewriting the above equation in terms of molar mass (M) yields:

$$M = \frac{(K_f)(g_{solute})}{(\Delta T_f)(kg_{solvent})} = \frac{(5.12)(2.50)}{(5.5-4.3)(0.025)}$$

Note that you had to convert the 25.0 g of benzene solvent to 0.025 kg before substituting into the expression.

5.   C

0.20 M CH₃OH < 0.20 M CH₃COOH < 0.20 M NaOH < 0.20 M MgCl₂

The number of particles in solution determines the boiling point elevation of the solution. The more particles, the higher the boiling point. All the solutions have the same molarity but the degree of dissociation varies. *CH₃OH*, an alcohol, does not ionize at all so there are 0.20 mol of particles. *CH₃COOH* ionizes slightly since it is a weak acid and there are slightly more than 0.20 mol of particles. $CH_3COOH \Leftrightarrow CH_3COO^- + H^+$
NaOH is a strong electrolyte and dissociates completely. There are 0.40 mol of particles. $NaOH \rightarrow Na^+ + OH^-$
MgCl₂ is a soluble salt (strong electrolyte) producing three ions per molecule. There are 0.6 mol of particles. $MgCl_2 \rightarrow Mg^{2+} + 2Cl^-$

6.   E

Na_2SO_4 ionizes in water and produces three ions: $Na_2SO_4 \rightarrow 2Na^+ + SO_4^{2-}$

$$14 \text{ g Na}_2\text{SO}_4 \times \frac{1 \text{ mol Na}_2\text{SO}_4}{142 \text{ g Na}_2\text{SO}_4} \times \frac{3 \text{ particles}}{1 \text{ mol Na}_2\text{SO}_4} = 0.30 \text{ mol of particles}$$

$\Delta T_f = K_f m = (1.86 \text{ °C/}m)(0.30 \text{ mol/0.1 kg}) = 5.6$

$T_{f, \text{ solution}} = T_{f, \text{ pure solvent}} - \Delta T_f = 0 \text{ °C} - 5.6 = -5.6 \text{ °C}$

7.   C

The moles of solvent: $180 \text{ g H}_2\text{O} \times \frac{1 \text{ mol H}_2\text{O}}{18 \text{ g H}_2\text{O}} = 10. \text{ mol H}_2\text{O}$

The moles of solute particles:

$$58 \text{ g NaCl} \times \frac{1 \text{ mol NaCl}}{58 \text{ g NaCl}} \times \frac{2 \text{ mol ions}}{1 \text{ mol NaCl}} = 2 \text{ mol ions}$$

$$P_{\text{solution}} = VP_{\text{solvent}} X_{\text{solvent}} = \frac{(50 \text{ torr})(10 \text{ mol H}_2\text{O})}{(10 \text{ mol H}_2\text{O} + 2 \text{ mol ions})}$$

8.   A

Henry's law cannot be used to calculate the solubility of $SO_2(g)$ in water because $SO_2(g)$ is highly soluble in water, forming H_2SO_3. Henry's law applies only to gases with low solubility in the solvent.

9.   B

$$\% \text{ sugar} = \frac{\text{mass of sugar}}{\text{total mass of solution}} = \frac{18 \text{ g} \times 100\%}{(180+18)\text{g}}$$

10.   B

There are two volatile components in this solution and Raoult's law applies:
$$P_{total} = P_A + P_M = VP_A X_A + VP_M X_M$$
$$= 270\,(0.5) + 140\,(0.5) = 135 + 70 = 205 \text{ torr}$$

The mole fraction of acetone in the vapor is: $\dfrac{P_A}{P_{TOTAL}} = \dfrac{135}{205}$

Answers to Comprehension Questions

1)   a)   To solve this type of problem, it's easiest to choose a convenient volume to study, perhaps 1 L (exactly). Then the calculation simply becomes one of determining the mass of solute in 1 L of solution (through the use of moles and molar mass of solute) and the mass of the solution (through the use of its density).

mass of solute = 1 L solution × (18.2 mol H_2SO_4 / liter of solution) × (98.0 g H_2SO_4 /mol) = 1784 g H_2SO_4

mass of solution = 1000 mL solution × 1.83 g/mL = 1830 g of solution per liter

mass % = (mass of solute (or component) / mass of solution) × 100% = (1784 g/1830 g) × 100% = **97.5%**

b)   Molality is defined as moles of solute per kilogram of solvent. In the above example, in 1 L of solution there were 18.2 mol, and the mass of the solvent was 1830 g solution – 1784 g of solute = 46 g of solvent.

molality = 18.2 mol/0.046 kg solvent = **396 molal**

c)   i)   The molality and molarity differ by more than an order of magnitude.

ii)   This is an unusually large difference between molarity and molality. These two values are usually closer together for common lab solutions. The reason is that this solution is unusually concentrated, 98% solute by mass or 18.2 M. Solution concentrations of 0.1 – 2 are much more common in routine lab work. In these cases the molarity and molality values are commonly quite similar.

2)   a)   As in Question 1, we will choose a convenient volume or mass to work with, 1 L in this case because we're working our way toward molarity so a volume of solution would be helpful. Once this has been established, the mass (and then the moles) of solute contained in that volume will be determined.

moles of solute in 1 L = 1000 mL of solution × (1.18 g solution / mL) × (37 g of solute per 100 g of solution) × (1 mol of solute/36.45 g of solute) = 12 mol

molarity of solution = moles of solute/liter of solution = 12 mol /liter = **12 molar**

b)   The "hot day" is a good clue here. Since HCl is a gas, its solubility is likely to be indirectly proportional to temperature, that is, as temperature increases, the solubility decreases. If the saturated solution were made on a day cooler than the bottle was opened, then excess HCl would be in the bottle in the form of a gas. This accounts for the buildup of pressure.

3)   a)   To answer this question we must find the molality of the solution and then multiply it by the K_f for benzene. This number, which is a change in freezing point, must then be subtracted from the normal freezing point.

moles of solute = 16.4 g triethylamine × (1 mol/101.19 g of triethylamine) = 0.162 mol of triethylamine

mass of solvent = 455 mL benzene × (0.879 g benzene/mL) × (1 kg/1000 g) = 0.400 kg benzene

molality = 0.162 mol of triethylamine/0.400 kg solvent = 0.405 molal

freezing point depression = 0.405 m × (5.12 °C / m) = 2.07 °C

freezing point of the solution = normal freezing point – freezing point depression = 5.5 °C – 2.07 °C = **3.4 °C**

b)   For this problem, we will use the boiling point change and the given K_b to determine the molality of the solution. From there we will multiply the molality by the known mass of solvent to arrive at the moles of solute. Finding the molar mass then becomes the simple matter of dividing the mass of the solute by the number of moles.

increase in boiling point = 83.8 °C – 80.1 °C = 3.7 °C

molality of solution = 3.7 °C × (1 molal/2.53 °C) = 1.5 molal

moles of solute = 0.0879 kg benzene × (1.5 mol solute/kg benzene) = 0.13 mol of solute

molar mass = 7.45 g solute/0.13 mol = **58 g/mol**

4)   a)   As temperature increases, the solubility increases. This must mean that the reaction is endothermic, that is, heat is a reactant. That would explain why the addition of heat (by raising the temperature) causes the reaction to shift right, or towards more dissolved products. Endothermic reactions have positive ΔH^0 values.

        b)   Generally solids become more soluble in water as temperature increases. The opposite is true for gases, that is, they become less soluble at higher temperatures.

        c)   Solving this part of the problem involves determining the molality of the solution and the change in freezing point, and then division of the change in freezing point by the molality. Since pure water normally freezes at 0 °C, the change in freezing point must be 1.91 °C.

molality of solution = moles of solute/mass of solvent (kg) = (10.74 g $NaNO_3$ × (1 mol $NaNO_3$ / 85.00 g)) / 0.2461 kg solvent = 0.5134 molal

In this case we actually want the moles of particles, and, since $NaNO_3$ is an ionic compound that produces 2 mol of particles upon dissolving and dissociating, the actual molarity that we're interested in is 2 × 0.5134 m or 1.027 m

K_f for water = 1.91 °C/1.027 molal = **1.86 °C / m**

        d)   For this calculation we can use our newly determined K_f for water and the molality of the solution. Remember to take into account the fact that 1 mol of magnesium nitrate yields 3 mol of particles, ions, in solution so we will have to multiply the molality number by three. This issue is actually addressed in something called the van't Hoff factor, or i. In this case i = 3, and the important formula is $\Delta T_{freezing} = i \, K_f \, m$.

molality of solution = (5.00 g Mg(NO$_3$)$_2$ × (1 mol Mg(NO$_3$)$_2$/148.3 g)) / 0.100 kg solvent = 0.337 m

$\Delta T_{freezing}$ = i K$_f$ m = 3 × 1.86 °C/m × 0.337 m = 1.88 °C

Since pure water freezes at 0 °C, then the new freezing point would be depressed or lowered to **–1.88 °C.**

CHAPTER 13
CHEMICAL KINETICS

This chapter reviews the basic topics associated with the rates of chemical reactions. The topics include:

- The rate of a reaction
- Rate laws
- Integrated rate laws and the half-lives of reactions
- Collision theory and the rate of reaction
- Why every collision does not result in a reaction
- The Arrhenius equation
- Reaction mechanisms
- Catalysts

The laws of thermodynamics (see Chapter 18 of this review book) determine whether a particular reaction will occur. The laws of thermodynamics do not, however, predict how long the reaction will take. **Chemical kinetics** is *concerned with the speeds, or rates, at which chemical reactions occur.*

The rate of a reaction

Some reactions are fast and some are slow. The *speed at which the reaction proceeds* is called the **reaction rate**. There are three common clues as to whether a reaction is fast or slow:

- If the reaction involves a bimolecular collision of molecules rather than a trimolecular collision, it will probably be a faster reaction.
- If the reaction proceeds by *a series of steps,* **a reaction mechanism**, it is probably a slower reaction.
- If the reaction involves the rupture of many bonds such as those found in polyatomic ions, it will probably be a slower reaction.

These three clues often indicate whether a reaction is faster or slower. To know the speed of a reaction with certainty, however, specific experiments must be performed.

Reaction rates are usually determined by following **the change in concentration (M) of a reactant or product over time**. For example, for the simple reaction

$$H_2 (g) \quad + \quad Br_2 (g) \rightarrow \quad 2HBr (g)$$

the rate can be determined by following the disappearance of the reactants H_2 or Br_2 over time or by following the rate of appearance of HBr over time:

$$\text{Rate} = -\frac{[\Delta H_2]}{\Delta t} \qquad \text{or} \qquad \text{Rate} = \frac{\Delta[HBr]}{\Delta t}$$

This rate can be calculated at any time during the reaction and is referred to as the **instantaneous rate**. Further, if the rate of disappearance of the reactant is known, then the rate of appearance of a product can be calculated by the normal stoichiometric relationships:

$$-\frac{[\Delta H_2]}{\Delta t} \times \frac{2\,mol\,HBr}{1\,mol\,H_2} = \frac{\Delta[HBr]}{\Delta t} = -2\frac{[\Delta H_2]}{\Delta t}$$

Notice that there is a negative sign before the rate of disappearance of reactants since their concentration is decreasing. Product rates have a positive sign since their concentration is increasing.

Rate laws

Examination of the data taken from rate experiments demonstrates that a direct relationship exists between rate and concentration. The **rate law** *expresses the relationship of the rate of reaction to the rate constant (k) and the concentrations of the reactants raised to some powers*. For a general reaction

$$aA \;+\; bB \;\rightarrow\; cC \;+\; dD$$

the **rate law** takes the form

$$\textbf{Rate} = \textbf{k[A]}^x\textbf{[B]}^y$$

where the numbers *x and y must be experimentally determined.*

It is important to note that *x and y are <u>not</u> equal to the stoichiometric relationships in the balanced equation.* The exponents x and y in the rate law represent the relationship between the concentrations of the reactants A and B and the rate of the reaction. The **order of the reaction** *refers to the exponents to which the reactant concentrations are raised.* The reaction is xth order with respect to A and yth order with respect to B. The overall order of the reaction $= x + y$.

> **Take Note:** *When a kinetics problem appears on the Free Response section of the AP exam, you will frequently be asked to assess experimental data and to determine the correct rate law.*

Example 1. Determination of the rate law.

Using the following rate data for four experiments involving the reaction between F_2 and ClO_2 at a given temperature, determine a) the order of the reaction with respect to F_2 and ClO_2, b) the rate law for the reaction, and c) the rate constant, k, for the reaction.

[F₂] (M)	[ClO₂] (M)	Initial Rate (M/sec)
1. 0.10	0.010	1.2×10^{-3}
2. 0.10	0.040	4.8×10^{-3}
3. 0.20	0.010	4.8×10^{-3}

General Strategy	Solution to Example 1a, 1b, and 1c
Examine the rate data to determine the order of ClO_2. Select two experiments where one reactant has the same concentration in both experiments. The change in the rate is determined solely by the reactant whose concentration is changing.	a) Notice that in experiments 1 and 2, the concentration of F_2 is held constant and does not affect the reaction rate. The concentration of ClO_2 is increased by a factor of four and the rate increases by a factor of four. There is a simple, direct relationship between the rate and the concentration. $$Rate = [ClO_2]^1 \quad First\ order$$
Determine the order with respect to another reactant. Again, find two experiments in which only one concentration is changing.	Notice that in experiments 1 and 3, the concentration of ClO_2 is held constant and does not affect the rate of the reaction. The concentration of F_2 is increased by a factor of two and the rate increases by a factor of four. If the factor by which the concentration increases is squared, that is equal to the effect on the rate. $$Rate = [F_2]^2 \quad Second\ order$$
The superscripts are the order of the reaction.	b) $$Rate = k[F_2]^2[ClO_2]$$ *The **rate law** is 1st order with respect to ClO_2 and 2nd order with respect to F_2.*
From the reactant concentrations and the initial rate, the specific rate constant, **k**, can be calculated. When determining the value of k, you can take the data from any one of the three experiments and substitute it into the rate law.	c) $$Rate = k[F_2]^2[ClO_2]$$ $$k = \frac{Rate}{[F_2]^2[ClO_2]}$$ $$= \frac{1.2 \times 10^{-3}\,M/sec}{(0.10\,M)^2\,(0.010\,M)}$$ $$= 12\,\frac{L^2}{mol^2 \cdot sec}$$

Integrated rate laws and the half-lives of reactions

Rate laws allow the rate of the reaction to be calculated from the specific rate constant and the reactant concentrations. The integrated form of the rate law can be used to calculate the concentration at any time during the course of the reaction. The integrated rate law also allows data to be analyzed graphically to determine the order of the reaction and the specific rate constant k. Most reactions are first or second order.

For **first-order reactions,** *reactions whose rate depends on the reactant concentrations raised to the first power*, the integrated form of the rate law is:

$$\ln \frac{[A]_t}{[A]_0} = -kt$$

or

$$\ln[A]_t = -kt + \ln[A]_0 \quad \text{where}$$

$[A]_0$ is the initial concentration of the reactant
$[A]_t$ is the concentration at some time t in the reaction
k is the rate constant at a specific temperature

Two key characteristics of first-order reactions are that the reactant decreases with time and that a plot of the $\ln[A]_t$ vs. time will yield a straight line graph whose slope is –k.

Another important aspect of first-order reactions is the **half-life of the reaction,** *the time required for the amount of a reactant to decrease to half of its initial amount.* The substitution of ½[A]$_0$ for [A]$_t$ yields the following equation for a first-order reaction:

$$t_{\frac{1}{2}} = \frac{0.692}{k}$$

Second order reactions, *reactions whose rate depends on the amount of one reactant raised to the second power,* yield the following integrated form of the rate law:

$$\frac{1}{[A]_t} = kt + \frac{1}{[A]_0}$$

A plot of $\dfrac{1}{[A]_t}$ vs. time will yield a straight line graph whose slope is k. The half-life equation for the second order reaction is:

$$t_{\frac{1}{2}} = \dfrac{1}{k[A]_0}$$

There are reactions that are **zero order**, *reactions whose rates are constant and independent of amount*, but they are rare. The rate law for zero-order reactions is given by the expression:

$$\textbf{Rate} = \textbf{k[A]}^{\textbf{0}}$$

The following table summarizes the kinetics of first-order, second-order, and zero-order reactions:

Order	Rate Law	Concentration-Time Equation	Half-life Equation	Graphical Analysis	Graph
1	Rate = k[A]	$\ln\dfrac{[A]_t}{[A]_0} = -kt$	$t_{\frac{1}{2}} = \dfrac{0.692}{k}$	$\ln[A]_t$ vs. time slope $= -k$	
2	Rate = k[A]^2	$\dfrac{1}{[A]_t} = kt + \dfrac{1}{[A]_0}$	$t_{\frac{1}{2}} = \dfrac{1}{k[A]_0}$	plot of $\dfrac{1}{[A]_t}$ vs. time slope $= k$	
0	Rate = k[A]^0	$[A] = -kt + [A]_0$	$t_{\frac{1}{2}} = \dfrac{[A]_0}{2k}$	plot of [A] vs. t slope $= -k$	

Take Note: *A common kinetics problem on the Free Response section of the AP exam asks you to determine the order of the reaction using primary rate data. You can use graphical analysis to obtain the answer.*

Another common question is to ask you to determine the amount of reactant or product present at some time, t, during the reaction. You can use the integrated rate law to find that answer. Don't forget that first-order and second-order rates are the most common and the equations are given to you in the equation list.

Example 2. Determination of the order of reaction.

From the following experimental data, determine the order of the reaction and the rate constant for the reaction:

$$C_4H_8\ (g) \rightarrow 2C_2H_4\ (g)$$

The reaction was carried out in a constant volume container at 532 °C.

Time (sec)	$P_{C_4H_8}$ (mm Hg)
0	800
200	732
400	496
600	392
800	310
1000	244

General Strategy	Solution to Example 2
First plot ln[A] vs time. (If this does not give a straight line, then you can proceed to plot $\dfrac{1}{[A]}$ vs. t, etc.)	
Examine the resulting graph.	The resulting graph is a straight line with a negative slope of –k.
Determine the order of the reaction.	Since the graph is a straight line, the reaction data indicate a first-order reaction with respect to [A].
Calculate the rate constant from the slope of the graph.	$-1.55 \times 10^{-3}\ \dfrac{mm\ Hg}{sec}$

Collision theory and the rate of reaction

For reactions to occur, the reactants must collide with sufficient energy. Rate laws show a direct relationship between the concentration of reactants and the rate of the reaction. This is because *greater concentrations of reactants increase the probability of molecular collisions and, therefore, increase the rate of reaction.*

Increasing the temperature, pressure, or surface area of the reactants also increases the collision frequency and, therefore, *increases the rate* of reaction. The addition of a catalyst also increases the rate of reaction (see catalyst discussion at the end of this chapter).

Why every collision does not result in a reaction

Even when the concentration, temperature, pressure, and surface area are increased, resulting in increased collision frequency, reactions are not always instantaneous. Not every collision results in a reaction. Two other key factors also influence whether a particular collision results in a reaction: *activation energy* and the *orientation of the molecules*.

Examination of the reaction process reveals that bond rupture is the first step toward the formation of product. *Reactant molecules need a minimum amount of energy to break the bonds of the reactants and initiate the reaction.* This minimum energy barrier is called the **activation energy**, E_a. Molecules move at different speeds. Fast moving molecules will most likely collide with sufficient energy to begin bond rupture and succeed in reacting. Slower molecules that collide with less than this minimum energy do _not_ react. They simply bounce off each other.

When molecules collide with sufficient energy to reach the activation energy, they form **the transition state** (also called the **activated complex**), *a temporary species formed by the reactant molecules before they form products.* If the products formed are at a lower energy than the reactants, the reaction will release heat, that is, it is an exothermic reaction (see diagram a). If the products are at a higher energy than the reactants, the reaction will absorb heat, that is, it is an endothermic reaction (see diagram b).

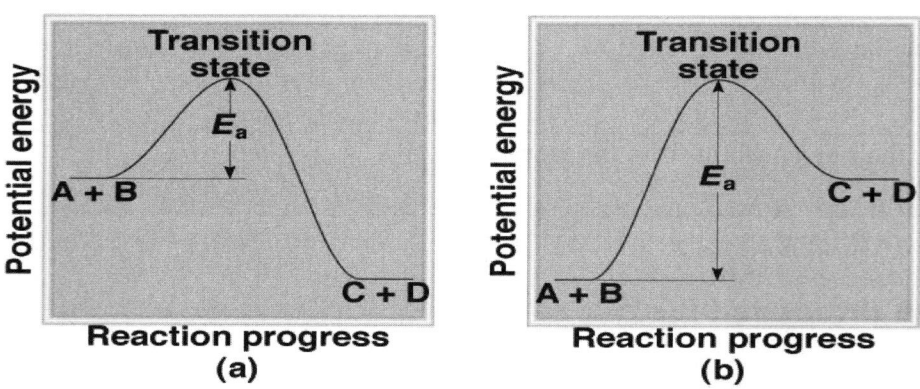

Molecules must also collide with the correct **orientation** for the reaction to occur. Molecules that do not collide with the correct orientation will not form product (see the diagram).

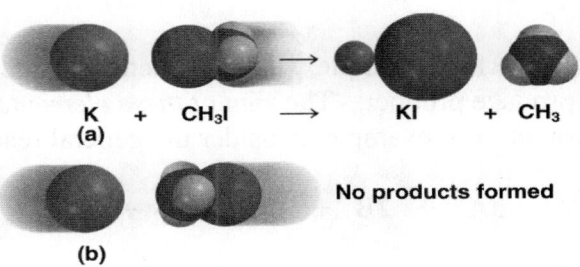

K + CH₃I ⟶ KI + CH₃
(a)

No products formed

(b)

The Arrhenius equation

The Arrhenius equation shows the dependence of the rate constant, k, on temperature and is expressed by the following equation:

$$k = Ae^{-\frac{E_a}{RT}}$$

or the more useful form

$$\ln k = \left(\frac{-E_a}{R}\right)\left(\frac{1}{T}\right) + \ln A \quad \text{where}$$

E_a is the activation energy for the reaction

$$R \text{ is } 8.31 \frac{J}{mol \cdot K}$$

T is the temperature in Kelvin

A plot of ln k vs. $\frac{1}{T}$ produces a slope of $\frac{-E_a}{R}$. *This allows the activation energy for the reaction to be calculated.*

The Arrhenius equation also allows the rate constant to be calculated at various temperatures:

$$\ln\frac{k_1}{k_2} = \frac{E_a}{R}\left(\frac{1}{T_2} - \frac{1}{T_1}\right)$$

Reaction mechanisms

Many chemical reactions are fairly complex. These reactions usually proceed by a series of elementary steps to produce product. The *sum of these elementary steps* is referred to as the **reaction mechanism**. For example, consider the general reaction:

$$3A + 2B + C \rightarrow 2X + Y$$

Examination of the reactants would indicate that six molecules must collide for the reaction to occur. A simultaneous six-bodied collision is unlikely. A reaction mechanism for the reaction might be:

$A + C \rightarrow X$		elementary step 1, bimolecular collision
$2A + B \rightarrow Y + \cancel{Z}$		elementary step 2, termolecular collision
$+ \quad B + \cancel{Z} \rightarrow X$		elementary step 3, bimolecular collision
$3A + 2B + C \rightarrow 2X + Y$		

Since bimolecular collisions are more likely than termolecular collisions, elementary step 2 is probably *the slowest step in the sequence of steps leading to the product,* or the **rate-determining step**. The rate law for this reaction can be written from the rate-determining step. The coefficients of the rate-determining step are the exponents for the rate law. For the reaction above, the rate law is rate = $k[A]^2[B]$.

Notice that the product Z is made in step 2 and consumed in step 3 so that it does not appear in the overall equation. *Substances that are made in one step and consumed in another step* are called **reaction intermediates**. If one of the reactants in the rate-determining step is a reaction intermediate, you must substitute for it into the rate law (see Multiple Choice Question 6 for an example).

The elementary steps of a reaction mechanism must satisfy two requirements:

- The sum of the elementary steps must add up to the overall equation.
- The rate-determining step should predict the same rate law as the experimental evidence.

Example 3. Determination of reaction mechanism.

The reaction $2N_2O \rightarrow 2N_2 + O_2$ has the experimentally determined rate law: rate = $k[N_2O]$. The proposed mechanism for the reaction is:

$N_2O \rightarrow N_2 + O$		(slow)
$N_2O + O \rightarrow N_2 + O_2$		(fast)

Determine whether this is a reasonable mechanism for the reaction.

General Strategy	Solution to Example 3
Determine if the elementary steps of the mechanism add up to the overall equation.	$N_2O \rightarrow N_2 + \cancel{O}$ $\underline{+ \quad N_2O + \cancel{O} \rightarrow N_2 + O_2}$ $2N_2O \rightarrow 2N_2 + O_2$ (overall equation)
Determine if the rate-determining step predicted by the mechanism coincides with the experimentally determined rate law.	The slow step in the mechanism is the rate-determining step. Therefore, the rate law from the mechanism is $rate = k[N_2O]$ This coincides with the experimentally determined rate law.
Draw your conclusion.	The two criteria for a mechanism have been met. This is a reasonable mechanism.

Take Note: *A common kinetics question on the AP exam will ask you to assess some kinetics data for a particular reaction. Usually, you will then be asked to choose the correct mechanism from among several possibilities.*

Catalysts

A **catalyst** is *a substance that increases the rate of a chemical reaction without itself being consumed.* Catalysts increase the rate of reaction by *lowering the activation energy* for the reaction. Diagram **a** shows the reaction path with no catalyst, while diagram **b** shows the activation energy interval with a catalyst. Thus, a greater fraction of the molecules that collide will succeed in achieving the activation energy and producing product.

If the reaction is an equilibrium, the catalyst lowers the activation energy for both the forward and the reverse reaction by the same amount. There is no effect on the value of the equilibrium constant, K.

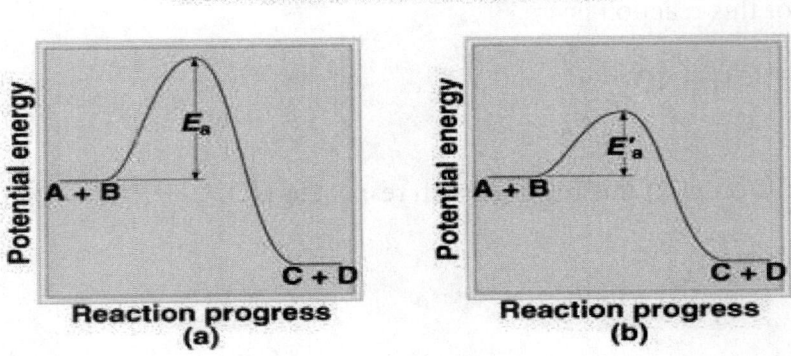

Take Note: *A common Multiple Choice question on the AP exam will ask you to assess the impact of a catalyst on an equilibrium system. Remember that a catalyst has no effect on the value of the equilibrium constant since the activation energy for the forward and for the reverse reaction are lowered to the same degree.*

SAMPLE MULTIPLE CHOICE QUESTIONS

1. The rate law for the reaction $2A + B \rightarrow C$ is: rate $= k[A]^2$. Which graph will produce a straight line for this reaction?

    A. [A] vs. time

    B. ln[A] vs. time

    C. [A] vs. $\dfrac{1}{time}$

    D. $\dfrac{1}{[A]}$ vs. time

    E. $\dfrac{1}{[A]}$ vs. $\dfrac{1}{time}$

2. A first-order reaction has a rate constant of 3.00×10^{-3} mol $L^{-1}sec^{-1}$. The original sample contained 20. g of reactant. How long will it take for only 10. g of reactant to remain?

    A. 20 sec
    B. 1500 sec
    C. 230 sec
    D. 5.0 sec
    E. 100 sec

For questions 3 – 6, consider the following reaction:

$$2H_2\,(g) + \quad 2NO\,(g) \quad \rightarrow \quad 2H_2O\,(g) \quad + \quad N_2\,(g)$$

The rate law for this reaction is

$$\text{rate} = [H_2][NO]^2$$

3. What is the order of this reaction with respect to NO?

    A. 0
    B. 1
    C. 2
    D. 3
    E. 4

4. What is the overall order of the reaction?

    A. 0
    B. 1
    C. 2
    D. 3
    E. 4

5. If the concentration of NO is tripled, the reaction rate will increase by a factor of

    A. 3
    B. 9
    C. 27
    D. 54
    E. 81

6. The following are three mechanisms by which the reaction can proceed. On the basis of the observed rate law, which mechanisms can be ruled out?

    Mechanism I

H_2 + NO	$\rightarrow$	H_2O + N	(slow)
N + NO	$\rightarrow$	N_2 + O	(fast)
O + H_2	$\rightarrow$	H_2O	(fast)

    Mechanism II

H_2 + 2NO	$\rightarrow$	N_2O + H_2O	(slow)
N_2O + H_2	$\rightarrow$	N_2 + H_2O	(fast)

    Mechanism III

2NO	$\Leftrightarrow$	N_2O_2	(fast equilibrium)
N_2O_2 + H_2	$\rightarrow$	N_2O + H_2O	(slow)
N_2O + H_2	$\rightarrow$	N_2 + H_2O	(fast)

    A. I only
    B. II only
    C. III only
    D. I and III
    E. II and III

7.   Consider the following energy profile for a reaction:

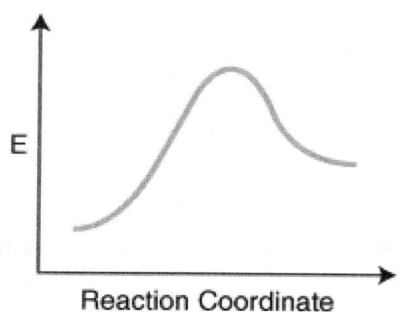

Reaction Coordinate

The addition of a catalyst:

    I.   increases the amount of product formed
    II.   has no effect on the reaction since it is endothermic
    III. lowers the activation energy for the reaction
    IV. increases the rate at which product is formed

A. I only
B. II only
C. I and III
D. II and III
E. III and IV

8.   Consider the following equilibrium at 25 °C:

$$PCl_5 \ (g) \ \Leftrightarrow \ PCl_3 \ (g) \ + \ Cl_2 \ (g)$$

When a catalyst is added to this equilibrium, it will:

    I.   shift the reaction to the right
    II.   shift the reaction to the left
    III. alter the value of K_p
    IV. have no effect on the value of K_p
    V.   have no effect on the partial pressures

A. I and III
B. II and III
C. I and V
D. III and IV
E. IV and V

Questions 9 and 10 refer to the following reaction and data.

$$CH_3COCH_3\ (l)\ +\ Br_2\ (g)\ \xrightarrow{\ H^+\ }\ CH_3COCH_2Br\ +\ H^+\ +\ Br$$

Experiment	$[CH_3COCH_3]$	$[Br_2]$	$[H^+]$	Disappearance of Br_2 $Msec^{-1}$
1	0.300	0.050	0.050	5.7×10^{-5}
2	0.300	0.100	0.050	5.7×10^{-5}
3	0.300	0.100	0.100	1.2×10^{-4}
4	0.600	0.500	0.100	4.8×10^{-4}

9. The correct rate law for this reaction is:

   A. Rate = k $[CH_3COCH_3]^2$ $[H^+]$
   B. Rate = k $[Br_2]$
   C. Rate = k $[CH_3COCH_3]^2$ $[Br_2]$ $[H^+]$
   D. Rate = k $[CH_3COCH_3]$ $[Br_2]$
   E. Rate = k $[CH_3COCH_3]$ $[H^+]$

10. Bromine, Br_2:

   A. is a catalyst
   B. is part of the rate-determining step
   C. is not part of the rate-determining step
   D. is a reaction intermediate
   E. is part of a fast equilibrium step

Comprehension Questions

1) Initial rate data were collected for the following reaction in which iodide ion is oxidized to triiodide by peroxydisulfate ion:

$$S_2O_8^{2-}\ (aq) + 3I^-\ (aq) \rightarrow 2SO_4^{2-}\ (aq) + I_3^-\ (aq)$$

Experiment	$[S_2O_8^{2-}]$ (M)	$[I^-]$ (M)	Initial rate (M/sec)
1	0.080	0.034	2.2×10^{-4}
2	0.080	0.017	1.1×10^{-4}
3	0.16	0.017	2.2×10^{-4}

a)   Write rate expressions for the reaction in terms of all reactants and products.

b)   Determine the order of the reaction with respect to $S_2O_8^{2-}$ and I^- from the data provided, and write a rate law for this reaction.

c)   Determine the value of the rate constant. Include proper units.

d)   What is the overall order of this reaction? What does this tell us about the slow step of the reaction mechanism?

2)   Hydrogen peroxide, H_2O_2, decomposes according to the following reaction:

$$2H_2O_2\,(aq) \rightarrow 2H_2O\,(l) + O_2\,(g)$$

The following data were collected during a study of its decomposition at a certain temperature:

$[H_2O_2]$ (M)	Time (sec)
0.100	0
0.088	120
0.070	300
0.051	600
0.025	1200
0.006	2400

a)   Graph the concentration and time data to determine the order of the reaction with respect to hydrogen peroxide. Explain your reasoning and work.

b)   Write a rate law for this reaction, assuming that the products do not affect the rate of the reaction, and determine a rate constant with appropriate units.

c)   Calculate the half-life of the reaction.

d)   Predict the concentration of hydrogen peroxide at a time of 8 min.

3)   Initial rate data were generated for the following (balanced) reaction at a temperature of 298 K:

$A + B \rightarrow X + Y$

Experiment	[A] (M)	[B] (M)	Initial rate (M/sec)
1	0.050	0.025	6.8×10^{-3}
2	0.050	0.075	2.0×10^{-2}
3	0.10	0.025	2.7×10^{-2}

a)   Determine the order of the reaction with respect to A and B, and the overall order.

b)   Write the rate law, and determine the value of the rate constant (with appropriate units).

c)   Several mechanisms were proposed for this reaction. Of the following choices, determine which, if any, are plausible based on the data (justify your choice(s) using the data and the rate law determined):

Mechanism A

$$A \leftrightarrow C \qquad\qquad \text{fast}$$

$$C + B \rightarrow X + Y \qquad\qquad \text{slow}$$

Mechanism B

$$2A \rightarrow D \qquad\qquad \text{fast}$$

$$D + B \rightarrow X + Y + Z \qquad\qquad \text{slow}$$

$$Z \leftrightarrow A \qquad\qquad \text{fast}$$

Mechanism C

$$2A \rightarrow D \qquad\qquad \text{fast}$$

$$D + B \rightarrow X + Y \qquad\qquad \text{slow}$$

ANSWERS TO SAMPLE MULTIPLE CHOICE QUESTIONS

1. D

    The reaction is a second-order rate law. Therefore, a plot of $\dfrac{1}{[A]}$ vs. time will give a straight line, the slope of which will yield the specific rate constant for this reaction.

2. C

    Since half of the reactant has been converted to product, the time it takes can be calculated by the half-life equation for a first-order rate law:

    $$t_{\frac{1}{2}} = \frac{0.693}{k}$$
    $$= \frac{0.693}{3.00 \times 10^{-3}}$$
    $$= 230 \ sec$$

3. C

    The exponent on [NO] is 2. Therefore, the reaction is second order with respect to NO.

4. D

    The overall order of the reaction is the sum of the exponents in the rate law.
    Since the rate law is rate = $[H_2][NO]^2$, $1 + 2 = 3$.

5. B

    Since the rate law is rate = $[H_2][NO]^2$, $[3]^2 = 9$. The rate will increase by a factor of 9.

6. A

Only mechanism I does not coincide with the observed rate law.
Nor does the sum of the steps in mechanism I yield the overall equation.

Mechanism I yields the rate law, rate = $k[H_2][NO]$ based on the slow step.
Mechanism II yields the rate law, rate = $[H_2][NO]^2$ based on the slow step.

Mechanism III yields the rate law, rate = $[H_2][NO]^2$ based on the slow step and the substitution of 2NO for N_2O_2 from the equilibrium step:

Slow step: rate = $[N_2O_2][H_2]$, but $[N_2O_2]$ depends on 2NO from the first step.
Therefore, substituting 2NO for N_2O_2, the rate law is: rate = $[H_2][NO]^2$

7. E

The presence of a catalyst lowers the activation energy for reaction (III). The fraction of molecules possessing the activation energy increases and, therefore, the rate at which product is formed increases (IV).

8. E

The presence of the catalyst lowers the activation energy for the reaction. The forward and reverse reactions both occur at a faster rate on the microscopic level. The macroscopic level is unchanged. Therefore, the value of K_p is unchanged since the temperature is fixed at 25 °C and the partial pressures remain the same.

9. A

Examination of Experiments 1 and 2 shows that doubling the concentration of Br_2, while keeping CH_3COCH_3 and H^+ constant, has no effect on the rate. Br_2 is not included in the rate law. Rate α $[Br_2]^0$.

Examination of Experiments 2 and 3 shows that doubling the concentration of H^+, while keeping the concentration of CH_3COCH_3 constant, causes the rate to double. Therefore, rate α $[H^+]$.

Examination of Experiments 3 and 4 shows that doubling the concentration of CH_3COCH_3, while keeping the concentration of H^+ constant, causes the rate to increase by a factor of four. Therefore, rate α $[CH_3COCH_3]^2$.

Thus the overall rate law for this reaction is: rate $= k [CH_3COCH_3]^2 [H^+]$.

10. C

Br$_2$ is not part of the rate-determining step since it is not included in the rate law.

Answers to Comprehension Questions

1)   a)   Rate expressions are simply ratios comparing the change in concentration of a given reactant to the change in time over which that change in concentration occurs.

$$\text{rate} = -\frac{\Delta[S_2O_8^{2-}]}{\Delta t} = -\frac{\Delta[I^-]}{3\Delta t} = \frac{\Delta[SO_4^{2-}]}{2\Delta t} = \frac{\Delta[I_3^-]}{\Delta t}$$

The negative signs preceding the ratios for the two reactants cause the observed rate to be a positive quantity, by convention, as the concentrations of these substances are decreasing over time. The ratios are divided by their coefficient in the balanced equation so that each of these expressions gives the same overall rate for the reaction.

b)   In this example, we are given data from a series of experiments each of which started with different concentrations of reactants. The idea is to compare the initial rates (literally the rate of the reaction as it just begins, and the concentrations are approximately equal to what you started with) between the various experiments and discover how changing one of the concentrations affects the rate of the reaction. The trick is to pick two experiments in which only a single concentration has been changed so that you can discover the effect produced by changing that reactant's concentration.

In this case, comparing experiment 2, in which the initial concentration of I^- is 0.17 M with experiment 1, in which its concentration is 0.34 M (the concentration of the other reactant remains unchanged between these two experiments) allows us to determine that the order of the reaction with respect to I^- is 1. This can be said because when the concentration of I^- was doubled, the rate of the reaction doubled; the effect of the change had a first-order effect on the rate. Likewise, comparing the data from experiments 2 and 3, you see that as the concentration of $S_2O_8^{2-}$ is doubled while leaving the concentration of I^- unchanged, the rate doubles. The order of the reaction with respect to $S_2O_8^{2-}$ is also 1.

The rate law for this reaction must be **rate** $= k[S_2O_8^{2-}][I^-]$

c)   The rate constant for the reaction can be determined by substituting reactant concentration values from any of the experiments into the rate law determined in b). The only variable left in the equation is the constant, k. You may determine a constant for all three experiments and average them if desired or requested. Therefore,

$$k = \text{rate} / ([S_2O_8^{2-}][I^-]) = 2.2 \times 10^{-4} \text{ Msec}^{-1} / (0.080 \text{ M} \times 0.034 \text{ M}) = 0.081 \text{ M}^{-1} \text{sec}^{-1}$$

d)   The overall order of the reaction is the sum of the individual orders associated with each reactant or two. Since the reaction is second order overall, the rate of the reaction is dependant on the concentrations of both reactants to the first power. This means that the slow step of the mechanism must be dependent on a molecule of each of these substances. They do not, however, have to be reacting directly with each other. For example, one of the reactants could undergo some sort of rapid, that is, non-rate limiting, decomposition or rearrangement, and the product of that fast step might react with the other reactant in the slow step. This would cause the concentrations of both reactants to be linked to the slow step of the reaction.

2)   a)   When the concentration, or some function of the concentration, of a reactant is graphed versus time, it can give insight on the order of the reaction with respect to that reactant. If a plot of concentration versus time yields a linear graph, then the reaction is zero order with respect to that substance. If a plot of the natural log of the concentration versus time yields a straight line, then the reaction is first order with respect to that reactant. And, if a plot of 1 / concentration of a reactant versus time yields a linear graph, then the reaction is second order with respect to that reactant.

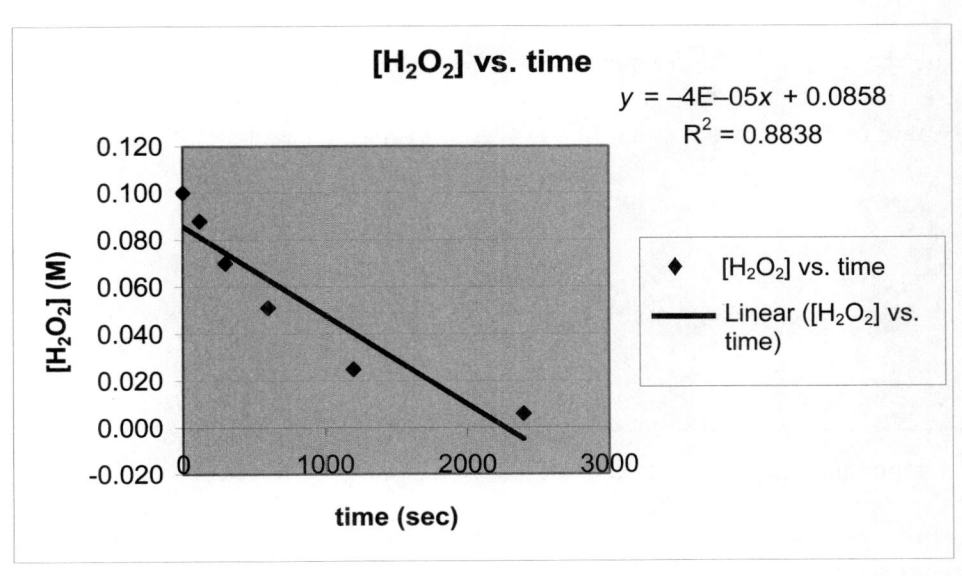

[H₂O₂] vs. time

$y = -4\text{E}{-}05x + 0.0858$
$R^2 = 0.8838$

- ♦ [H₂O₂] vs. time
- —— Linear ([H₂O₂] vs. time)

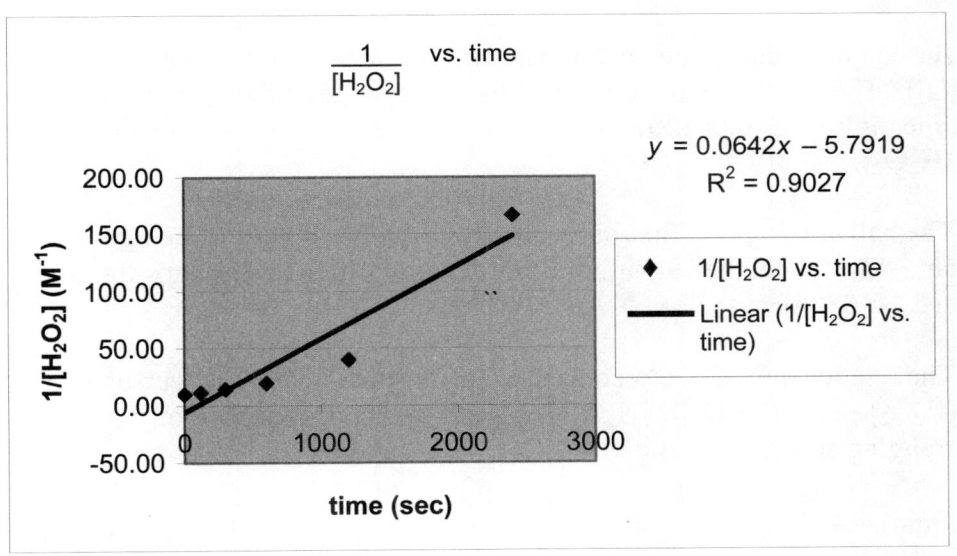

$\dfrac{1}{[\text{H}_2\text{O}_2]}$ vs. time

$y = 0.0642x - 5.7919$
$R^2 = 0.9027$

- ♦ 1/[H₂O₂] vs. time
- —— Linear (1/[H₂O₂] vs. time)

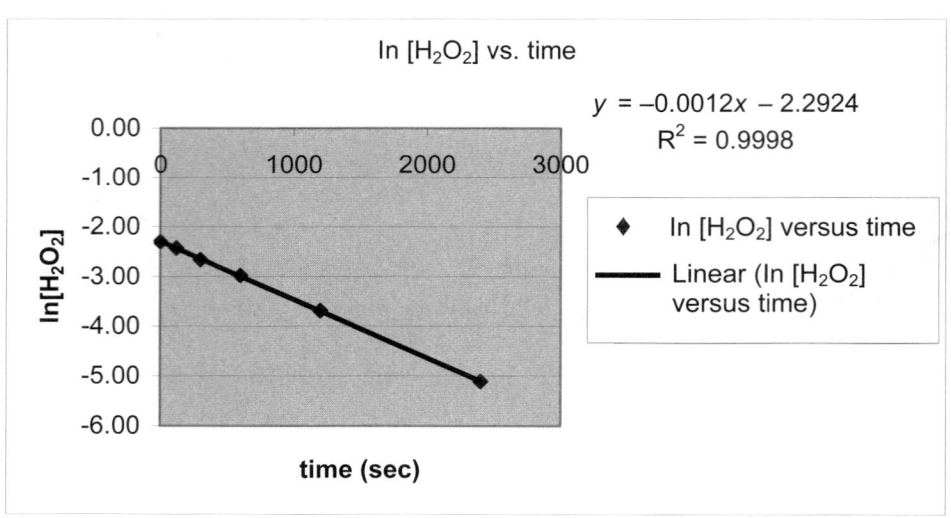

In [H₂O₂] vs. time

$y = -0.0012x - 2.2924$
$R^2 = 0.9998$

As can be seen by the preceding figures, only the plot of ln [H₂O₂] yields a straight line. Therefore, the reaction is first order with respect to H_2O_2.

b)   Based on the data generated in part a), the rate law for this reaction would be, rate = k [H₂O₂]. The rate constant, k, is equal to the negative of the slope of plot of ln [H₂O₂] versus time, or $k = -(-0.00120) = 0.0012 \text{ sec}^{-1}$. The rate law then becomes **rate = 0.0012 sec⁻¹ [H₂O₂].**

c)   The half-life, $t_{1/2}$, of a first-order reaction equals 0.693 / k, k being the first-order rate constant. Therefore, the half-life for this decomposition at this temperature is $0.693 / .00012 \text{ sec}^{-1} = \textbf{580 sec}$.

d)   The relationship between concentration at time = t, $[H_2O_2]_t$, versus initial concentration, $[H_2O_2]_0$, is $\ln([H_2O_2]_t / [H_2O_2]_0) = -kt$, ($t$ being the amount of time in question). Rearranging and substituting,

8 min × (60 sec/min) = 480 sec

$([H_2O_2]_t / [H_2O_2]_0) = e^{-kt}$

$[H_2O_2]_t = [H_2O_2]_0 \times e^{-kt} = (0.100 \text{ M}) \times e^{-(0.0012 \text{ sec}^{-1} \times 480 \text{ sec})} = \textbf{0.056 M}$

This answer seems reasonable as it is between the concentration values for 5 and 10 min.

3)   a)   The reaction orders can be determined in the same way as for part 1b). Comparing experiments 1 and 2 we see that the concentration of A has remained unchanged while the concentration of B has tripled. The relative rates of experiments 1 and 2 show a one to three ratio, so changing the concentration of B has had a first-order effect. Comparing experiments 1 and 3 we see that the concentration of B remains constant while the concentration of A doubles. The initial rate of experiment 3 is four times as fast as experiment 1, therefore the concentration of A has a second-order effect on rate. **The order of the reaction with respect to A is 2 and with respect to B is 1.** Again as in 1d), we need only add the individual reactant orders to determine **the overall order, which in this case is 3.**

        b)   The rate law for this reaction would be as follows:

$$\text{rate} = k\,[A]^2[B]$$

As in part 1c) substituting the values from one of the experiments will allow us to calculate the rate constant,

$$k = \text{rate}\ /\ ([A]^2[B])\ = 6.8 \times 10^{-3}\ M\,sec^{-1}\ /\ ((0.050\ M)^2 \times 0.025\ M) = \mathbf{110\ M^{-2}\ sec^{-1}}$$

So the rate law can be expressed as:

$$\text{rate} = 110\ M^{-2}\ sec^{-1}\ [A]^2[B]$$

c) Answering this question involves comparing the rate law and individual reactant orders with the slow step of each mechanism (and the mechanism in general). The first mechanism proposed, A, cannot be correct as its slow step is dependent on one molecule of A, ultimately through the intermediate C, and one molecule of B. This would correspond to reactant orders of 1 for both species. Mechanism B could be consistent with the observed rate law as the slow step is dependent on one molecule of B and two molecules of A, through the dimeric intermediate D. Mechanism C fulfills the need for a second-order relationship with substance A and a first-order relationship with compound B, but the overall stoichiometry of the reaction is wrong. This mechanism would have A and B reacting in a two to one ratio. **Therefore, only proposed mechanism B is a possibility.**

CHAPTER 14
CHEMICAL EQUILIBRIA

This chapter reviews the basic concepts associated with chemical equilibria, the main component of the AP curriculum. The basic equilibrium concepts presented are:

- The nature of chemical equilibria
- Factors that affect chemical equilibria
- The equilibrium expression and the equilibrium constant
- General strategy for solving equilibrium problems
- Examples of equilibrium problems

Chemical equilibrium is a complex subject. It is addressed by three chapters: Chapter 14, Chemical Equilibria; Chapter 15, Acids and Bases; and Chapter 16, Acid-Base Equilibria and Solubility Equilibria.

To understand the more sophisticated equilibria presented in Chapters 15 and 16, you must have a strong foundation in the essential equilibria concepts presented in this chapter.

Take Note: *The first question in the Free Response section of the AP exam is always an equilibrium question. This question is weighted as 20% of the Free Response exam grade.*

The nature of chemical equilibria

Most chemical reactions do not go to completion, but are reversible and achieve an equilibrium. At **equilibrium**, *the rate of the forward reaction is equal to the rate of the reverse reaction and thus the concentrations of reactants and products are unchanging.* On the *macroscopic level it looks as if no change is occurring*, a steady state has been achieved. On the *molecular level, reactants are continuously forming products (forward reaction) and products are continuously forming reactants (reverse reaction).* Equilibrium is a **dynamic** process and is symbolized by a reversible arrow. The conditions necessary for a chemical equilibrium to exist are: a closed system, constant conditions of temperature and pressure, and a reversible chemical reaction.

Factors that affect chemical equilibria

Le Chatelier's principle *states that when a stress is placed on a system in equilibrium, the system reacts in such a way as to counteract the stress.* A stress, such as concentration, temperature, pressure, or a catalyst, can affect an equilibrium in three ways:

1. It has no effect;
2. The products are favored and the equilibrium shifts to the right; or
3. The reactants are favored and the equilibrium shifts to the left.

Example 1. Applying Le Chatelier's principle.

$$3H_2 + N_2 \Leftrightarrow 2NH_3 \quad \Delta H^0 = -80 \text{ kJ/mol}$$

How will the equilibrium be affected by the following?

a. Hydrogen gas is added to the reaction vessel.
b. Ammonia is added to the reaction vessel.
c. The reaction vessel is heated.
d. The volume of the reaction vessel is halved.
e. Helium gas is added to the reaction vessel.

Stress placed on the equilibrium reaction	Solution to Example 1
a. Hydrogen gas is added to the reaction vessel.	*Net shift to the right.* This shift reduces the amount of added H_2.
b. Ammonia is added to the reaction vessel.	*Net shift to the left.* This shift reduces the amount of added NH_3.
c. Reaction vessel is heated.	*Net shift to the left.* The endothermic direction is favored so that the added heat is consumed. The forward direction is exothermic since the ΔH^o value is a negative number. Another approach is to think of heat as a product in an exothermic reaction and thus adding heat will drive the reaction to the left, absorbing some of the added heat.
d. Volume of the reaction vessel is halved.	*Net shift to the right.* Decreasing the volume of the container increases the partial pressures of the gases and thus the reaction shifts to the side with fewer gas moles to restore equilibrium.
e. Helium gas is added to the reaction vessel.	*No change.* Adding a noble gas increases the total pressure inside the container but does not affect the individual partial pressures of the gases that appear in the K_p expression.

The equilibrium expression and the equilibrium constant

The **equilibrium expression** for the reaction $aA + bB \Leftrightarrow cC + dD$ can be written as a K_c or K_p expression.

$$K_c = \frac{[C]^c[D]^d}{[A]^a[B]^b}$$

In K_c expressions, the brackets [ ] indicate concentration. The concentration units are moles per liter, molarity (M). The superscripts are the coefficients of the balanced reaction.
Solids and solvents (such as water in an aqueous reaction) do not appear in the equilibrium expression because these entities *do not affect the equilibrium.*

$$K_p = \frac{P_C^c P_D^d}{P_A^a P_B^b}$$

In the K_p expression, the P refers to the partial

pressure of the gases where the units of pressure are in atmospheres.

K is a dimensionless quantity that *indicates the extent to which the reaction proceeds toward products* before the equilibrium is established. A large K value (greater than 1) indicates that the products are favored. A small K (less than 1) implies that the reactants are favored. Small K values are often associated with weak acids and weak bases (see Chapter 15 of this review guide) and precipitation reactions (see Chapter 16 of this review guide) where the forward reaction is very small.

To *convert between K_p and K_c* use the formula:

$$K_p = K_c(RT)^{\Delta n} \quad \text{where}$$
R is the ideal gas law constant 0.0821 L·atm/mol·K
T is temperature in K
Δn = moles of product gas – moles of reactant gas

The *magnitude of K, the equilibrium constant, is temperature dependent.*

Example 2. Calculating the equilibrium constant.

Consider the Haber processs for making ammonia:

$$3H_2\,(g) + N_2\,(g) \Leftrightarrow 2NH_3\,(g)$$

The $K_c = 8.7$ at a given temperature, T for this reaction.

a. Write the equilibrium expression for the equation.
b. Calculate the equilibrium constant if the reaction is written as:
    I. $3/2 H_2 + \frac{1}{2}N_2 \Leftrightarrow NH_3$
    II. $2NH_3 \Leftrightarrow 3H_2 + N_2$

Solution to Example 2.

a. $Kc = \dfrac{[NH_3]^2}{[H_2]^3[N_2]}$

b.   I. Halving the coefficients of the balanced reaction means taking the square root of K_c to obtain K_c'

$$3/2 H_2 + \frac{1}{2}N_2 \Leftrightarrow NH_3 \qquad K_c' = \dfrac{[NH_3]}{[H_2]^{3/2}[N_2]^{1/2}} = (8.7)^{1/2} = 2.9$$

II. Reversing the reaction means taking the inverse of K_c to obtain K_c''

$$2NH_3 \Leftrightarrow 3H_2 + N_2 \qquad K_c'' = \dfrac{[H_2]^3[N_2]}{[NH_3]^2} = \dfrac{1}{8.7} = 0.11$$

General strategy for solving equilibrium problems

A general strategy for solving equilibrium problems involves the following steps:

1. Write the balanced equation.
2. Write the equilibrium expression.
3. Place the data from the problem underneath the balanced equation and identify the quantity to be calculated.
4. Determine which way the reaction will shift:

    a) If initially there are only reactants present, the reaction will shift to the right.**
    b) If initially there are only products present, the reaction will shift to the left.**
        ** Note: Statements (a)and (b) are true as long as the reactant and product molecules appear in the equilibrium expression (see Example 6).

c) If initially there is a mixture of reactants and products present, you must determine the reaction quotient, Q. The reaction quotient is calculated by substituting the *initial* concentrations into the equilibrium constant expression.

> If Q = **K,** the reaction *is at equilibrium* and *no net change* will occur.
>
> If Q > **K**, the reaction *shifts to the left* (reducing the product concentrations and increasing the reactant concentrations) until equilibrium is established, at which point Q = K.
>
> If Q < **K**, the reaction *shifts to the right* (increasing the product concentrations and decreasing reactant concentrations) until equilibrium is established, at which point Q = K.

Note that if you are given the initial amounts of molecules and the amount of one of the molecules at equilibrium, you can calculate the other equilibrium concentrations based on the coefficients of the balanced equation (see to Example 3).

Examples of equilibrium problems

Example 3. Determining the equilibrium constant.

3.0 mol of iodine and 4.0 mol of bromine are placed in a 2.0-L reaction vessel at 150 °C. The reaction comes to equilibrium, at which point 3.2 mol of iodine bromide are present. Determine the equilibrium constant for the reaction:

$$I_2 (g) + Br_2 (g) \Leftrightarrow 2IBr (g)$$

General Strategy	Solution to Example 3
Write the balanced equation.	$I_2 (g) + Br_2 (g) \Leftrightarrow 2IBr (g)$

Write the equilibrium expression.	$K_c = \dfrac{[IBr]^2}{[I_2][Br_2]}$
Place the data from the problem underneath the balanced equation. The initial conditions, *(i)* and equilibrium concentrations *(eq)* must be expressed in $\dfrac{mol}{L}$ for K_c.	$\begin{array}{lccc} & I_2\,(g) & + \; Br_2\,(g) & \Leftrightarrow \; 2IBr\,(g) \\ (i) & \underline{3.0\text{ mol}} & \underline{4.0\text{ mol}} & 0 \\ & 2.0\text{ L} & 2.0\text{ L} & \\ (eq) & & & \underline{3.2\text{ mol}} \\ & & & 2.0\text{ L} \end{array}$
Use the stoichiometric relationships from the balanced equation to determine the equilibrium concentrations of all species. Initially there was no product present, so the reaction shifts to the right. This means that the reactant molecules decrease in amount. The amount is determined by the coefficients of the balanced equation.	Initially there was zero IBr present. At equilibrium there was 3.2 mol of IBr present. From the stoichiometric relationships 1.6 mol of I_2 and 1.6 mol Br_2 reacted to produce the 3.2 mol of IBr. $3.2\text{ mol IBr} \times \dfrac{1\text{ mol }I_2}{2\text{ mol IBr}} = 1.6\text{ mol }I_2$ reacted $\begin{array}{l} 3.0\text{ mol }I_2 \text{ present initially} \\ \underline{-\,1.6\text{ mol }I_2 \text{ reacted}} \\ 1.4\text{ mol }I_2 \text{ remaining at equilibrium} \end{array}$ $[I_2] = \dfrac{1.4\text{ mol}}{2.0\text{ L}} = 0.70\text{ M }I_2$ $3.2\text{ mol IBr} \times \dfrac{1\text{ mol }Br_2}{2\text{ mol IBr}} = 1.6\text{ mol }Br_2$ reacted $\begin{array}{l} 4.0\text{ mol }Br_2 \text{ present initially} \\ \underline{-\,1.6\text{ mol }Br_2 \text{ reacted}} \\ 2.4\text{ mol }Br_2 \text{ remaining at equilibrium} \end{array}$ $[Br_2] = \dfrac{2.4\text{ mol}}{2.0\text{ L}} = 1.2\text{ M }Br_2$
Solve for the *desired* quantity by substituting the equilibrium concentrations into the equilibrium expression.	$K_c = \dfrac{[IBr]^2}{[I_2][Br_2]} = \dfrac{[3.2\text{ mol/2 L}]^2}{[0.70\text{ M}][1.2\text{ M}]}$ $K_c = 3.0$

Example 4. Calculating equilibrium concentrations using K_c.

Consider the equilibrium: $H_2(g) + I_2(g) \Leftrightarrow 2HI(g)$

If 2.0 mol of hydrogen gas and 2.0 mol of iodine gas are introduced into a 2.0-L reaction vessel at 400 °C, determine the concentrations of all species at equilibrium ($K_c = 64$ at 400 °C).

General Strategy	Solution to Example 4
Write the balanced equation.	$H_2(g) + I_2(g) \Leftrightarrow 2HI(g)$
Write the equilibrium expression.	$K_c = \dfrac{[HI]^2}{[H_2][I_2]}$
Place the data from the problem underneath the equation. These represent the initial conditions *(i)*. Units must be expressed in mol/L for K_c.	$$\begin{array}{cccc} & H_2(g) & + I_2(g) & \Leftrightarrow 2HI(g) \\ (i) & 2.0 \text{ mol} & 2.0 \text{ mol} & 0 \\ & 2.0 \text{ L} & 2.0 \text{ L} & \end{array}$$
Use the stoichiometric relationships from the balanced equation to express the *change*s that occur in going from the initial to the equilibrium *(eq)* condition. Initially there was no product present, so the reaction shifts to the right. This means that the reactant molecules decrease in amount (*–nx*, where *n* refers to the coefficient in the balanced reaction; $n = 1$ in this reaction). Product is formed so product molecules increase in amount (*+ nx*, where *n* refers to the coefficient in the balanced reaction; $n = 2$ in this reaction). *Equilibrium concentrations (eq)* are given by adding the *changes* to the *initial (i)* concentrations.	$$\begin{array}{cccc} & H_2(g) & + I_2(g) & \Leftrightarrow 2HI(g) \\ (i) & 1.0\text{M} & 1.0\text{ M} & 0 \\[4pt] changes & -x & -x & +2x \\[4pt] (eq) & 1.0\text{ M} -x & 1.00\text{ M} - x & 0 + 2x \end{array}$$
Solve for the *desired* quantity by substituting the equilibrium concentrations into the equilibrium expression.	$K_c = 64 = \dfrac{[HI]^2}{[H_2][I_2]} = \dfrac{[2x]^2}{[1.0\text{ M} -x]^2}$ The square root of both sides can be taken

| | to solve for x.

$$8 = \frac{2x}{1.0-x}$$
$$x = 0.80$$

$[H_2] = 1.0\ M - x = 0.20\ M$
$[I_2]\ = 1.0\ M - x = 0.20\ M$
$[HI]\ = 2x = 1.60\ M$ |

Note the large extent to which the reaction proceeds forward in Example 4. This is expected when the K value is large.

Example 5. K_p and partial pressures at equilibrium.

Consider the equilibrium: $H_2(g) + I_2(g) \Leftrightarrow 2HI(g)$
At 400 °C, $K_c = 64$ for this reaction.

(a) Calculate K_p for this reaction at 400 °C.

(b) Determine the equilibrium partial pressure of HI at 400 °C if initially the partial pressures of H_2 and I_2 are 2.0 and 4.0 atm, respectively, at 400 °C.

General Strategy	Solution to Example 5a
Recall the equation : $K_p = K_c(RT)^{\Delta n}$ where Δn = (moles of product gas – moles of reactant gas). Substitute into the equation and solve.	$\Delta n = 0$ since there are 2 mol of gas on each side of the equation. Substituting into the equation: $K_p = 64$ $(RT)^0 = 64$.

Note that $K_p = K_c$ when the number of moles of *gas* on each side of the equation is equal.

General Strategy	Solution to Example 5b
Write the balanced equation.	$H_2(g) + I_2(g) \Leftrightarrow 2HI(g)$
Write the equilibrium expression.	$$K_p = \frac{P^2_{HI}}{P_{H2}P_{I2}}$$

Place the data from the problem underneath the equation. These are the initial conditions, *(i)*. Units must be expressed in atm for K_p.	$\quad\quad$ H$_2$ (g) $\quad$ + I$_2$ (g) $\quad$ ⇔ 2HI (g) *(i)* $\quad$ 2.0 atm $\quad\quad$ 4.0 atm $\quad\quad$ 0
Use the stoichiometric relationships from the balanced equation to express the *changes* that occur in going from the initial to the equilibrium condition *(eq)*. Initially there was no product present so the reaction shifts to the right. This means that the reactant molecules decrease in amount (−*n*x, where *n* refers to the coefficient in the balanced reaction, *n* = 1 in this reaction). Product is formed. Therefore, using the coefficient for the product, the increase is +2*x*. The *equilibrium pressures (eq)* are given by adding the *changes* to the *initial pressures (i)*.	$\quad\quad$ H$_2$ (g) $\quad$ + I$_2$ (g) $\quad$ ⇔ 2HI (g) *(i)* $\quad$ 2.0 atm $\quad\quad$ 4.0 atm $\quad\quad$ 0 *changes* $\;$ −*x* $\quad\quad\quad$ −*x* $\quad\quad\quad$ +2*x* *(eq)* $\;$ 2.0 −*x* $\quad\quad$ 4.0 − *x* $\quad\quad$ 0 + 2*x*
Solve for the *desired* quantity by substituting the equilibrium concentrations into the equilibrium expression.	$64 = K_p = \dfrac{P_{HI}^2}{P_{H2}P_{I2}} = \dfrac{(2x)^2}{(2.0-x)(4.0-x)}$ Using the quadratic equation $x = 1.9$ and 4.5 atm. Only 1.9 is reasonable since the $P_{H2} = 2.0 - x$ and *x* cannot be larger than 2.0. $P_{HI} = 2x = 2\,(1.9) = 3.8\ atm$

Example 6. Predicting the direction of a reaction.

Consider the equilibrium: $CaCO_3$ (s) ⇔ CaO (s) + CO_2 (g).

A 1.0-L reaction vessel contains 10.0 g of CaO at 1000 °C. At this temperature, 0.20 atm of carbon dioxide gas is introduced into the reaction vessel. Will any calcium carbonate form? Justify your answer. At 1000 °C, $K_c = 0.0370$ for this reaction.

General Strategy	Solution to Example 6
Write the balanced equation.	$CaCO_3 (s) \Leftrightarrow CaO (s) + CO_2 (g)$
Write the equilibrium expression. The K_c is given but the data are provided in atmospheres, so using K_p is also possible. Remember that solids do not appear in an equilibrium expression.	$K_c = [CO_2]$ $K_p = P_{CO2}$
Place the data from the problem underneath the equation. These are the initial conditions, *(i)*. Notice that the units, g and atm, are not consistent.	$CaCO_3 (s) \Leftrightarrow CaO (s) + CO_2 (g)$ *(i)* 0 10.0 g 0.20 atm

Initially there is no reactant present so you might expect the reaction to shift to the left. However, because solids do not affect the equilibrium condition, CO_2 is the *only* factor that affects the equilibrium in this reaction. The question is about the CO_2 pressure: is it at equilibrium, does it exceed equilibrium pressure, or is it not yet at equilibrium pressure?	$K_p = K_c(RT)^{\Delta n}$
	    $= 0.0370\ [0.0821)(1273)^1$
	    $= 3.87$
	
	$K_p = P_{CO2} = 3.87$
	
	The carbon dioxide equilibrium pressure is 3.87 atm. The pressure of the CO_2 is 0.020 atm. CO_2 is not yet at equilibrium and therefore *no $CaCO_3 (s)$ will form.*

Note: Another method of solving this problem is to use the K_C expression and compare the moles of CO_2 initially with the moles of CO_2 at equilibrium.	Using $PV = nRT$ and solving for moles of CO_2 :
	   $n = \dfrac{PV}{RT} = \dfrac{(0.20\ atm)\ (1.0\ L)}{(0.0821)\ (1273\ K)}$
	   $n = 1.9 \times 10^{-3}$ mol CO_2 gas present
	
	The moles of CO_2 gas is also the reaction quotient, *Q*, since solids do not appear in the *Q* or K expression.
	
	$K_c = [CO_2] = 0.037$
	$Q = [CO_2] = 0.0019$
	Q < K. The reaction is not yet at equilibrium and no calcium carbonate will form.

Example 7. Applying Le Chatelier's principle.

$$CaCO_3\,(s) \Leftrightarrow CaO\,(s) + CO_2\,(g) \qquad \Delta H^\circ = 175\ kJ/mol$$

How will the equilibrium be affected by each of the following?

    a. Adding $CaCO_3\,(s)$.
    b. Removing $CO_2\,(g)$.
    c. Halving the volume of the reaction vessel.
    d. Adding helium gas to the reaction vessel.
    e. Heating the reaction vessel.
    f. Adding a catalyst to the reaction vessel.

Solution to Example 7.

Le Chatelier's principle provides the basis for answering these questions. Whatever stress is placed on a system at equilibrium, the system reacts in such a way as to counteract the stress.

a. *No effect.* Adding or removing a solid does not affect the equilibrium. Some solid needs to be present but how much of the solid is present does not influence the equilibrium. Solids never appear in the equilibrium expression.

b. *A net shift to the right.* Removing $CO_2\,(g)$ is the stress; the system acts to replace the missing $CO_2\,(g)$.

c. *A net shift to the left.* By halving the volume of the reaction vessel all the partial pressures of the gases are doubled. To counteract this stress, the equilibrium shifts to the side with less gas moles. In this case there is only 1 mol of gas, CO_2, on the product side, so by shifting to the left the partial pressure of $CO_2\,(g)$ is reduced.

d. *No effect.* Adding helium gas increases the total pressure inside the container but does not affect the partial pressure of the gas, CO_2, and thus does not affect the equilibrium.

e. *A net shift to the right.* The stress is the additional heat. The forward reaction is endothermic (a positive ΔH° value) and absorbs some of the additional heat. Another approach is to think of heat in an endothermic reaction as a reactant and thus adding heat will drive the reaction forward according to Le Chatelier's principle.

f. *No effect.* A catalyst only speeds up the reaction, that is, the equilibrium is attained faster. A catalyst does not affect equilibrium concentrations.

SAMPLE MULTIPLE CHOICE QUESTIONS

Questions 1–6 refer to the following reaction.

$$2NaHCO_3 (s) \Leftrightarrow Na_2CO_3 (s) + H_2O (g) + CO_2 (g) \qquad \Delta H = 128 \text{ kJ/mol}$$

For each question determine if:

    A. The amount of CO_2 gas increases;
    B. The amount of CO_2 gas decreases; or
    C. The amount of CO_2 gas remains unchanged.

1.   Sodium carbonate is added to the reaction vessel.
2.   Water vapor is removed from the reaction vessel.
3.   The volume of the reaction vessel is halved.
4.   Neon gas is introduced into the reaction vessel.
5.   The reaction vessel is heated.
6.   A catalyst is added to the reaction.

7.   Consider the reaction: $N_2 (g) + O_2 (g) \Leftrightarrow 2NO (g)$

A reaction mixture at 400 $^{\circ}$C has the following concentrations: [NO] = 0.01 M, [N_2] = 0.001 M, and [O_2] = 0.001 M. (K_c = 0.0156 at 400 $^{\circ}$C.)
At 400 $^{\circ}$C the reaction will:

    A. Shift to the right
    B. Shift to the left
    C. Remain unchanged
    D. Not enough information is given to predict the outcome
    E. Shift to the right if a catalyst is added

8.   Consider the equilibrium: $2NOCl (g) \Leftrightarrow 2NO (g) + Cl_2 (g)$

0.50 mol of NOCl were introduced into a 2.0-L container at 25 $^{\circ}$C. After equilibrium was established it was found that the molar concentration of chlorine gas was 0.10 M. What is the molar concentration of NOCl at equilibrium?

    A. 5.0×10^{-1} M
    B. 1.0 M
    C. 1.0×10^{-1} M
    D. 5.0×10^{-2} M
    E. None of these

9. Consider the reaction: $Ca(OH)_2 (s) + H_2O (l) \Leftrightarrow Ca^{2+}(aq) + 2OH^- (aq)$

For this reaction $K = 6 \times 10^{-5}$ at $0\ ^\circ C$ and $K = 1 \times 10^{-6}$ at $100\ ^\circ C$. Which of the following statement(s) is true?

    A. Adding HCl to the reaction will shift the equilibrium to the right.
    B. The forward reaction is exothermic.
    C. If heat is added to the system, reactants are favored.
    D. A and B are true.
    E. All statements are true.

10. Consider the reaction $H_2 (g) + S (s) \Leftrightarrow H_2S (g)$ at $25\ ^\circ C$. ($\Delta H^\circ = -21$ kJ/mol for this reaction.) Which statement(s) apply?

    A. $K_p = K_c$
    B. Adding heat favors the formation of hydrogen gas.
    C. Adding heat to the system will raise the K value.
    D. Statements A and B are true.
    E. All statements are true.

Comprehension Questions

1.   Write equilibrium constant expressions, Kc and Kp (where appropriate), for the following reactions:

    a) $2N_2 (g) + 5O_2 (g) \leftrightarrow 2N_2O_5 (g)$

    b) $CaCO_3 (s) + H_2SO_4 (aq) \leftrightarrow CaSO_4 (s) + H_2O (l) + CO_2 (g)$

    c) $C_2H_4 (g) + H_2 (g) \leftrightarrow C_2H_6 (g)$

    d) $HNO_2 (aq) \leftrightarrow H+(aq) + NO_2^- (aq)$

    e) $2A_2 (aq) + 2B_2D_3 (aq) \leftrightarrow 4AB (aq) + 3D_2 (aq)$

2)   Provide equilibrium constant expressions for the following representations of aqueous phase reactions. Write a balanced summary reaction that accounts for both of these reactions occurring simultaneously, or consecutively (eliminating formulas for compounds that are produced in one reaction only to be consumed in another), and write an equilibrium constant expression for the summary reaction.

$AB + B \leftrightarrow 2C$

$2C \leftrightarrow A_2 + D$

What is the relationship between the equilibrium constants for the reactions provided and the summary reaction?

3)   The equilibrium constant for the following reaction

$2SO_2 (g) + O_2 (g) \leftrightarrow 2SO_3 (g)$

is 5.6×10^4 at 350 °C. Calculate the partial pressure of SO_2 for a closed system at equilibrium if the partial pressures of O_2 and SO_3 are 0.55 atm and 10.8 atm, respectively.

4)   Consider the following reaction with a $K_c = 55.6$ at a temperature of 698 K:

$$H_2 (g) + I_2 (g) \leftrightarrow 2HI (g)$$

a)   Write the equilibrium constant expression, K_c, for this reaction.

b)   Suppose that the reaction was set up at 698 K with initial concentrations as follows: $[H_2] = 0.12$ M, $[I_2] = 0.041$ M, and $[HI] = 2.6$ M. Is the system at equilibrium? If not, in what direction will the system shift to reach equilibrium?

c)   Starting with 0.50 mol of $H_2 (g)$ and 0.88 mol of $I_2 (g)$ in a closed 3.0-L vessel at 698 K, calculate the concentrations of all species present once equilibrium is reached.

d)   Would the equilibrium mixture in part c) be described as reactant or product favored?

e)   Describe the observed effect on the above system at equilibrium caused by making the following changes:

     i)   Addition of HI (g)

     ii)   Addition of a catalyst

     iii)   Decreasing the volume of the container to 1.5 L

     iv)   Selective removal of $I_2 (g)$

     v)   Addition of 0.50 mol of argon gas

f)   Calculate the value of the equilibrium constant, K_c, for the following reactions:

i) $2H_2\,(g)\ +\ 2I_2\,(g)\ \leftrightarrow\ 4HI\,(g)$

ii) $HI\,(g)\ \leftrightarrow\ \frac{1}{2}H_2\,(g)\ +\ \frac{1}{2}I_2\,(g)$

ANSWERS TO SAMPLE MULTIPLE CHOICE QUESTIONS

1.   C
The amount of CO_2 gas remains unchanged.
Adding a solid reactant or product to a reaction that is at equilibrium does not affect that equilibrium.

2.   A
The amount of CO_2 gas increases. Removing the water vapor is the stress placed on the system. According to Le Chatelier's principle, the system will react to counteract the stress and thus shift forward to produce more water vapor.

3.   B
The amount of CO_2 gas decreases. Halving the volume of the container increases the partial pressure of the gases. To reduce the partial pressures the equilibrium shifts to the side with fewer moles of gas. In this case, the equilibrium shifts to the left, favoring the reactants.

4.   C
The amount of CO_2 gas remains unchanged. Adding a noble gas to the reaction increases the total pressure inside the container but does not affect the equilibrium since it does not affect the partial pressure of the gases that determine the equilibrium.

5.   A
The amount of CO_2 gas increases. Adding heat to the system favors the direction that absorbs heat, that is, the direction that is endothermic. ΔH is a positive number, indicating that the forward direction is endothermic, so the amount of carbon dioxide gas is increased. Another approach is to think of heat as a reactant in an endothermic reaction and thus adding heat will drive the reaction forward according to Le Chatelier's principle.

6.   C
The amount of CO_2 gas remains unchanged. A catalyst speeds up a reaction so that equilibrium is established more quickly. A catalyst does not affect the equilibrium concentrations.

7.   B

When a mixture of reactants and products exist, Q must be determined. If $Q = K$, the reaction is at equilibrium and no net shift occurs; if $Q > K$ the reaction will shift to the left; if $Q < K$ the reaction will shift to the right.

$$Q = \frac{[NO]^2}{[N_2][O_2]} = \frac{(0.01)^2}{(0.001)^2} = 100$$ Since $Q > K$, *the reaction shifts to the left,* decreasing the amount of NO.

8.   D

General Strategy	Solution to Question 8
Write the balanced equation.	$2NOCl\,(g) \Leftrightarrow 2NO\,(g) + Cl_2\,(g)$
Write the equilibrium expression.	$K_c = \dfrac{[NO]^2\,[Cl_2]}{[NOCl]^2}$
Place the data from the problem underneath the balanced equation. The initial conditions, *(i)* and equilibrium concentrations, *(eq),* must be expressed in $\dfrac{mol}{L}$ for K_c. Identify the quantity to be calculated.	$\begin{array}{lccc} & 2NOCl\,(g) \Leftrightarrow & 2NO\,(g) & + Cl_2\,(g) \\ (i) & 0.50 \text{ mol} & 0 & 0 \\ & 2.0 \text{ L} & & \\ (eq) & ? & & 0.10 \text{ M} \end{array}$
Use the stoichiometric relationships from the balanced equation to determine the equilibrium concentration of NOCl. Initially, there was no product present so the reaction shifts to the right. This means that the reactant molecule decreases in amount. The amount is determined by the coefficients of the balanced equation.	Initially there was zero NO and Cl_2 present. At equilibrium the Cl_2 gas has a 0.10 M concentration. In a 2.0-L container that means 0.20 mol of Cl_2 are present. There exists a 1 mol Cl_2 to 2 mol NOCl relationship so that if 0.20 mol of chlorine are formed, then 0.40 mol of NOCl reacted. $0.20 \text{ mol } Cl_2 \times \dfrac{2 \text{ mol NOCl}}{1 \text{ mol } Cl_2} = 0.40 \text{ mol NOCl}$ $\begin{array}{l} 0.50 \text{ mol NOCl present initially} \\ \underline{-\ 0.40 \text{ mol NOCl reacted}} \\ \ \ \ 0.10 \text{ mol NOCl remaining at equilibrium} \end{array}$
Solve for the *desired* quantity.	$[NOCl] = \dfrac{0.10 \text{ mol}}{2.0 \text{ L}} = \textit{0.050 M NOCl}$

9. E

All statements are true.
If HCl is added, the HCl will neutralize the OH⁻ ions to form water. As the OH⁻ ion concentration is reduced, the system reacts to replenish the OH⁻ concentration and more $Ca(OH)_2 (s)$ dissolves: that is, the equilibrium shifts to the right.

As the temperature goes from 0 °C to 100 °C, K gets smaller; this means that reactants are favored as heat is added to the system. In accordance with Le Chatelier's principle, as heat is added to the system, the system reacts to counteract this stress (heat), and the reaction shifts in the endothermic direction. The reverse reaction is endothermic, so the forward reaction is exothermic.

10. D

Statement (A) is true, as can be seen by applying the formula $K_p = K_c(RT)^{\Delta n}$. In this reaction $\Delta n = 0$, since there is 1 mol of reactant gas and 1 mol of product gas and therefore $K_p = K_c$. *Statement (B) is also true*; Le Chatelier's principle applies here. The reaction is exothermic in the forward direction since ΔH is negative. Adding heat to the reaction drives the reaction in the endothermic direction. In this case, that is the reverse direction, and the amount of hydrogen gas is increased.

Answers to Comprehension Questions

1) a) $$K_c = \frac{[N_2O_5]^2}{[N_2]^2[O_2]^5} \qquad K_p = \frac{(P_{N_2O_5})^2}{(P_{N_2})^2(P_{O_2})^5}$$

    b) $$K_c = \frac{[CO_2]}{[H_2SO_4]}$$

c) $\quad K_c = \dfrac{[C_2H_6]}{[C_2H_4][H_2]} \qquad K_p = \dfrac{(P_{C_2H_6})}{(P_{C_2H_4})(P_{H_2})}$

d) $\quad K_c = \dfrac{[NO_2^-][H^+]}{[HNO_2]}$

e) $\quad K_c = \dfrac{[AB]^4[D_2]^3}{[A_2]^2[B_2D_3]^2}$

2) $\qquad AB + B \leftrightarrow 2C \qquad K_c = \dfrac{[C]^2}{[AB][B]}$

$\qquad 2C \leftrightarrow A_2 + D \qquad K_c' = \dfrac{[A_2][D]}{[C]^2}$

The summary reaction for this two-step process is

$\qquad AB + B \leftrightarrow A_2 + D$

And its equilibrium constant expression is

$\qquad K_c'' = \dfrac{[A_2][D]}{[AB][B]}$

The relationship between this expression and the others, as well as the relationship between K_c, K_c', and K_c'', is that K_c'' is the mathematical product of $K_c \times K_c'$.

3)   In this situation all but one of the values in the equilibrium constant expression have been provided. Simply substitute and solve for the desired variable. The equilibrium constant expression for this reaction is

$$K_{eq} = \frac{(P_{SO_3})^2}{(P_{SO_2})^2(P_{SO_2})}$$

Rearranging to solve for the partial pressure of SO_2

$$P_{SO_2} = [\ (P_{SO_3})^2 / (K_{eq} \times (P_{O_2}))\ ]^{\frac{1}{2}}$$

And substituting

0.062atm

$$P_{SO_2} = [\ (10.8\ \text{atm})^2 / (5.6 \times 10^4 \times (0.55\ \text{atm}))\ ]^{\frac{1}{2}} = \text{0.0062 atm}$$

4)   a) $K_c = \dfrac{[HI]^2}{[H_2][I_2]}$

      b)   Given a set of concentrations or pressures for a reaction, you need only compare the value of the reaction quotient, Q, to the equilibrium constant to determine whether the reaction is at equilibrium. The reaction quotient is simply the equilibrium constant expression substituted with nonequilibrium values. Substituting the concentrations given,

$$Q = \frac{[2.6]^2}{[0.12][0.041]}$$

The value of Q does not equal the equilibrium constant at this temperature, therefore, **the system is not at equilibrium.** Since Q is greater than K in this particular case, **the reaction must run in reverse according to the way it is written** (HI must decompose to form H_2 and I_2).

c)   Given initial concentrations or partial pressures and the value of the equilibrium constant, equilibrium values may be calculated. A mathematical gimmick called the "ICE-box" has been created to help with assigning variables to represent equilibrium values. Since we only have one equation relating the equilibrium constant and the equilibrium values, a single variable must be used to represent all of these values. The "ICE-box" helps simplify and clarify this procedure. Only concentrations or partial pressures, not moles or any other quantity, should be plugged into the "ICE-box." The initials ICE stand for initial, change, and equilibrium, respectively.

First we must divide the numbers of moles given by the volume of the reaction vessel to calculate initial concentrations in moles/liter.

[HI] = 0 M

[H_2] = 0.50 mol/3.0 L = 0.17 M

[I_2] = 0.88 mol/3.0 L = 0.29 M

$$H_2 \ (g) \ + \quad I_2 \ (g) \ \leftrightarrow \quad 2HI \ (g)$$

	H_2 (g)	I_2 (g)	2HI (g)
Initial	0.17 M	0.29 M	0 M
Change	$-x$	$-x$	$+2x$
Equilibrium	0.17 M $- x$	0.29 M $- x$	$2x$

Notice that the coefficients of the balanced equation have been used to assign the variables in terms of x. The concentrations, which will be going down, have negative signs in front of them to distinguish between these values and those of the products, which will be increasing. The rate of change of hydrogen and iodine will be equal and in the same direction. The change of hydrogen iodide's concentration will be opposite in direction and changing twice as fast as the rate of change of hydrogen and iodine. These equilibrium values, in terms of the variable x, may be substituted into the equilibrium constant expression and the value of x determined.

$$K_c = \frac{[HI]^2}{[H_2][I_2]} = \frac{[2x]^2}{[0.17-x][0.29-x]} = \frac{4x^2}{0.049 - 0.46x + x^2}$$

$$4x^2 = 2.7 - 26x + 55.6x^2$$

$$0 = 51.6x^2 - 26x + 2.7$$

Solving the quadratic equation for x yields values of 0.36 and 0.15. Only one value is ever meaningful when substituted back into the equilibrium values (i.e., does not lead to negative concentrations, etc.), in this case 0.15 works whereas 0.36 leads to negative concentration values for both hydrogen and iodine. Substituting this value of x yields the following values for equilibrium concentrations:

$[HI] = 2 \times 0.15 = $ **0.30 M**

$[H_2] = 0.17 - 0.15 = $ **0.02 M**

$[I_2] = 0.29 - 0.15 = $ **0.14 M**

d)   Since this product mixture contains more hydrogen iodide than hydrogen and iodine, it would be considered **product favored.**

e)   This question involves application of Le Chatelier's principle to a system that is at equilibrium. Le Chatelier's principle states that a chemical reaction at equilibrium will react to an external stress in a manner that will re-establish a new equilibrium position.

i)   Addition of a product to an equilibrium mixture causes the reaction to shift in the direction of formation of additional reactants from the products, that is, **to the left.**

ii)   Catalysts speed up a reaction by lowering the activation energy of the reaction intermediates. This has the effect of speeding up the forward and reverse reactions equally, and therefore does not affect the position of the equilibrium, that is, **no change.**

iii)   Decreasing the volume of the container has the effect of increasing the concentrations and partial pressures of the reactants. If a gas phase reaction can shift left or right to produce a smaller number of moles of gaseous substances, it will. This will allow the reaction to take up less space, since equal moles of a gas and volumes are directly proportional. In this particular case, the reaction will not adjust since there are 2 mol of gas on either side of the reaction as written, that is, **no effect.**

iv)   Removal of a substance from an equilibrium reaction mixture will cause the reaction to shift toward the side that has lost matter. In this case the reaction will **shift left** to replace, partially, the iodine that has been removed.

v)   Addition of an inert gas, or anything that does not appear in the equilibrium constant expression or react with any of the substances involved, will have **no effect** on the position of the equilibrium. The total pressure of the system will increase, but the partial pressures, or concentrations, of the reactants and products will not be affected, therefore the reaction will still be at equilibrium.

f)   The value of an equilibrium constant is dependent on the manner in which a chemical equation is written. The actual amounts of reactants and products in an equilibrium mixture, however, are obviously not dependent on how the equation is written. If you reverse the written chemical equation, the new equilibrium constant will be the reciprocal of the original value. If the coefficients in the balanced equation are multiplied by a constant, the new equilibrium constant will be the old one raised to the power of the value by which the coefficients were multiplied. In part i) of this question, the coefficients have been multiplied by 2, so the new equilibrium constant, $K_c' = (55.6)^2 = $ **3090**. In part ii) of the problem, the reaction is written in reverse, and the coefficients multiplied by ½ . The new value for the equilibrium constant will then be $K_c'' = (1/55.6)^{1/2} = $ **0.134**.

CHAPTER 15
ACIDS AND BASES

This chapter reviews the following concepts of acid-base chemistry:

- Definitions of acids and bases
- Relative strengths of acids and bases
- K_a and K_b expressions
- pH and pOH
- Interrelationships between K_a, K_b, pH, and pOH
- Calculating pH: strong and weak acids
- Acidic, basic, and neutral salts

Both this chapter and Chapter 16 address acid-base chemistry. To master these concepts, you must thoroughly understand the equilibria concepts described in Chapter 14.

Take Note: *Equilibria concepts are the major topic on the AP exam. To do well on the AP exam, you <u>must</u> understand the concept of equilibrium and apply it to a variety of phenomena!*
In the Free Response section of the exam, the first question is an equilibrium question. This problem is 20 % of the Free Response exam grade. Numerous questions on equilibrium are also asked in the Multiple Choice section of the exam.

Definitions of acids and bases

Generally, **acidic solutions** have a *sour taste, dissolve certain metals, react with carbonates to form CO_2 gas, turn blue litmus paper red, and have a pH of less than 7.* In contrast, **basic solutions** generally have a *bitter taste, feel slippery, turn red litmus paper blue, and have a pH greater than 7.*

There are three major theories used to describe acids and bases: Arrhenius theory, Brønsted theory (also called Brønsted-Lowry theory), and Lewis theory. Which definition is appropriate depends upon the particular acid or base. The *Arrhenius theory* can be used when a *strong acid and strong base* are involved. The *Brønsted theory is most generally applicable to describe reactions in aqueous solution.* The *Lewis theory* is often used in the context of *nonaqueous reactions.*

Summary of acid-base theories:

Theory	*Acid definition* **and examples**	*Base definition* **and examples**
Arrhenius	*Produces H^+ ions* Examples: HCl, H_2SO_4	*Produces OH^- ions* Examples: $NaOH$, $Ca(OH)_2$
Brønsted** Note that by looking at the products you can determine how the reactant behaved.	*Proton (H^+) donor* Example: H_2O in the reaction: $$NH_3 + H_2O \Leftrightarrow NH_4^+ + OH^-$$ Water *donates* an H^+ to NH_3 and forms OH^-. By donating a proton, H_2O has acted as a Brønsted acid.	*Proton acceptor* Example: NH_3 in the reaction: $$NH_3 + H_2O \Leftrightarrow NH_4^+ + OH^-$$ Ammonia *accepts* an H^+ from H_2O and forms NH_4^+. By accepting a proton, NH_3 has acted as a Brønsted base.
Lewis	*Electron pair acceptor* Examples: BF_3, Zn^{2+} BF_3 has an empty orbital and thus can accept an electron pair. To use the Lewis definition, draw a Lewis dot diagram. If there is an empty orbital, the species can behave as a Lewis acid.	*Electron pair donor* Example: NH_3 NH_3 has an unbonded electron pair (draw Lewis dot diagram). A Lewis acid and Lewis base combine when they react: $BF_3 + NH_3 \rightarrow BF_3\!\!-\!\!NH_3$

** When using the Brønsted theory, the conjugate acid-base pair reference may be used. A **conjugate acid-base pair** is an *acid with its conjugate base or a base with its conjugate acid.* For example:

$$NH_3 \quad + \quad H_2O \quad \Leftrightarrow \quad NH_4^+ \quad + \quad OH^-$$
$$\text{Base} \qquad \text{Acid} \qquad \text{Conjugate acid} \qquad \text{Conjugate base}$$

In the forward reaction, the NH_3 acts as the Brønsted base by accepting the proton from water. Once NH_3 has accepted the proton, it becomes an acid, NH_4^+, and is called the conjugate acid. The $NH_3 - NH_4^+$ pair is a conjugate acid-base pair.

Water can act as either a weak acid or a weak base. Substances that can *act as either an acid or base* are known as **amphoteric** substances.

- When water reacts with an acid, water behaves as a base. For example, when HCl is dissolved in water the water accepts a proton: $HCl + H_2O \rightarrow H_3O^+ + Cl^-$

- When water reacts with a base, water behaves as an acid. For example, when ammonia is dissolved in water, the water donates a proton: $NH_3 + H_2O \Leftrightarrow NH_4^+ + OH^-$

Relative strengths of acids and bases

Strong acids dissociate 100% in water; weak acids usually dissociate less than 5%. Weak acids have K_a values associated with them and weak bases have K_b values.

- Common strong acids are: $HClO_4$, $HClO_3$, HI, HBr, HCl, HNO_3, H_2SO_4, HSO_4^-, H_3PO_4 (HSO_4^- and H_3PO_4 are generally considered strong acids even though they have K_a values. Their K_a values are large.)

- Common weak acids are: CH_3COOH, HCOOH, HF, $HClO_2$, HClO, H_3PO_3, HNO_2

- Common weak bases are: NH_3 and the conjugate bases of weak acids

Table 15.2 lists common acids arranged in order from strong to weak acids.

TABLE 15.2

Relative Strengths of Conjugate Acid-Base Pairs

Acid	Conjugate Base
$HClO_4$ (perchloric acid)	ClO_4^- (perchlorate ion)
HI (hydroiodic acid)	I^- (iodide ion)
HBr (hydrobromic acid)	Br^- (bromide ion)
HCl (hydrochloric acid)	Cl^- (chloride ion)
H_2SO_4 (sulfuric acid)	HSO_4^- (hydrogen sulfate ion)
HNO_3 (nitric acid)	NO_3^- (nitrate ion)
H_3O^+ (hydronium ion)	H_2O (water)
HSO_4^- (hydrogen sulfate ion)	SO_4^{2-} (sulfate ion)
HF (hydrofluoric acid)	F^- (fluoride ion)
HNO_2 (nitrous acid)	NO_2^- (nitrite ion)
HCOOH (formic acid)	$HCOO^-$ (formate ion)
CH_3COOH (acetic acid)	CH_3COO^- (acetate ion)
NH_4^+ (ammonium ion)	NH_3 (ammonia)
HCN (hydrocyanic acid)	CN^- (cyanide ion)
H_2O (water)	OH^- (hydroxide ion)
NH_3 (ammonia)	NH_2^- (amide ion)

(Left margin: Acid strength increases; Strong acids; Weak acids. Right margin: Base strength increases.)

The conjugate bases of weak acids are weak bases. For example, CH_3COO^- is the conjugate base of the weak acid CH_3COOH. The CH_3COO^- anion is a weak base; it undergoes a hydrolysis reaction with water to form CH_3COOH and OH^- according to the reaction: $CH_3COO^- + H_2O \Leftrightarrow CH_3COOH + OH^-$. *Reactions that produce OH^- ions form basic solutions.*

On the other hand, the *conjugate bases of strong acids do not exhibit basic behavior.* For example, when considering the HCl-Cl$^-$ conjugate acid-base pair in the reaction $HCl + H_2O \rightarrow H_3O^+ + Cl^-$, the Cl$^-$ ion does not act as a base. There is no reverse reaction when strong acids dissociate in water. The Cl$^-$ ion does not accept a proton from H_3O^+ to form HCl.

In conclusion, *the stronger the acid, the weaker its conjugate base and the weaker the acid, the stronger its conjugate base.*

Summary of strong acid vs. weak acid behavior:

Acid	Concentration before dissociation	Concentration at equilibrium	Comments
Strong acid, HA	Only HA	Only H$^+$ and A$^-$ ~~HA~~ $\rightarrow$ H$^+$ + A$^-$	HA is 100% dissociated.
Weak acid, HA	Only HA	Lots of HA, very little H$^+$ and A$^-$ HA $\Leftrightarrow$ H$^+$ + A$^-$	HA is less than 5% dissociated. A dynamic equilibrium exists.

K_a and K_b expressions

A **weak acid** *dissociates to a very limited extent in water, producing H_3O^+ ions* and has a K_a **expression.** There is more than one way to represent the equilibrium. Three common and interchangeable equations for representing the ionization of the weak acid acetic acid are:

1. $CH_3COOH + H_2O \Leftrightarrow H_3O^+ + CH_3COO^-$ $\quad K_a = \dfrac{[H_3O^+][CH_3COO-]}{[CH_3COOH]} = 1.8 \times 10^{-5}$

2. $CH_3COOH \Leftrightarrow H^+ + CH_3COO^-$ $\quad K_a = \dfrac{[H^+][CH_3COO^-]}{[CH_3COOH]} = 1.8 \times 10^{-5}$

3. $HA \Leftrightarrow H^+ + A^-$ $\quad K_a = \dfrac{[H^+][A^-]}{[HA]} = 1.8 \times 10^{-5}$

Notice that equation 3 is a generic representation of any acid, HA. (Recall from Chapter 14 of this review book that water, the solvent, is not included in a K_a or K_b expression.)

A **weak base** undergoes *a hydrolysis reaction in water, producing OH⁻ ions and has a* K_b expression. Two common and interchangeable representations for the weak base equilibrium are given for the reaction between the weak base ammonia and water:

1. $NH_3 + H_2O \Leftrightarrow NH_4^+ + OH^-$ $K_b = \dfrac{[NH_4^+][OH^-]}{[NH_3]} = 1.8 \times 10^{-5}$

or more generally:

2. $B + H_2O \Leftrightarrow BH^+ + OH^-$ $K_b = \dfrac{[BH^+][OH^-]}{[B]} = 1.8 \times 10^{-5}$

The relationship between K_a and K_b for a weak acid and its conjugate base and for a weak base and its conjugate acid is given by the equation: $\mathbf{K_a \times K_b = [H^+][OH^-]}$. To illustrate this relationship, examine the NH_3 - NH_4^+ conjugate base-acid pair and write the appropriate K_a and K_b expression for each species:

$NH_3 + H_2O \Leftrightarrow NH_4^+ + OH^-$ $NH_4^+ + H_2O \Leftrightarrow NH_3 + H_3O^+$

$K_b = \dfrac{[NH_4^+][OH^-]}{[NH_3]}$ $K_a = \dfrac{[NH_3][H_3O^+]}{[NH_4^+]}$

When you multiply the K_a by the K_b expressions, all the species cancel except for H_3O^+ and OH^-. Thus $K_a \times K_b = [H^+][OH^-]$. Some examples of ionization constants for weak acids and their conjugate bases at 25 °C are given in Table 15.3.

TABLE 15.3

Ionization Constants of Some Weak Acids and Their Conjugate Bases at 25°C

Name of Acid	Formula	Structure	K_a	Conjugate Base	K_b
Hydrofluoric acid	HF	H—F	7.1×10^{-4}	F^-	1.4×10^{-11}
Nitrous acid	HNO_2	O=N—O—H	4.5×10^{-4}	NO_2^-	2.2×10^{-11}
Acetylsalicylic acid (aspirin)	$C_9H_8O_4$	(structure)	3.0×10^{-4}	$C_9H_7O_4^-$	3.3×10^{-11}
Formic acid	HCOOH	(structure)	1.7×10^{-4}	$HCOO^-$	5.9×10^{-11}
Ascorbic acid*	$C_6H_8O_6$	(structure)	8.0×10^{-5}	$C_6H_7O_6^-$	1.3×10^{-10}

* For ascorbic acid it is the upper left hydroxyl group that is associated with this ionization constant.

The acids listed in Table 15.3 are **monoprotic acids:** they have *one ionizable hydrogen ion*. Acids with *two ionizable hydrogen ions*, such as H_2SO_4, are **diprotic acids.** Acids with *more than two ionizable hydrogen ions*, such as H_3PO_4, are **polyprotic acids.**

The ionization reactions that occur with diprotic and polyprotic acids happen in a stepwise fashion. Usually the first ionization (K_{a1}) occurs to a much larger extent than the second ionization (K_{a2}). For example, hydrosulfuric acid, H_2S, ionizes in two steps. The first reaction has a much greater forward reaction (larger K value) and thus K_{a1} determines the pH in the beaker.

$$H_2S \Leftrightarrow H^+ + HS^- \qquad\qquad K_{a1} = 1 \times 10^{-7}$$
$$HS^- \Leftrightarrow H+ + S^{2-} \qquad\qquad K_{a2} = 1 \times 10^{-19}$$

pH and pOH

Pure water ionizes to give H^+ and OH^- concentrations of 1.0×10^{-7} M at 25 °C.

$$H_2O \Leftrightarrow \qquad H^+ \qquad + \qquad OH^-$$
$$1.0 \times 10^{-7}\,M \qquad 1.0 \times 10^{-7}\,M$$

$[H^+]\,[OH^-] = (1.0 \times 10^{-7})^2 = 1.0 \times 10^{-14}$ at 25 °C = K_w, known as the **ion-product constant**. The addition of an acid or base to the water will change the concentrations of $[H^+]$ and $[OH^-]$, but the product of their concentrations must always equal 1×10^{-14} at 25 °C.

The concentration of $H^+(aq)$ can be expressed in terms of the pH scale: **pH = – log $[H^+]$.** An acidic solution has a pH less than 7, a neutral solution has a pH equal to 7, and a basic solution has a pH greater than 7.

The concentration of $OH^-(aq)$ can be expressed in terms of the pOH scale: **pOH = – log $[OH^-]$.**

Interrelationships between K_a, K_b, pH, and pOH

When taking the negative log of the ion product expression: **$[H^+][OH^-]$ = 1.0 × 10^{-14}**, the equation becomes: **pH + pOH = 14** at 25 °C.

Summary of equations:

- pH = – log $[H^+]$, which can also be written as $[H^+] = 10^{-pH}$
- pOH = – log $[OH^-]$, which can also be written as $[OH^-] = 10^{-pOH}$
- pH + pOH = 14
- $[H^+][OH^-] = 1.0 \times 10^{-14}$
- $K_a \times K_b = K_w = 1.0 \times 10^{-14}$

Calculating pH: strong and weak acids

A strong acid ionizes 100% in water. This is not an equilibrium since there is no reverse reaction. Therefore, the concentration of the strong acid is the concentration of the H^+ ion, as illustrated in Example 1. In contrast, a weak acid in water ionizes only a little. An equilibrium exists and the K_a expression is used to calculate the pH, as illustrated in Example 2.

Example 1. pH determination of a strong acid.

Determine the pH of a 0.10 M HCl solution.

General Strategy	Solution to Example 1
Write the equation. Two common representations of the reaction are given.	$HCl + H_2O \rightarrow H_3O^+ + Cl^-$ $HCl \rightarrow H^+ + Cl^-$
Place the data from the problem underneath the equation. HCl is a strong acid and thus completely dissociates. There is no reverse reaction.	$HCl + H_2O \rightarrow H_3O^+ \ + \ Cl^-$ 0.10 M 0.010 M
Solve for the unknown quantity.	$pH = -\log [H_3O^+] = -\log(1.0 \times 10^{-1}) = 1.00$

Example 2. pH determination of a weak acid.

Determine the pH of a 0.10 M solution of CH_3COOH. K_a for acetic acid is 1.8×10^{-5}.

General Strategy	Solution to Example 2
Write the equation. Two commonly used representations are given.	$CH_3COOH + H_2O \Leftrightarrow H_3O^+ + CH_3COO^-$ $CH_3COOH \Leftrightarrow H^+ + CH_3COO^-$
Write the appropriate K expression. CH_3COOH is a weak acid. Therefore, the K_a expression is used to calculate H^+ concentration. H_2O does not appear in a K_a or K_b expression.	$K_a = \dfrac{[H_3O][CH_3COO]}{[CH_3COOH]} = 1.8 \times 10^{-5}$

Place the data underneath the equation. These are the initial conditions, (*i*).	$CH_3COOH + H_2O \Leftrightarrow H_3O^+ + CH_3COO^-$ (*i*) 0.010 M 0 0
Then express the *changes (+x and –x)* that occur as the reaction moves toward equilibrium.	*changes* –*x* +*x* +*x*
At equilibrium (*eq*), the concentrations of the species are equal to the *initial concentrations + changes*.	(*eq*) 0.10 M – *x* *x* *x*
Substitute into the K_a expression.	$1.8 \times 10^{-5} = \dfrac{[H_3O^+][CH_3COO^-]}{[CH_3COOH]} = \dfrac{x^2}{(0.10 \text{ M} - x)}$
Make assumptions to simplify the equilibrium expression when possible. Since K_a is small, the amount of acid that dissociates is negligible when compared to the concentration of the acid initially.	Assume $0.10 - x = 0.10$ $1.8 \times 10^{-5} = \dfrac{x^2}{(0.10 \text{ M})}$
Solve for *x*. Validate the accuracy of any assumptions made.	$x = 1.3 \times 10^{-3}$ Check assumption: $0.10 - 0.0013 = 0.10$ M. Assumption is valid.
Solve for *desired* unit (pH in this case). $pH = -\log [H_3O^+]$	$pH = -\log (1.3 \times 10^{-3}) = 2.89$

In general, you can assume the amount of acid that dissociates is negligible compared to the original amount of the acid if the K_a of the acid is less than or equal to 1×10^{-5}. The forward reaction, that is, the dissociation of the acid, is very small when the K_a is small. The acid concentration is changed to such a small degree that it can be ignored. It is important to state this assumption and then test its validity.

The percent ionization of an acid is the amount of acid that dissociates at equilibrium divided by the initial amount of acid multiplied by 100:

$$\textbf{\% ionization} = \frac{\textbf{concentration of acid ionized at equilibrium}}{\textbf{initial concentration of acid}} \times \textbf{100}$$

Example 3. Percent ionization determination.

Determine the percent ionization of acetic acid in Example 2.

$$\text{\% ionization} = \frac{\text{concentration of acid ionized at equilibrium}}{\text{initial concentration of acid}} \times 100 = \frac{1.3 \times 10^{-3}}{0.10} \times 100$$

% ionization = *1.3%.* This means that 98.7% of the weak acid CH_3COOH is left in the undissociated form.

Acidic, basic, and neutral salts

When solid salts are dissolved in water they can form acidic solutions, basic solutions, or neutral solutions. To determine the type of salt, you should write two equations:

1. the dissociation equation; and
2. the reaction, if any, that the dissociated ions undergo with water. *H^+ is produced by acidic salts and OH^- is produced by basic salts.*

Type of salt	Example	Reaction with water	Comments
Acidic	$NaHSO_4$	$NaHSO_4\,(s) + H_2O\,(l) \rightarrow Na^+\,(aq) +$ $HSO_4^-\,(aq)$ $HSO_4^-\,(aq) \Leftrightarrow H^+ + SO_4^{2-}$	H^+ is produced so $NaHSO_4$ is an acidic salt. Another common acidic salt is NH_4Cl. Notice that the NH_4^+ is the conjugate acid of the weak base NH_3. *Conjugate acids of weak bases form acidic salts.*
Basic	NaF	$NaF\,(s) + H_2O\,(l) \rightarrow Na^+\,(aq) + F^-\,(aq)$ $F^-\,(aq) + H_2O \Leftrightarrow HF\,(aq) + OH^-\,(aq)$	OH^- is produced, so NaF is a basic salt. F^- is the conjugate base of the weak acid HF. Another common basic salt is CH_3COONa. The CH_3COO^- ion is the

			conjugate base of the weak acid CH_3COOH.

In general, the conjugate bases of weak acids form basic solutions. |
| **Neutral** | KCl | $KCl\ (s) + H_2O\ (l) \rightarrow K^+\ (aq) + Cl^-\ (aq)$ | Neither H^+ nor OH^- is formed, so KCl is a neutral salt.

Usually neutral salts are formed when a strong acid neutralizes a strong base, such as: $KOH + HCl \rightarrow H_2O + KCl$ |

Table 15.7 gives you examples of the three types of salts.

TABLE 15.7

Acid-Base Properties of Salts

Type of Salt	Examples	Ions That Undergo Hydrolysis	pH of Solution
Cation from strong base; anion from strong acid	NaCl, KI, KNO_3, RbBr, $BaCl_2$	None	≈ 7
Cation from strong base; anion from weak acid	CH_3COONa, KNO_2	Anion	> 7
Cation from weak base; anion from strong acid	NH_4Cl, NH_4NO_3	Cation	< 7
Cation from weak base; anion from weak acid	NH_4NO_2, CH_3COONH_4, NH_4CN	Anion and cation	< 7 if $K_b < K_a$

≈ 7 if $K_b \approx K_a$

> 7 if $K_b > K_a$ |
| Small, highly charged cation; anion from strong acid | $AlCl_3$, $Fe(NO_3)_3$ | Hydrated cation | < 7 |

To calculate the pH of a solution made from an acidic, basic, or neutral salt, the first step is to write the dissociation equation and then the reaction(s) that the dissociated ions may have with water. If it is an acidic salt solution, the K_a expression is used to solve for pH. If it is a basic solution, the K_b expression is used. If the solution is neutral, then the pH = 7 since $[H^+] = [OH^-] = 1.0 \times 10^{-7}$. See Example 4.

Example 4. The pH determination of an acidic salt.

Determine the pH of a 0.10 M solution of NH_4Cl. (K_b for NH_3 is 1.8×10^{-5}.)

General Strategy	Solution to Example 4

Write the equation. Since it is a salt there are two equations:
  a. the dissociation equation for the salt; and
  b. the reaction of the ion(s) with water.
	a. $NH_4Cl\ (s) + H_2O\ (l) \rightarrow NH_4^+\ (aq)\ +\ Cl^-\ (aq)$
	
	b. $NH_4^+\ (aq) \Leftrightarrow H^+\ (aq)\ +\ NH_3\ (aq)$
	
	Since H^+ is produced, $NH_4Cl\ (s)$ is an acidic salt.

Write the appropriate K expression. NH_4^+ is a weak acid; therefore, the K_a expression is used. $K_a \times K_b = K_w = 1.0 \times 10^{-14}$ and therefore $K_a = \dfrac{1.0 \times 10^{-14}}{K_b}$	$K_a = \dfrac{[H^+][NH_3]}{[NH_4^+]}$ $K_a = \dfrac{K_w}{K_b} = \dfrac{1.0 \times 10^{-14}}{1.8 \times 10^{-5}} = 5.5 \times 10^{-10}$
Place the data underneath the equation. These are the initial conditions, (i). Express the *changes* (−x and +x) that occur as the reaction moves toward equilibrium. At equilibrium (*eq*), the concentrations of the species are equal to the *initial concentrations + changes*.	$\qquad\quad NH_4^+\ (aq) \Leftrightarrow H^+\ (aq)\ +\ NH_3\ (aq)$ (*i*) $\qquad\quad$ 0.010 M $\qquad$ 0 $\qquad\quad$ 0 *changes* −x $\qquad\qquad$ +x $\qquad$ +x (*eq*) $\qquad$ 0.10 M − x $\quad$ x $\qquad\quad$ x
Substitute into the K_a expression.	$5.5 \times 10^{-10} = \dfrac{[H^+][NH_3]}{[NH_4^+]} = \dfrac{x^2}{(0.10\ M - x)}$
Make assumptions to simplify the equilibrium expression when possible. Since K_a is small, the amount of acid that dissociates is negligible when compared to the concentration of the acid initially.	Assume $0.10 - x = 0.10$ $5.5 \times 10^{-10} = \dfrac{x^2}{(0.10\ M)}$
Solve for x. Validate the accuracy of any assumptions made.	$x = 7.4 \times 10^{-6}\ M = [H^+] = [NH_3]$ *Check assumption: $0.10 - 0.0000074 = 0.10$ M
Solve for *desired* unit (pH in this case). pH $= -\log [H^+]$	pH $= -\log (7.4 \times 10^{-6}) = 5.13$

SAMPLE MULTIPLE CHOICE QUESTIONS

1. Consider the conjugate bases: CN^-, CH_3COO^-, F^-, NO_2^- and SO_4^{2-}. Which arrangement is in order of *increasing* base strength?

Acid	K_a value
HSO_4^-	1.1×10^{-2}
HNO_2	4.5×10^{-4}
CH_3COOH	1.8×10^{-5}
HF	7.1×10^{-4}
HCN	4.9×10^{-10}

    A. SO_4^{2-}, F^-, NO_2^-, CH_3COO^-, CN^-
    B. CN^-, CH_3COO^-, NO_2^-, SO_4^{2-}, F^-
    C. SO_4^{2-}, NO_2^-, F^-, CH_3COO^-, CN^-
    D. CN^-, CH_3COO^-, F^-, NO_2^-, SO_4^{2-}
    E.   None of these

2. Which of the following salts will form an acidic solution?

    NH_4Cl, CaO, $NaNO_2$, NaI, KH_2PO_4, KBr

    A. NaI, KBr and KH_2PO_4
    B. CaO, $NaNO_2$, and NaI
    C. NH_4Cl, $NaNO_2$, NaI, and KH_2PO_4
    D. NH_4Cl and KH_2PO_4
    E.   All of them

3. What is the $[H_3O^+]$ concentration in a 1.0 M CH_3COONa solution? (The K_a for CH_3COOH is 2×10^{-5}.)

    A. 2×10^{-5}
    B. 5×10^{-9}
    C. 5×10^{-10}
    D. 2×10^{-11}
    E. 5×10^{-12}

4. Consider the reaction: $H_2PO_4^- + H_2O \Leftrightarrow H_3O^+ + HPO_4^{2-}$

   Which of the following statements are true?

I.	$H_2PO_4^- - H_3O^+$ are a conjugate acid-base pair
II.	$H_2PO_4^- - HPO_4^{2-}$ are a conjugate acid-base pair
III.	$H_2PO_4^-$ is a stronger acid than HPO_4^{2-}
IV.	Water acts as a base in this reaction
V.	The above reaction has a pH greater than seven

   A. All the statements are true
   B. Only statements I, III, IV, and V are true
   C. Only statements II, III, IV, and V are true
   D. Only statements II, III, and IV are true
   E. Only statements II and III are true

5. What is the $[OH^-]$ concentration when 36.5 g of HCl are dissolved in enough water to make 500.0 mL of solution?

   A. 5.00×10^{-15}
   B. $\times 10^{-2}$
   C. 1.13×10^{-14}
   D. 1.56×10^{-3}
   E. 5.00×10^{-2}

6. Consider the following molecules and ions: BF_3, NH_3, Al^{3+}, Zn^{2+}, and HCl. Which species is a Lewis acid?

   A. BF_3
   B. BF_3 and Al^{3+}
   C. Al^{3+} and Zn^{2+}
   D. HCl
   E. BF_3, Al^3, and Zn^{2+}

7. The pOH of a 0.010 M solution of C_5H_5N is 5. What is the K_b for C_5H_5N?

   A. 1×10^{-8}
   B. 1×10^{-9}
   C. 5×10^{-5}
   D. 1×10^{-4}
   E. None of these

8. Consider the following molecules: HNO_2, HNO_3, HCl, HI, and NaF. The correct arrangement in terms of *increasing* acid strength is?

   A. HNO_2, HNO_3, HCl, HI, NaF
   B. NaF, HNO_2, HNO_3, HCl, HI

C. NaF, HNO_2, HNO_3, HI, HCl
D. NaF, HCl, HI, HNO_2, HNO_3
E. NaF, HI, HCl, HNO_2, HNO_3

9. What is the pH of a 0.10 M solution of H_2S?

K_a for $H_2S = 1 \times 10^{-7}$ $H_2S \Leftrightarrow HS^- + H^+$

K_a for $HS^- = 1 \times 10^{-19}$ $HS^- \Leftrightarrow S^{2-} + H^+$

A. 1.5
B. 2.0
C. 3.0
D. 4.0
E. 5.0

10. A 0.10 M solution of H_2S has an adjusted pH of 3. What is the $[S^{2-}]$ concentration?

K_a for $H_2S = 1 \times 10^{-7}$ $H_2S \Leftrightarrow HS^- + H^+$

K_a for $HS^- = 1 \times 10^{-19}$ $HS^- \Leftrightarrow S^{2-} + H^+$

A. 1×10^{-21}
B. 1×10^{-4}
C. 1×10^{-10}
D. 1×10^{-12}
E. 1×10^{-13}

Comprehension Questions

1)   Identify the conjugate acid-base pairs for the following chemical equations (all species are in aqueous solution unless otherwise noted):

a) $2HNO_3 + Ba(OH)_2 \rightarrow BaNO_3 + 2H_2O$ (l)

b) $HCl + NH_3 \rightarrow NH_4Cl$

c) $CH_3CO_2^- + H_2O$ (l) $\leftrightarrow CH_3CO_2H + OH^-$

d) $HF + H_2O \leftrightarrow H_3O^+ + F^-$

2)   Calculate the pH of the following aqueous solutions:

a) 0.15 M HBr

b) A solution with pOH of 4.5

c) A solution with $[H^+] = 3.1 \times 10^{-9}$

d) 1×10^{-3} M $Sr(OH)_2$

3)   For the following chemical species, a) identify each as a weak acid or base, b) provide balanced equations in aqueous solution for their reaction with water, and c) write equilibrium constant expressions, K_a or K_b, for each.

a) H_3PO_4

b) CH_3NH_2

c) NO_2^-

d)   HX

4)   Consider the following reaction of formic acid in aqueous solution:

$$HCO_2H \ (aq) + H_2O \ (l) \leftrightarrow H_3O^+ \ (aq) + HCO_2^- \ (aq) \qquad K_a = 1.7 \times 10^{-4}$$

a)   Write the equilibrium constant expression for the above reaction.

b)   Calculate the pH and pOH of a 0.215 M solution of formic acid in water. Provide a list of all species present at equilibrium along with their concentrations in moles per liter.

c)   Identify the formate ion and write a balanced equation for its reaction with water along with an equilibrium constant expression for this species. What is the value of the equilibrium constant, K_b, for formate's reaction with water?

ANSWERS TO SAMPLE MULTIPLE CHOICE QUESTIONS

1.   A

     The stronger the acid, the weaker its conjugate base is. A strong acid dissociates 100%. The conjugate base will not accept a proton; it is a very, very weak base. The strongest acid listed is HSO_4^- since it has the largest K_a value. Its conjugate base, SO_4^{2-}, is the weakest base. The weakest acid listed is HCN and its conjugate base CN^- is the strongest base. Arranging the conjugate bases from weakest to the strongest base means arranging the corresponding acids from strongest acid (largest K_a) to weakest acid: $SO_4^{2-} < F^- < NO_2^- < CH_3COO^- < CN^-$.

2.   D

     The salts that form acidic solutions are *NH_4Cl and KH_2PO_4*. The solid ionizes in water. The ion then undergoes a hydrolysis reaction with water, producing H_3O^+ ions.

$$NH_4Cl \ (s) + H_2O \rightarrow NH_4^+ \ (aq) + Cl^- \ (aq)$$
$$NH_4^+ + H_2O \Leftrightarrow H_3O^+ + NH_3$$

$$KH_2PO_4 \ (s) + H_2O \rightarrow H_2PO_4^- + K^+$$
$$H_2PO_4^- + H_2O \Leftrightarrow H_3O^+ + HPO_4^{2-}$$

     CaO and $NaNO_2$ form basic solutions. NaI and KBr form neutral solutions.

3.   C

General Strategy	Solution to Question 3
Write the equation.	CH_3COONa is a soluble salt that exists as Na^+ and CH_3COO^- ions in aqueous solution. The CH_3COO^- ion undergoes a hydrolysis reaction with water, forming the OH^- ion, which results in a basic solution: $$CH_3COO^- + H_2O \Leftrightarrow CH_3COOH + OH^-$$ Recognize that CH_3COO^- is the conjugate base of the weak acid CH_3COOH.
Write the appropriate K expression. CH_3COO^- is a weak base; therefore, the K_b expression is used. $K_a \times K_b = K_w = 1.0 \times 10^{-14}$ and therefore $K_b = \dfrac{1.0 \times 10^{-14}}{K_a}$	$$K_b = \frac{[OH^-]\,[CH_3COOH]}{[CH_3COO^-]}$$ $$K_b = \frac{1.0 \times 10^{-14}}{2 \times 10^{-5}} = 5 \times 10^{-10}$$
Place the data underneath the equation. These are the initial conditions, (*i*). Then express the *changes* (–*x* and +*x*) that occur as the reaction moves toward equilibrium. At equilibrium (*eq*), the concentrations of the species are equal to the *initial concentrations + changes*.	$$CH_3COO^- + H_2O \Leftrightarrow CH_3COOH + OH^-$$ (*i*) 1 M *changes* –*x* +*x* +*x* (*eq*) 1 M – *x* *x* *x*
Substitute into the K_b expression.	$K_b = \dfrac{[OH^-]\,[CH_3COOH]}{[CH_3COO^-]}$ $$K_b = \frac{x^2}{1-x}$$
Make assumptions to simplify the equilibrium expression when possible. Since K is $\leq 10^{-5}$, it is reasonable to assume the dissociated amount is negligible when compared to the initial amount of the base.	Assume $1 - x = 1$ $$K_b = 5 \times 10^{-10} = \frac{x^2}{1}$$
Solve for *x*. Validate the accuracy of any assumptions made.	$x = [OH^-] = 2.2 \times 10^{-5} = 2 \times 10^{-5}$ Assumption: $1 - (2 \times 10^{-5}) = 1$. The assumption is valid.

Calculate *desired* unit, $[H_3O^+]$ in this case. $[H_3O^+][OH^-] = 1 \times 10^{-14}$	$[H_3O] = \dfrac{1 \times 10^{-14}}{[OH]} = \dfrac{1 \times 10^{-14}}{2 \times 10^{-5}} = 5 \times 10^{-10}$

4.   D

Statement II is true. The acid is $H_2PO_4^-$ and its conjugate base is HPO_4^{2-}. Once the acid donates its proton, it becomes the conjugate base.

Statement III is true since $H_2PO_4^-$ donates a proton to water and is a stronger acid than water. Once the $H_2PO_4^-$ has donated the proton, it becomes a conjugate base, HPO_4^{2-}. $H_2PO_4^-$ is a stronger acid than HPO_4^{2-}.

Statement IV is true since water accepts a proton from $H_2PO_4^-$ and acts as a Brønsted base.

Statement I is false. $H_2PO_4^-$ and HPO_4^{2-} form a conjugate acid-base pair.
Statement V is false. The solution has a pH below 7 since it is an acidic solution.

5.   A

HCl is a strong acid and dissociates 100%. $HCl \rightarrow H^+ (aq) + Cl^- (aq)$. Use dimensional analysis to calculate molarity of HCl. The molarity of H^+ is 2.0 M.

$$\frac{36.5\,g\,HCl}{0.50\,L} \times \frac{1\,mol\,HCl}{36.5\,g\,HCl} \times \frac{1\,mol\,H}{1\,mol\,HCl} = 2.0\,M\,H^+$$

$[H^+][OH^-] = 1.0 \times 10^{-14}$ and therefore $[OH^-] = \dfrac{1.0 \times 10^{-14}}{2.0} = 5.0 \times 10^{-15}$

6.   E

A Lewis acid is an electron pair acceptor, which means it has an empty electron orbital. BF_3 is sp^2 hybridized and there is an empty orbital. The metal ions (Al^{3+} and Zn^{2+}) have lost electrons so they too have empty electron orbitals. NH_3 is a Lewis base since it has an unbonded electron pair. HCl is an Arrhenius acid.

7.   A

General Strategy	Solution to Question 7
Write the equation.	$C_6H_5N + H_2O \Leftrightarrow C_6H_5NH^+ + OH^-$
Write the K_b expression.	$K_b = \dfrac{[OH^-][C_6H_5NH^+]}{[C_6H_5N]}$
Place the data from the problem underneath the equation.	pOH is given as 5. $[OH^-] = 10^{-pH} = 1 \times 10^{-5}$ The equation shows a 1 to 1 ratio between $C_6H_5NH^+$ and OH^-. Thus the $C_6H_5NH^+$ concentration is also 1×10^{-5}. $\quad\quad C_6H_5N + H_2O \Leftrightarrow C_6H_5NH^+ + OH^-$ *(eq)* 0.010 M $\quad\quad\quad 1 \times 10^{-5}\,M \quad 1 \times 10^{-5}M$
Substitute into the K_b expression.	$K_b = \dfrac{[OH^-][C_6H_5NH^+]}{[C_6H_5N]} = \dfrac{(1 \times 10^{-5})^2}{0.010}$
Solve the K_b expression	$K_b = 1 \times 10^{-8}$

8.   B

NaF is a basic salt. $F^- + H_2O \Leftrightarrow HF + OH^-$, so NaF is the least acidic.
HNO_2 is a weaker acid than HNO_3 because oxoacids are weaker when they have fewer oxygens.
HCl and HI are both considered strong acids, but HI is the stronger. The hydrogen ion is furthest from the nucleus in HI and more easily ionized.
The arrangement from least acidic to strongest acid is:

$$NaF < HNO_2 < HNO_3 < HCl < HI$$

9.   D

General Strategy	Solution to Question 9
Write the equations.	(1) $H_2S \Leftrightarrow HS^- + H^+$ $\qquad K_{a1} = 1 \times 10^{-7}$ (2) $HS^- \Leftrightarrow S^{2-} + H^+$ $\qquad K_{a2} = 1 \times 10^{-19}$
Write the relevant K expression. $K_{a1} \gg K_{a2}$ and thus K_{a1} determines the pH in the solution. In general, when there is a very much stronger acid with a weak acid in the same solution, the stronger acid determines the pH. The weaker acid's contribution of H^+ is negligible.	$K_{a1} = \dfrac{[H^+][HS^-]}{[H_2S]} = 1 \times 10^{-1}$
Place the data underneath the equation. These are the initial conditions, (*i*). Express the *changes* (*–x and +x*) that occur as the reaction moves toward equilibrium. At equilibrium (*eq*), the concentrations of the species are equal to the *initial concentrations + changes*.	$\qquad\qquad H_2S \quad \Leftrightarrow HS^- + H^+$ (*i*) $\qquad\quad$ 0.10 changes $\qquad -x \qquad +x \quad +x$ (*eq*) $\qquad$ 0.10 $-x \quad\;\; x \qquad x$
Substitute into the K_{a1} expression.	$K_{a1} = \dfrac{[H^+][HS^-]}{[H_2S]} = 1 \times 10^{-7} = \dfrac{x^2}{(0.10-x)}$

Make assumptions to simplify the equilibrium expression when possible.	Since K is small ($\leq 10^{-5}$), you can assume $0.10 - x = 0.10$. $$K_{a1} = \frac{[H^+][HS^-]}{[H_2S]} = 1 \times 10^{-7} = \frac{x^2}{(0.10)}$$
Solve for x. Validate the accuracy of any assumptions made.	$[x] = [H^+] = 1 \times 10^{-4}$ Assumption is valid: $0.10 - 0.0001 = 0.10$
Solve for *desired* quantity, pH.	$pH = -\log[H^+] = -\log[1 \times 10^{-4}] = 4.0$

10.   A

Since the H^+ concentration is adjusted to $pH = 3$, $[H^+] = 1 \times 10^{-3}$ and you can use equation (3) in the chart, which combines the two equilibria, to solve for the S^{2-} concentration.

General Strategy	**Solution to Question 10**
Write the equations. All the species are in the same solution and are in equilibrium with each other. Each species can only have one concentration at equilibrium. For example, the H^+ concentration is 1×10^{-3}.	(1) $H_2S \Leftrightarrow HS^- + H^+$ (2) $HS^- \Leftrightarrow S^{2-} + H^+$ (3) $H_2S \Leftrightarrow S^{2-} + 2H^+$
Write the appropriate K expression. $K_{a3} = K_{a1} \times K_{a2}$	$$K_{a3} = \frac{[H^+]^2}{[H_2S]} = 1 \times 10^{-26}$$ $K_{a3} = K_{a1} \times K_{a2} = (1 \times 10^{-7})(1 \times 10^{-19}) = 1 \times 10^{-26}$
Place the data underneath the equation.	$H_2S \Leftrightarrow S^{2-} + 2H^+$ *(eq)* $\;0.10\,M\quad ?\quad 1 \times 10^{-3}\,M$
Substitute into the K_{a3} expression.	$$K_{a3} = \frac{[H^+]^2[S^{2-}]}{[H_2S]} = 1 \times 10^{-26} = \frac{(1 \times 10^{-3})^2[S^{2-}]}{(0.10)}$$
Solve K_{a3} expression for x.	$[S^{2-}] = 1 \times 10^{-21}$

Notice the very small concentration of $[S^{2-}]$. By lowering the pH (adding a strong acid), the ionization of the weak acid becomes very, very small.

Answers to Comprehension Questions

1)   a)   This is the reaction of a strong acid with a strong base, which is essentially irreversible. Both the acid and the base will be completely ionized in aqueous solution so the reaction is more appropriately represented with a net ionic equation as follows;

$$H_3O^+ + OH^- \rightarrow 2H_2O$$

In this case, H_3O^+ is the acid and its conjugate base is H_2O. The base is OH^-, and its conjugate acid is also H_2O.

    b)   This is the reaction of a strong acid with a weak base. The acid is HCl and its conjugate is Cl^-. The base is NH_3 and its conjugate acid is NH_4^+.

    c)   This is a reaction of a weak base, acetate ion, with the amphoteric substance water. The base, in the forward direction is $CH_3CO_2^-$, and its conjugate acid is CH_3CO_2H. Water is the acid in the forward reaction and its conjugate base is OH^-.

    d)   This is the reaction of a weak acid, hydrofluoric acid, with the amphoteric substance water. In the forward direction, the acid is HF and its conjugate base is F^-. The base in the forward direction is water itself, and its conjugate acid is H_3O^+.

2)   a)   HBr is a strong acid, and therefore completely ionized in water. The result is that the final $[H^+]$ will be equal to whatever the concentration of acid was initially, that is, 0.15 M. Since pH is simply $- \log_{10} [H^+]$, substituting the value 0.15 yields a **pH = 0.82**.

    b)   The product of $[H^+]$ and $[OH^-]$ ions in aqueous solutions at 25 °C always equals 1.0×10^{-14}. Since pH and pOH are the negative logs of these concentrations, respectively, the sum of pH and pOH for aqueous solutions at 25 °C always equals 14. This question then becomes the one of simply subtracting the pOH from 14, $14 - 4.5 = $ **9.5 = pH**, to arrive at the pH.

    c)   You must simply take the negative log of the given $[H^+]$ here to arrive at pH. **$pH = - \log_{10} [3.1 \times 10^{-9}] = 8.51$.**

d) $Sr(OH)_2$ is a strong base that yields 2 mol of OH^- upon dissolving and dissociating in water. As stated above, the product of $[H^+]$ and $[OH^-]$ ions in aqueous solutions at 25 °C always equals 1.0×10^{-14}. To find pH, you must have $[H^+]$, so this problem becomes one of first determining the $[OH^-]$ from the initial base concentration, then dividing 1.0×10^{-14} by $[OH^-]$ to arrive at $[H^+]$. Finally the negative log of this value may be taken to arrive at pH.

$[OH^-] = 1.0 \times 10^{-3}$ M $Sr(OH)_2 \times$ (2 mol OH^-/mol $Sr(OH)_2$)$= 2.0 \times 10^{-3}$ M $[OH^-]$

$[H^+] = 1.0 \times 10^{-14} / [OH^-] = 1.0 \times 10^{-14} / [2.0 \times 10^{-3}] = 5.0 \times 10^{-12}$

pH $= -\log_{10} [5.0 \times 10^{-12}] = 11.30$

3)   a)   This is the weak acid phosphoric acid.

$$H_3PO_4\,(aq) + H_2O\,(l) \leftrightarrow H_3O^+\,(aq) + H_2PO_4^-\,(aq) \quad K_a = \frac{[H_3O^+][H_2PO_4^-]}{[H_3PO_4]}$$

b)   This is the weak base methylamine (a close relative to ammonia).

$$CH_3NH_2\,(aq) + H_2O\,(l) \leftrightarrow OH^-\,(aq) + CH_3NH_3^+\,(aq) \quad \frac{[OH^-][CH_3NH_3^+]}{[CH_3NH_2]}$$

c)   This is the weak base nitrite anion (conjugate base of the weak acid nitrous acid).

$$NO_2^-\,(aq) + H_2O\,(l) \leftrightarrow OH^-\,(aq) + HNO_2\,(aq) \quad K_b = \frac{[OH^-][HNO_2]}{[NO_2^-]}$$

d)   This is one of the ways in which a weak acid can be generally represented. Others are HB or HA.

$$HX(aq) + H_2O(l) \leftrightarrow H_3O^+(aq) + X^-(aq) \quad K_a = \frac{[H_3O^+][X^-]}{[HX]}$$

4) a)

$$K_a = 1.7 \times 10^{-4} = \frac{[H_3O^+][HCO_2^-]}{[HCO_2H]}$$

b) This is a time to use the **ICE-box** since we are given an Initial concentration and asked for the pH that is related to the Equilibrium concentration of $[H^+]$.

	HCO$_2$H (aq) +	H$_2$O (l)	$\leftrightarrow$	H$_3$O$^+$ (aq) +	HCO$_2^-$ (aq)
Initial	0.215 M	NA		0	0
Change	$-x$	NA		$+x$	$+x$
Equilibrium	$0.215 - x$	NA		x	x

Notice that the concentration of water is not included as it is neither in the aqueous nor gas phases and therefore is not a variable associated with the equilibrium constant expression.

$$K_a = 1.7 \times 10^{-4} = \frac{[H_3O^+][HCO_2^-]}{[HCO_2H]} = \frac{[x][x]}{[0.215 - x]}$$

$$x^2 = 3.66 \times 10^{-5} - 1.7 \times 10^{-4}x$$

$$0 = x^2 + 1.7 \times 10^{-4}x - 3.66 \times 10^{-5}$$

Solving the quadratic provides values of 0.00597 and –0.00614. Only the positive value makes sense as the negative would lead to negative, and meaningless, equilibrium concentrations of products.

The **pH** is equal to $-\log_{10}[H^+] = -\log_{10}[0.00597] = \mathbf{2.22}$. Subtracting from 14 we get a **pOH** of $14 - 2.22 = \mathbf{11.78}$. The species present at equilibrium, apart from water, are H_3O^+, HCO_2^-, HCO_2H, and OH^-. Their concentrations are as follows:

$[H_3O^+] = \mathbf{0.00597\ M}$

$[HCO_2^-] = \mathbf{0.00597\ M}$

$[HCO_2H] = 0.215 - 0.00597 = \mathbf{0.209\ M}$

$[OH^-] = 1.0 \times 10^{-14} / [H^+] = 1.0 \times 10^{-14} / [0.00597] = \mathbf{1.68 \times 10^{-12}\ M}$

c)   The **formate ion**, conjugate base of formic acid, is $\mathbf{HCO_2^-}$, and its reaction with water and equilibrium constant expressions are:

$$HCO_2^-(aq) + H_2O(l) \leftrightarrow OH^-(aq) + HCO_2H(aq) \quad K_b = \frac{[OH^-][HCO_2H]}{[HCO_2^-]} = 1.7 \times 10^{-4}$$

The relationship between K_a and K_b for any acid-base conjugate pair at 25 °C is $K_a \times K_b = 1.0 \times 10^{-14}$. Therefore,

$\mathbf{K_{b\ formate}} = 1.0 \times 10^{-14} / K_a = 1.0 \times 10^{-14} / 1.7 \times 10^{-4} = \mathbf{5.88 \times 10^{-11}}$

CHAPTER 16
ACID-BASE EQUILIBRIA AND SOLUBILITY EQUILIBRIA

The principles you learned in Chapter 14 and Chapter 15 are applied to more complex equilibria situations in this chapter. *To understand the topics presented in Chapter 16, you must know the material in the previous two chapters.*

Chapter 16 reviews the following concepts of acid-base and solubility chemistry:

- The common ion effect
- Buffer solutions
- Acid and base titration curves and titration calculations
    Strong acid and strong base titration
    Weak acid and strong base titration
    Weak base and strong acid titration
- Acid and base indicators
- Solubility equilibria
    K_{sp} expression and molar solubility calculations
    Predicting formation of a precipitate and ion concentration calculations
    Complex ion formation

> **Take Note:** *The equilibrium concepts presented in Chapter 16 are a significant part of the AP exam. Question 1 in the Free Response section is always an equilibrium problem and is weighted as 20% of the Free Response grade. The Multiple Choice section also contains questions on equilibrium.*

The common ion effect

The **common ion effect** is the *shift in an equilibrium caused by the addition of an ion that is the same as one of the product ions* of a dissociation reaction. This product ion addition shifts the equilibrium to the left according to Le Chatelier's principle. For example, the weak acid CH_3COOH dissociates less than 5% in water as shown in the reaction:

$$CH_3COOH + H_2O \Leftrightarrow CH_3COO^- + H_3O^+$$

If sodium acetate, $NaCH_3COO$, is added to an acetic acid solution, the ionization of acetic acid is suppressed. CH_3COONa is a strong electrolye and dissociates 100%. The added CH_3COO^- ion is the stress placed on the acetic acid equilibrium and, according to Le Chatelier's principle, the system reacts to relieve the stress. To reduce the concentration of CH_3COO^-, the equilibrium shifts to the left.

Example 1. pH calculation involving the common ion effect.

41 g of sodium acetate, CH_3COONa, have been added to 500.0 mL of 0.50 M CH_3COOH solution. Assume no change in volume. Determine the pH of the solution. (The K_a for CH_3COOH is 1.8×10^{-5}.)

General Strategy	Solution to Example 1
Write the equation. Two commonly used representations are given.	$CH_3COOH + H_2O \Leftrightarrow H_3O^+ + CH_3COO^-$ $CH_3COOH \Leftrightarrow H^+ + CH_3COO^-$
Write the appropriate equilibrium expression. H_2O is not included in K_a or K_b expressions.	$K_a = \dfrac{[H_3O^+][CH_3COO^-]}{[CH_3COOH]} = 1.8 \times 10^{-5}$
Molarity is the appropriate unit with K_c expressions such as K_a and K_b.	$\dfrac{41 \text{ g NaCH}_3\text{COO}}{0.500 \text{ L}} \times \dfrac{1 \text{ mol NaCH}_3\text{COO}}{82 \text{ g NaCH}_3\text{COO}} = 1.0 \text{ M}$ 1.0 M $NaCH_3COO$ = 1.0 M CH_3COO^- since $NaCH_3COO$ is a strong electrolyte and dissociates 100%.
Place the data underneath the equation. These are the initial conditions, (*i*).	$\quad\quad CH_3COOH + H_2O \Leftrightarrow H_3O^+ \ + \ CH_3COO^-$ (*i*) $\quad\quad$ 0.50 M $\quad\quad\quad\quad\quad$ 0 $\quad\quad$ 1.0 M
Express the *changes* (–*x* and +*x*) that occur as the reaction moves toward equilibrium. At equilibrium (*eq*), the concentrations of the species are equal to the *initial concentrations + changes*.	*changes* $\ -x \quad\quad\quad\quad\quad\quad +x \quad\quad +x$ (*eq*) $\quad$ 0.50 M $- x \quad\quad\quad\quad x \quad\quad$ 1.0 M $+ x$
Substitute into the K_a expression.	$1.8 \times 10^{-5} = \dfrac{[H_3O^+][CH_3COO^-]}{[CH_3COOH]} = \dfrac{x(1.0 \text{ M} + x)}{(0.50 \text{ M} - x)}$
Make assumptions to simplify the equilibrium expression when possible. Since K_a is small, the amount of acid that dissociates is negligible when compared to the concentration of the acid initially.	Assume $1.0 + x = 1.0$ M Assume $0.50 - x = 0.50$ M $1.8 \times 10^{-5} = \dfrac{x(1.0 \text{ M})}{(0.50 \text{ M})}$
Solve for *x*. Validate the accuracy of any assumptions made.	$x = 9.0 \times 10^{-6}$ Check assumptions: $1.0 + (9.0 \times 10^{-6}) = 1.0$ M. $\quad\quad\quad\quad\quad\quad 0.50 - (9.0 \times 10^{-6}) = 0.50$ M Assumptions are valid.
Solve for the *desired* quantity, which is pH in this case.	pH $= -\log [H^+] = -\log (9.0 \times 10^{-6}) = 5.05$

Usually, when the $K_a \leq 1 \times 10^{-5}$, the amount of acid that dissociates is negligible. For example, if a weak acid has an initial concentration of 1.0 M and at equilibrium its concentration is $1.0 - x$, the x is usually so small that for all intents and purposes the acid concentration is still 1.0 M. *It is important both to state assumptions when using them in calculations and to test their validity at the end of the calculation (see Example 1).*

Buffer solutions

Buffer solutions are *solutions whose pH does not appreciably change when a strong acid or strong base is added.* An **acidic buffer** is made from a *weak acid plus its conjugate base.* An example of an acidic buffer is CH_3COOH and its conjugate base CH_3COO^-, where both species are present in significant amounts such as in Example 1. A **basic buffer** is made from a *weak base plus its conjugate acid.* An example of a basic buffer is a solution composed of equal volumes of 1.0 M NH_3 and 1.0 M NH_4Cl.

Example 2. Identifying buffer solutions and predicting pH.

Determine which of the following solutions form buffers and identify each solution as acidic (pH < 7.00) , basic (pH > 7.00), or neutral (pH = 7.00).

a. 50.0 mL of 1.0 M HCl + 50.0 mL of NaCl
b. 100.0 mL of 1.0 M HF + 100.0 mL of 1.0 M NaF
c. 100.0 mL of 1.0 M KOH + 50.0 mL of 2.0 M HBr
d. 30.0 mL of 1.0 M NH_3 + 15.0 mL of 1.0 M HCl
e. 30.0 mL of 1.0 M NH_3 + 45.0 mL of 1.0 M HCl

Solution to Example 2
a. *This is not a buffer solution; pH < 7.00.*   A strong acid and its conjugate base do not have buffering action. HCl is 100% dissociated and therefore the pH of the solution is acidic. NaCl is a neutral salt.
b. *This is an acidic buffer; pH < 7.00.*   HF, a weak acid, and F^-, its conjugate base, are present in significant amounts. $$HF \Leftrightarrow H^+ + F^-$$ $$F^- + H_2O \Leftrightarrow HF + OH^-$$ Since both HF and F^- are present in significant amounts, the HF neutralizes any added base and the F^- neutralizes any added acid. The solution is buffered.   It is an acidic buffer because the K_a for HF is 7.1×10^{-4} and is greater than the K_b for F^-, which is 1.4×10^{-11} (Recall that $K_b = K_w/K_a$.)

c. *This is not a buffer; pH =7.00.*

This is a neutral solution made by the complete neutralization of a strong acid with a strong base. K^+ and Br^- are spectator ions. KBr is a neutral salt. The net ionic equation is: $H^+ + OH^- \rightarrow H_2O$.

$$\frac{1.0 \, mol \, KOH}{L} \times 0.100 \, L = 0.100 \, mol \, KOH = 0.100 \, mol \, OH^-$$

$$\frac{2.0 \, mol \, HBr}{L} \times 0.050 \, L = 0.100 \, mol \, HBr = 0.100 \, mol \, H^+$$

d. *This is a basic buffer; pH > 7.00.*

After neutralization has occurred, there are 0.015 mol of NH_3 and 0.015 mol of NH_4^+ present.

$$\frac{1.0 \, mol \, NH_3}{L} \times 0.030 \, L = 0.030 \, mol \, NH_3 \text{ initially}$$

$$\frac{1.0 \, mol \, HCl}{L} \times 0.015 \, L = 0.015 \, mol \, HCl = 0.015 \, mol \, H^+ \text{ initially}$$

NH_3 is neutralized by H^+:

	NH_3	+	H^+	$\rightarrow$	NH_4^+
Initially	0.030 mol		0.015 mol		0
After neutralization	0.015 mol		0		0.015 mol

e. *This is not a buffer; pH < 7.00.*

The buffering capacity has been exhausted because all of the weak base has been neutralized. There are 0.015 mol of HCl (H^+) and 0.030 mol of NH_4^+ present after the reaction has taken place. When a strong acid and a weak acid are present, the strong acid determines the pH. The contribution of H^+ from the weak acid (NH_4^+ in this case) is negligible.

$$\frac{1.0 \, mol \, NH_3}{L} \times 0.030 \, L = 0.030 \, mol \, NH_3$$

$$\frac{1.0 \, mol \, HCl}{L} \times 0.045 \, L = 0.045 \, mol \, HCl = 0.045 \, mol \, H^+$$

NH_3 is neutralized by H^+:

	NH_3	+	H^+	$\rightarrow$	NH_4^+
Initially	0.030 mol		0.045 mol		0
After neutralization	0 mol		0.015		0.030 mol

Acid and base titration curves and titration calculations

Titration is a *laboratory technique, usually involving the use of burets, in which the concentration of an unknown solution is determined.* Titration is the gradual addition of a **standardized solution,** *a solution of known concentration*, to a solution of unknown concentration, until the reaction is complete. The **equivalence point** *in an acid-base titration is when the moles of acid are equal to the moles of base.* The pH at the equivalence point is determined by the ions present. It can be acidic, basic, or neutral. An *indicator*, whose color changes in the pH range at the steep rise of the titration curve, indicates when the equivalence point has been reached.

There are three types of acid-base titrations:
1. strong acid and strong base titration (Example 3)
2. weak acid and strong base titration (Example 4)
3. weak base and strong acid titration (Example 5)

There are four major areas for pH calculation along a titration curve. These four areas are addressed in Examples 3, 4, and 5. The calculations are:
1. before titration begins
2. during titration
3. at the equivalence point
4. after the equivalence point

Take Note: *The AP exam contains many questions in both the Multiple Choice section and the Free Response section on titration, titration curves, indicators, and buffers. Examples 1–5 illustrate the equilibrium concepts that are most often tested. To do well on the AP exam, you must thoroughly understand this material!*

Strong acid and strong base titration

A strong acid and strong base neutralize one another, forming a neutral salt and water. The pH at the *equivalence point is 7.00.* Two common indicators used in this type of titration are phenolphthalein (pH range 8.3–10) and methyl red (pH range 4.2–6.3). As you can see from the titration curve in Figure 16.4, any indicator that changes color in the pH range from 4 to 11 may be used.

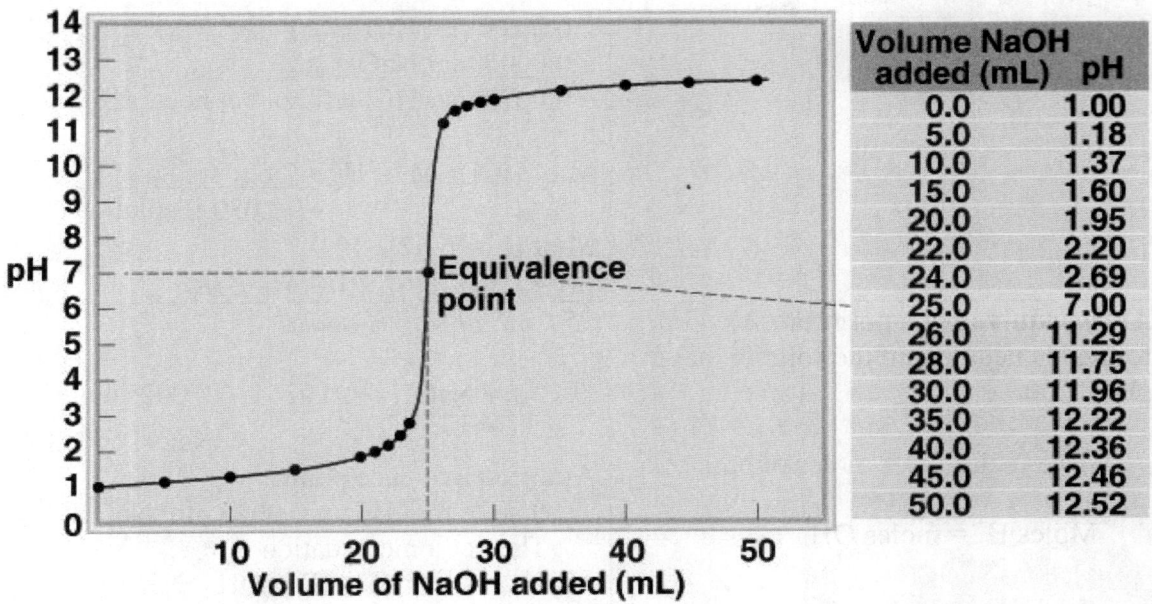

Figure 16.4

Example 3. Strong acid and strong base titration calculations.

Figure 16.4 follows the changes in pH as a strong acid is titrated with a strong base. 25.0 mL of 0.10 M HCl are titrated with 0.10 M NaOH. Determine the pH when:

      0.0 mL of 0.10 M NaOH have been added (before titration begins)
      5.0 mL of 0.10 M NaOH have been added (during titration)
      25.0 mL of 0.10 M NaOH have been added (equivalence point)
      30.0 mL of 0.10 M NaOH have been added (beyond equivalence point)

General Strategy	Solution to Example 3
Before titration the molarity of the H^+ in solution is equal to the molarity of the strong acid. Strong acids dissociate 100% in water. $pH = - \log [M_{\text{strong acid}}]$	*0.0 mL NaOH added* $0.10 \text{ M HCl} = 0.10 \text{ M H}^+$ $pH = -\log [H^+] = - \log (0.10) = 1.00$

During titration calculate the concentration of acid not yet neutralized:
 	
     moles of acid $_{\text{initial}}$
 $-$ <u>moles of base $_{\text{added}}$</u>
     moles of acid not yet neutralized
	5.0 mL of NaOH added
	
	 $\dfrac{0.10 \text{ mol HCl}}{L} \times 0.025 \text{ L} = 0.0025 \text{ mol HCl}$

$pH = -\log \dfrac{[\text{moles of acid not neutralized}]}{[\text{volume}_{total}]}$	$\dfrac{0.10 \text{ mol NaOH}}{\text{L}} \times 0.005 \text{ L} = 0.0005 \text{ mol NaOH}_{added}$ $\begin{array}{l} 0.0025 \text{ mol HCl}_{initially} \\ -\ \underline{0.0005 \text{ mol NaOH}_{added}} \\ 0.0020 \text{ mol HCl left not yet neutralized} \end{array}$ M of HCl = M of H^+ = $\dfrac{2.0 \times 10^{-3} \text{ mol } H^+}{0.030 \text{ L solution}}$ M of H^+ = 6.67×10^{-2} $pH = -\log(6.67 \times 10^{-2}) = 1.18$

At the equivalence point the salt formed is neutral and the solution has a pH = 7.00.
 	
Moles acid $_{initial}$ = moles of base $_{added}$
 	
Moles H^+ = moles OH^-
	25.0 mL of NaOH added
	
	$\dfrac{0.10 \text{ mol NaOH}}{\text{L}} \times 0.025 \text{ L} = 0.0025 \text{ mol NaOH added}$
	At the equivalence point:
	    0.0025 mol HCl = 0.0025 mol NaOH
	The net ionic equation is:
	$$H^+ + OH^- \rightarrow H_2O$$
	Moles of H^+ = moles OH^-
	The ions Na^+ and Cl^- do not interact
	    with water. NaCl is a neutral salt.
	
	pH = 7.00

After the equivalence point the moles of excess base determine the pH. $\dfrac{\text{Moles of } OH^-_{excess}}{\text{Volume}_{total}} = [OH^-]$ $pOH = -\log[OH^-]$ $pH = 14.00 - pOH$	*30.0 mL of NaOH added* 30.0 mL – 25.0 mL to reach equivalence point = 5.0 mL NaOH solution $_{excess}$ $\dfrac{0.10 \text{ mol NaOH}}{\text{L}} \times 0.0050 \text{ L} = 0.00050 \text{ mol NaOH excess}$ Volume $_{total}$ = 25.0 mL + 30.0 mL = 55.0 mL $\dfrac{5.0 \times 10^{-4} \text{ mol } OH^-}{5.5 \times 10^{-2} \text{ L}} = 9.1 \times 10^{-3} \text{ M } OH^-$ $pOH = -\log(9.1 \times 10^{-3} \text{ M}) = 2.04$ $pH = 14.00 - 2.04 = 11.96$

Weak acid and strong base titration

This type of reaction occurs when a weak acid such as HF is neutralized by a strong base such as NaOH: $HF + NaOH \rightarrow H_2O + F^-$. *When a weak acid is neutralized by a strong base,* the pH at the *equivalence point is greater than 7.00.* In the case of the HF + NaOH titration, the basic pH is due to the presence of the F^- ion at the equivalence point. F^- acts as a base and undergoes a hydrolysis reaction with water, producing OH^- ions: $F^- + H_2O \Leftrightarrow HF + OH^-$. Phenolphthalein (pH range 8.3–10) is the most commonly used indicator in this type of titration. As you can see from the titration curve in Figure 16.5, any indicator that changes color in the pH range from 6 to 11 may be used.

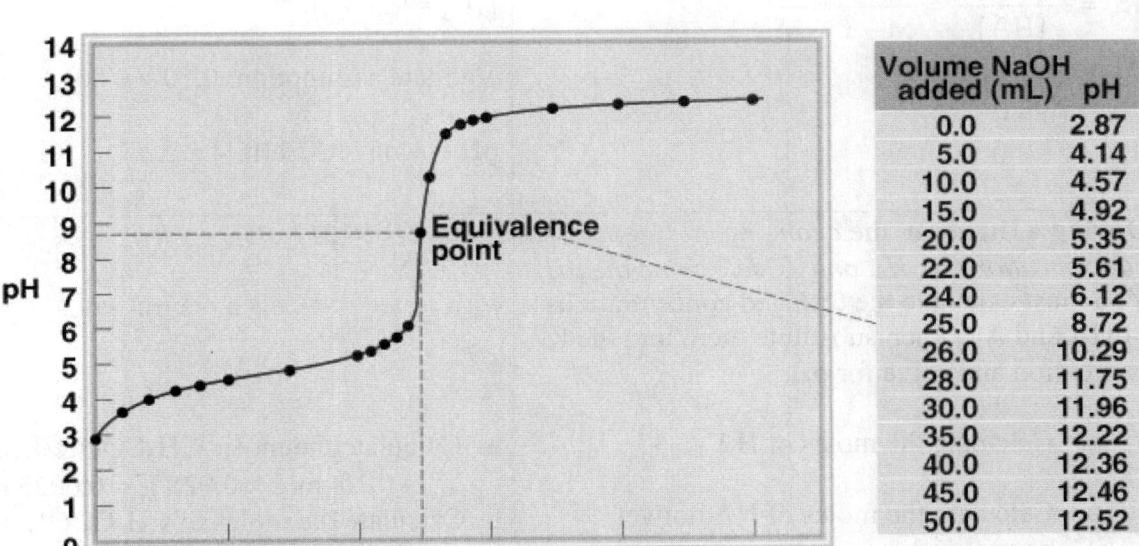

Volume NaOH added (mL)	pH
0.0	2.87
5.0	4.14
10.0	4.57
15.0	4.92
20.0	5.35
22.0	5.61
24.0	6.12
25.0	8.72
26.0	10.29
28.0	11.75
30.0	11.96
35.0	12.22
40.0	12.36
45.0	12.46
50.0	12.52

Figure 16.5

Example 4. Weak acid and strong base titration calculations.

Figure 16.5 follows the changes in pH as a weak acid is titrated with a strong base. 25.0 mL of 0.10 M CH_3COOH are titrated with 0.10 M NaOH. Determine the pH when:

0.0 mL of 0.10 M NaOH have been added (before titration begins)
5.0 mL of 0.10 M NaOH have been added (during titration)
25.0 mL of 0.10 M NaOH have been added (at the equivalence point)
30.0 mL of 0.10 M NaOH have been added (beyond the equivalence point)

General Strategy	Solution to Example 4
Before titration the dissociation of the *weak acid HA determines the pH* in solution. You use the K_a expression to calculate pH. $$HA \Leftrightarrow H^+ + A^-$$ *Initially* $\quad$ M* $\quad$ 0 $\quad$ 0 *Changes* $\quad$ $-x$ $\quad$ $+x$ $\quad$ $+x$ *Equilibrium* $\quad$ $M-x$ $\quad$ x $\quad$ x M* stands for the initial molarity of the acid solution. $$K_a = \frac{[H^+][A^-]}{[HA]} = \frac{x^2}{M-x} = \frac{x^2}{M}$$ $$[H^+] = x$$ $$pH = -\log [H^+]$$	*0.0 mL of 0.10 M NaOH added* $$CH_3COOH \Leftrightarrow CH_3COO^- + H^+$$ *Initial* $\quad$ 0.10 M $\quad$ 0 $\quad$ 0 *Changes* $\quad$ $-x$ $\quad$ $+x$ $\quad$ $+x$ *Equilibrium* $\quad$ 0.10 M–x $\quad$ x $\quad$ x Assume $0.10 \text{ M} - x = 0.10 \text{ M} = [CH_3COOH]$ $$K_a = \frac{[H^+][CH_3COO^-]}{[CH_3COOH]} = \frac{x^2}{0.10} = 1.8 \times 10^{-5}$$ $$[H^+] = 1.3 \times 10^{-3}$$ Validate assumption: $0.10 - 0.0013 = 0.10$. $$pH = -\log [1.3 \times 10^{-3}] = 2.87$$
During a titration the *changing concentrations of HA and A^- determine the pH*. You must calculate the changed concentrations of HA and A^-. Then substitute them into the K_a expression and solve for pH. a. Calculate the moles of HA ₍initially.₎ b. Calculate the moles of HA not yet neutralized. $\quad$ The neutralization reaction is: $\quad$ $HA + OH^- \rightarrow H_2O + A^-$. Note that this is NOT an equilibrium. $\quad$ The moles of *HA not yet neutralized* are equal to the moles of HA ₍initial₎ minus the moles of base added. c. Determine the moles of A^- formed: moles of A^- = moles of strong base ₍added₎. You can see from the neutralization equation that every mole of HA that is neutralized forms an A^- ion. d. Calculate the pH using the K_a	*5.0 mL 0.10 M NaOH added* a. Calculate the moles CH_3COOH ₍initially₎ $\quad = 0.10 \text{ mol} \times 0.025 \text{ L} = 0.0025 \text{ mol}$ b. Calculate the moles of CH_3COOH not yet neutralized. The neutralization reaction is: $\quad CH_3COOH + OH^- \rightarrow H_2O + CH_3COO^-$ The moles of NaOH ₍added₎ $= 0.10 \text{ mol} \times 0.0050 \text{ L}$ $= 0.00050 \text{ mol } OH^-$ ₍added₎ 0.0025 mol CH_3COOH ₍initially₎ $- 0.00050$ mol OH^- ₍added₎ $= 0.0020$ mol CH_3COOH not yet neutralized. c. moles of CH_3COO^- = moles of NaOH ₍added₎ $= 0.00050$ mol d. The *final* concentrations after neutralization has occurred are the *initial* concentrations for the new equilibrium that is established.

expression. The relevant equilibrium equation that *determines pH* is the dissociation of the weak acid: $HA \Leftrightarrow H^+ + A^-$. Substitute the HA and A^- concentrations, resulting from the neutralization, into the K_a expression and solve for H^+.

$$K_a = \frac{[H^+]\,[\text{mol } A^-/\text{vol}]}{[\text{mol } HA/\text{vol}]}$$

Note that the volume cancels out in the expression. This only happens *during* a titration.

The equilibrium reaction that determines pH:
$$CH_3COOH \Leftrightarrow CH_3COO^- + H^+$$

initial (M)	$\dfrac{0.0020 \text{ mol}}{0.030 \text{ L}}$	$\dfrac{0.00050 \text{ mol}}{0.030 \text{ L}}$	0
changes	$-x$	$+x$	$+x$
equilibrium(M)	$0.067 - x$	$0.017 + x$	x

**assume $-x$ and $+x$ are negligible

$$K_a = \frac{[H^+]\,[CH_3COO^-]}{[CH_3COOH]}$$
$$\frac{[x]\,[0.017 \text{ M}]}{[0.067 \text{ M}]} = 1.8 \times 10^{-5}$$
$$x = [H^+] = 7.1 \times 10^{-5}$$

Validate assumptions:
$$0.067 - (7.1 \times 10^{-5}) = 0.067$$
$$0.017 + (7.1 \times 10^{-5}) = 0.017$$

$$pH = -\log(7.1 \times 10^{-5}) = 4.15$$

At the equivalence point all the HA has been neutralized and is now in the form of A^- (i.e., the moles of HA present initially are equal to the moles of A^- at the equivalence point). The equilibrium that determines pH is: $A^- + H_2O \Leftrightarrow HA + OH^-$ and you need to calculate K_b.

a. *The hydrolysis of the anion, A^-, in water determines the pH.* Since A^- acts as a base, the K_b expression is appropriate and its value must be calculated. $K_b = \dfrac{K_w}{K_a}$

b. To calculate the molarity of A^-, divide the moles of A^- (which are equal to the moles of HA_{initial}) by the total volume of solution at the equivalence point.

c. *The relevant equilibrium equation that determines pH is: $A^- + H_2O \Leftrightarrow HA + OH^-$.*

25.0 ml of 0.10 M NaOH added

The 0.0025 mol of CH_3COOH that were initially present are now neutralized and are all in the form of CH_3COO^-.

a. The hydrolysis of water by the acetate ion determines the pH of the solution. The relevant equation is $A^- + H_2O \Leftrightarrow HA + OH^-$. You need to calculate the K_b value for this reaction.

$$K_b = \frac{K_w}{K_a} = \frac{1.0 \times 10^{-14}}{1.8 \times 10^{-5}} = 5.5 \times 10^{-10}$$

b. The molarity of the CH_3COO^- ion needs to be determined.
   25.0 mL CH_3COOH + 25.0 mL NaOH = 50.0 mL of solution at the equivalence point.
   0.0025 mol $CH_3COOH_{\text{initially}}$ = 0.0025 mol CH_3COO^- at the equivalence point.
$$\frac{0.0025 \text{ mol}}{0.050 \text{ L}} = 0.050 \text{ M } CH_3COO^-$$

c. The equilibrium reaction that determines pH is:
$$CH_3COO^- + H_2O \Leftrightarrow CH_3COOH + OH^-$$

initial (M)	0.050	0	0

	changes $\quad -x \qquad\qquad +x \qquad +x$ *equilibrium(M)* $0.050 -x \qquad x \qquad x$ $\qquad\qquad 0.050 \qquad\qquad x \qquad x$ ** assume $-x$ is negligible $$K_b = \frac{[OH^-][CH_3COOH]}{[CH_3COO^-]}$$ $$\frac{[x]^2}{[0.050]} = 5.5 \times 10^{-10}$$ $x = [OH^-] = 5.2 \times 10^{-6}$
Note that titrating a weak acid with a strong base results in a basic equivalence point due to the presence of the conjugate base A⁻.	Validate assumption: $0.050 -(5.2 \times 10^{-6}) = 0.05$ pOH $= -\log (5.2 \times 10^{-6}) = 5.28$ pH $= 14.00 - 5.28 = 8.72$
After the equivalence point all the HA has been neutralized. The *concentration of excess strong base determines the pH* of the solution. The amount of OH⁻ produced by the weak base A⁻ is insignificant and can be ignored. Calculate the volume of excess base added. Convert this volume to moles of excess base added. Divide the moles of excess base added by the total volume of the solution to determine [OH⁻] and then pH. $$[OH^-] = \frac{\text{moles strong base}_{excess}}{\text{volume}_{total}}$$ pOH $= -\log [OH^-$ pH $= 14.00 - $pOH	*30.0 ml of 0.10 M NaOH added* 30.0 mL of base added $-$ 25.0 mL to reach the equivalence point $=$ 5.0 mL of NaOH $_{excess}$ $0.10 \frac{\text{mol}}{L} \text{NaOH} \times 0.0050 \text{ L} = 0.00050$ mol NaOH $\qquad\qquad\qquad\qquad = 5.0 \times 10^{-4}$ mol OH⁻ $_{excess}$ Volume$_{total}$ = 25.0 mL acid + 30.0 mL NaOH = 55.0 mL of total solution = 0.055 L $$\frac{5.0 \times 10^{-4} \text{ mol OH}^-_{excess}}{0.055 \text{ L}} = [OH^-] = 9.1 \times 10^{-3}$$ pOH $= -\log (9.1 \times 10^{-3}) = 2.04$ pH $= 14.00 - 2.04 = 11.96$

Weak base and strong acid titration

When a weak base, B, is neutralized by a strong acid such as HCl, water and an acidic salt, BHCl, are produced. The *pH at the equivalence point is less than 7.00.* BHCl is a soluble salt and BH^+ dissociates according to the reaction: $BH^+ \Leftrightarrow B + H^+$. Methyl red (pH range 4.2–6.3) is a commonly used indicator in this type of titration. As you can see by the titration curve in Figure 16.6, any indicator that changes color in the pH range from 3 to 8 may be used.

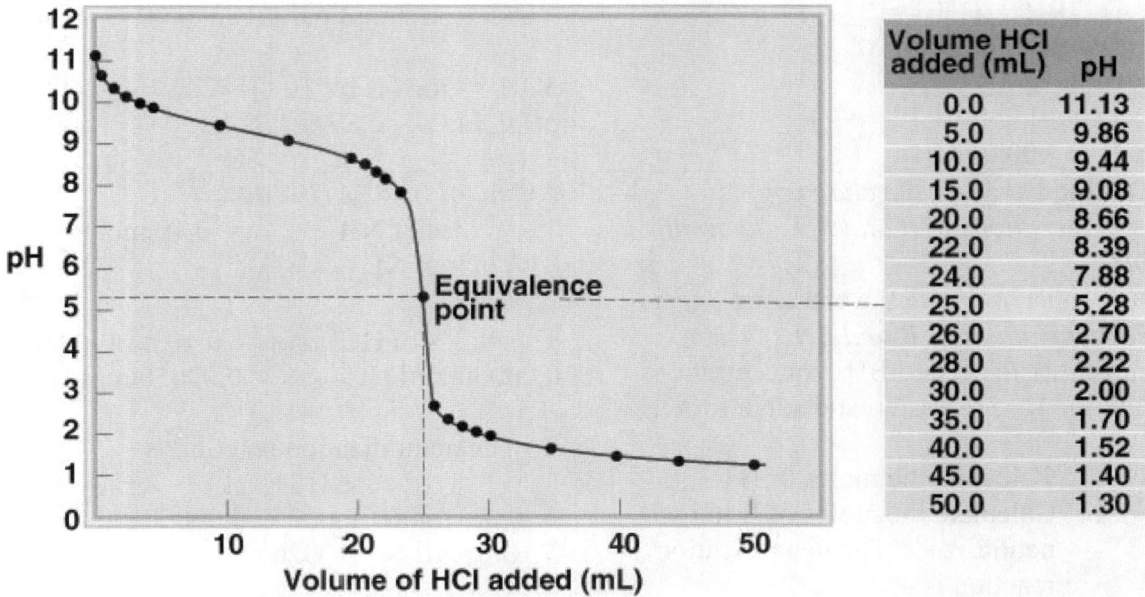

Volume HCl added (mL)	pH
0.0	11.13
5.0	9.86
10.0	9.44
15.0	9.08
20.0	8.66
22.0	8.39
24.0	7.88
25.0	5.28
26.0	2.70
28.0	2.22
30.0	2.00
35.0	1.70
40.0	1.52
45.0	1.40
50.0	1.30

Figure 16.6

Example 5. Weak base and strong acid titration calculations.

Figure 16.6 follows the changes in pH as a weak base is titrated with a strong acid. 25.0 mL of 0.10 M NH_3 are titrated with 0.10 M HCl solution. Determine the pH when:

  0.0 mL of 0.10 M HCl have been added (before titration begins)
  5.0 mL of 0.10 M HCl have been added (during titration)
  25.0 mL of 0.10 M HCl have been added (at the equivalence point)
  30.0 mL of 0.10 M HCl have been added (beyond the equivalence point)

General Strategy	Solution to Example 5
Before titration the weak base, B, undergoes a hydrolysis reaction with water. The pH depends on the extent of the hydrolysis.	*0.0 mL of 0.10 M HCl added*
$B + H_2O \Leftrightarrow BH^+ + OH^-$ *Initially* M^* $\quad$ 0 $\quad$ 0 *Changes* $-x$ $\quad +x$ $\quad +x$ *Equilibrium* $M-x$ $\quad x$ $\quad x$	$NH_3 + H_2O \Leftrightarrow NH_4^+ + OH^-$ *initially (M)* 0.10 $\quad$ 0 $\quad$ 0 *changes* $-x$ $\quad +x$ $\quad +x$ *equilibrium(M)* 0.10$-x$ $\quad x$ $\quad x$ Assume $-x$ is negligible
M^* refers to the initial molarity of the base.	$K_b = \dfrac{[NH_4^+][OH^-]}{[NH_3]} = \dfrac{x^2}{0.10} = 1.8 \times 10^{-5}$

- 293 -

Make assumptions to simplify the K expression, if possible. Validate any assumptions made. $K_b = \dfrac{[BH^+][OH^-]}{[B]} = \dfrac{x^2}{M-x} = \dfrac{x^2}{M}$ $[OH^-] = x$ $pOH = -\log[OH^-]$ $pH = 14.00 - pOH$	$[OH^-] = x = 1.3 \times 10^{-3}$ Validate assumption: $0.10 - 0.0013 = 0.10$ $pOH = -\log[1.3 \times 10^{-3}] = 2.89$ $pH = 14.00 - 2.89 = 11.11$

During titration *the changing concentrations of B and BH$^+$ determine the pH.*
You must calculate the changing *concentrations of B and BH$^+$*. Then substitute the B and BH$^+$ concentrations into the K$_b$ expression and solve for OH$^-$.
 	
  a. Calculate the moles of B $_{initially}$.
  b. Calculate the moles of B not yet neutralized. The neutralization reaction is:
      $B + H^+ \rightarrow BH^+$
Moles of *B not yet neutralized* = moles of B $_{initial}$ − moles of acid $_{added}$
 	
  c. Determine the moles of *BH$^+$* that are equal to the moles of strong acid$_{added}$.
 	
  d. Substitute into the K$_b$ expression.
$K_b = \dfrac{[OH^-][\text{mol } BH^+ / \text{vol}]}{[\text{mol } B / \text{vol}]}$
 	
Note that the volume cancels out in the K expression. This *only* happens *during* a titration.
	5.0 mL of 0.10 M HCl added
	The moles of NH$_3$ $_{initially}$ = 0.10 mol × 0.0250L = 0.0025 mol NH$_3$ $_{initially}$
	
	The moles of HCl $_{added}$ = 0.10 mol × 0.005 L = 0.00050 mol HCl $_{added}$ = 0.00050 mol H$^+$
	
	I. The neutralization reaction is:
	      $NH_3 + H^+ \rightarrow NH_4^+$
	Initially (mol) 0.0025 0.00050
	Final (mol) 0.0020 0 0.00050
	
	The *final* concentrations after neutralization has taken place are the *initial* concentrations of the new equilibrium that is established.
	
	II. Equilibrium reaction that determines pH:
	      $NH_3 + H_2O \Leftrightarrow NH_4^+ + OH^-$
	initial (M) $\dfrac{0.0020 \text{ mol}}{0.030 \text{ L}}$ $\dfrac{0.00050 \text{ mol}}{0.030 \text{ L}}$ 0
	changes $-x$ $+x$ $+x$
	equilibrium $0.067 - x$ $0.017 + x$ x
	assume $-x$ and $+x$ are negligible
	
	$K_b = \dfrac{[x][0.017]}{[0.067]} = 1.8 \times 10^{-5}$
	$x = [OH^-] = 7.1 \times 10^{-5}$
	
	Validate assumptions:
	  $0.067 - (7.1 \times 10^{-5}) = 0.067$
	  $0.017 + (7.1 \times 10^{-5}) = 0.017$
	
	$pOH = -\log(7.1 \times 10^{-5}) = 4.15$
	$pH = 14.00 - 4.15 = 9.85$

At the equivalence point all the B has been neutralized and is now in the form of BH^+. The moles of BH^+ at the equivalence point are equal to the moles of B initially present. The relevant *equilibrium equation that determines pH is:* $$BH^+ \Leftrightarrow B + H^+$$ The dissociation of BH^+ determines the pH. Since BH^+ acts as an acid, the K_a expression is appropriate and you must calculate its value: $$K_a = \frac{K_w}{K_b}$$ *When titrating a weak base with a strong acid, the equivalence point is acidic due to the presence of the conjugate acid, BH^+.*	*25.0 mL of 0.10 M HCl added* Initially there were 0.0025 mol of NH_3. These have all been neutralized by the HCl and all the NH_3 is now in the form of NH_4^+. A new equilibrium exists at the equivalence point: $$NH_4^+ \Leftrightarrow NH_3 + H^+$$ *initially(M)* $\dfrac{0.0025 \text{ mol}}{0.050 \text{ L}}$ 0 0 *equilibrium* 0.050 M $-x$ x x assume $-x$ is negligible $$K_a = \frac{1.0 \times 10^{-14}}{1.8 \times 10^{-5}} = 5.5 \times 10^{-10} = \frac{x^2}{0.050}$$ $$[H^+] = x = 5.2 \times 10^{-6}$$ Validate assumption: $$0.050 - (5.2 \times 10^{-6}) = 0.050$$ $$pH = -\log(5.2 \times 10^{-6}) = 5.24$$
After the equivalence point All the B has been neutralized. The amount of *excess* strong acid determines the pH of the solution. The amount of H^+ produced by the dissociation of the weak acid BH^+ is insignificant and can be ignored. $$[H^+] = \frac{\text{moles strong acid}_{excess}}{\text{volume}_{total}}$$ $$pH = -\log[H^+]$$	*30.0 mL of 0.10 M HCl added* 30.0 mL of HCl added – 25.0 mL HCl to reach the equivalence point = 5.0 mL HCl solution$_{excess}$ $$\frac{0.10 \text{ mol HCl}}{L} \times 0.0050 \text{ L} = 0.00050 \text{ mol HCl}_{excess}$$ $$= 0.00050 \text{ mol H}^+_{excess}$$ Volume$_{total}$ = 25.0 mL + 30.0 mL = 55.0 mL $$\frac{5.0 \times 10^{-4} \text{ mol H}^+}{5.5 \times 10^{-2} \text{ L}} = 9.1 \times 10^{-3} \text{ M H}^+$$ $$pH = -\log(9.1 \times 10^{-3} \text{ M}) = 2.04$$

The **midpoint (or half-way point)** of a titration is when half of the original solution is neutralized. The midpoint is significant because it allows you to determine K_a and K_b values using pH data. There are two cases to consider:

1. A weak acid HA is titrated with a strong base. The half-way point occurs when half of the initial HA has been converted to A^-. At the midpoint, $[HA] = [A^-]$, which causes these terms to cancel out of the K_a expression, yielding $K_a = [H^+]$. This can also be stated as $pK_a = pH$.

2.   A weak base B is titrated with a strong acid. The half-way point of titration occurs when half of the B has been converted to BH$^+$. At this point K_b = [OH$^-$] and pK_b = pOH.

Acid and base indicators

Figure 16.7 shows you that there may be more than one indicator appropriate for a particular titration. You can choose any indicator that changes color in the pH range that defines the steep part of the titration curve.

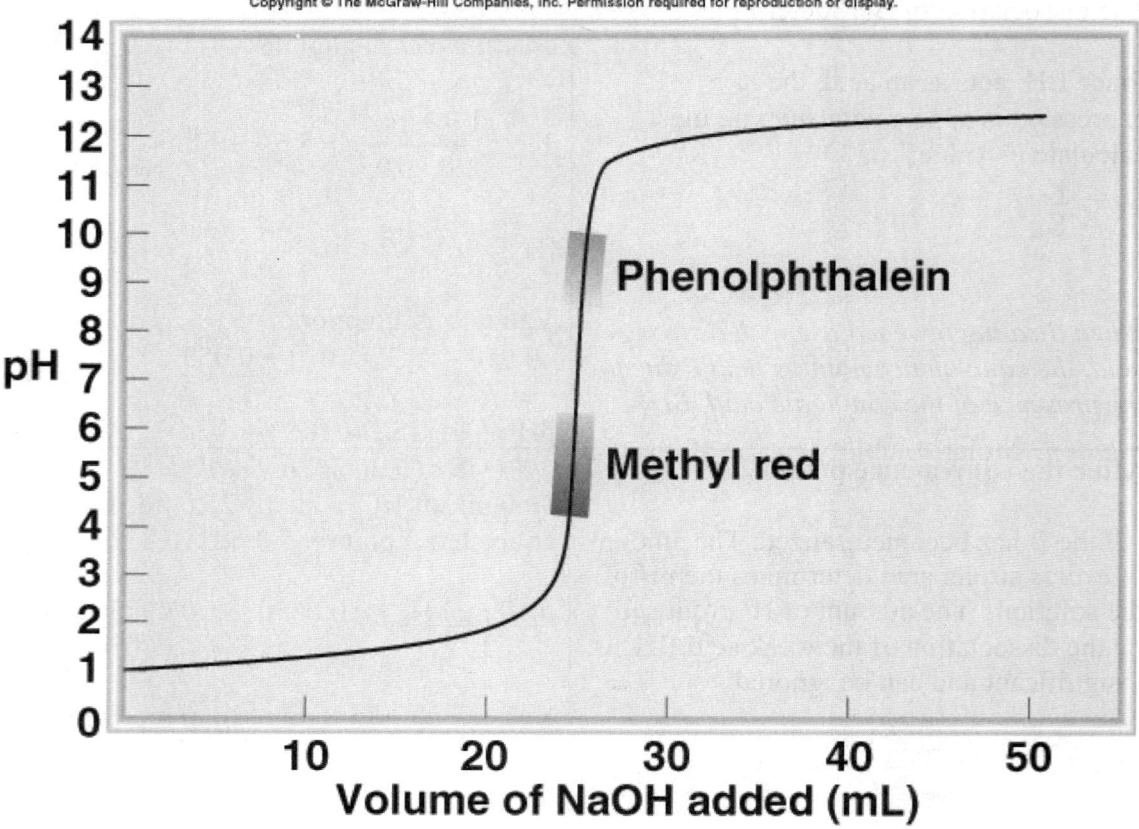

Figure 16.7

Indicators signal the equivalence point in a titration. **Indicators** *are weak organic acids or bases that change color in a specific pH range.* An indicator has one color in the undissociated form and another color when it is ionized. Depending upon the pH of the solution, either the forward or reverse reaction is favored, that is, color A or color B. When the concentration of HA$_\text{Indicator}$ equals the concentration of A$^-$, the **endpoint,** *the pH at which the indicator changes color* is reached. At this point, neither color A nor color B dominates; the in-between color, or endpoint color of the indicator, is observed and $pK_\text{a, Indicator}$ = pH ± 1. This pH range must fall somewhere in the steep part of the titration curve.

$$HA_\text{Indicator} \Leftrightarrow H^+ + A^-$$
Color A   Color B

$$K_\text{a, Indicator} = \frac{[H^+][A^-]}{[HA_\text{Indicator}]}$$

- 296 -

Solubility equilibria

K_{sp} expression and molar solubility calculations

Compounds that have low solubility in aqueous solutions are known as **precipitates**. For example, when a solution of silver nitrate is mixed with a solution of sodium chloride, the precipitate, silver chloride, forms. The nitrate and sodium ions are **spectator ions**, *ions that do not participate in the reaction*. The net ionic equation is:

$$Ag^+(aq) + Cl^-\ (aq) \Leftrightarrow AgCl\ (s)$$

Precipitates, compounds that *have low solubility in solution*, no matter how insoluble, are in equilibrium with the saturated solution. The **K_{sp} expression** is the *product of the soluble ions, expressed in molarity, raised to their coefficients* from the balanced dissociation equation. For example, the K_{sp} expression for $Ba_3(PO_4)_2$ is: $K_{sp} = [Ba^{2+}]3[PO_4^{3-}]^2$. The superscripts are obtained from the balanced dissociation equation: $Ba_3(PO_4)_2\ (s) \Leftrightarrow 3Ba^{2+}\ (aq) + 3PO_4^{3-}\ (aq)$. The *numerical value of K_{sp}* is known as the **solubility product constant.**

Molar solubility refers to *the number of moles of precipitate that dissolve per liter of solution*. Molar solubility can be affected in two ways:

1. Molar solubility can be reduced by adding a production; this is an example of the common ion effect. For example, adding NaCl solution to a saturated solution of AgCl will lower the solubility of AgCl.

2. The molar solubility of some precipitates can be affected by the *pH* of the solution. If a product ion interacts with H^+, the precipitate dissolves in acidic solution. For example, the precipitate AgF(s) dissolves in strong acid. The product ion, F^-, reacts with H^+ to form HF, a weak acid. As the F^- is removed from solution in the form of HF, the precipitation equilibrium shifts to the right (precipitate dissolves) according to Le Chatelier's principle.

Take Note: *Precipitation questions on the AP exam often involve the following precipitates: $BaSO_3$, $Ba_3(PO_4)_2$, $CaCO_3$, MgS, $Al(OH)_3$, and $Zn(OH)_2$. The common ion effect and the effect that pH has on molar solubilities are common topics for lab questions. You will also need to know the colors of some common precipitates. Most precipitates are white, but some are very distinctive, such as Ag_2CrO_4, which is red-brown.*

Table 16.2 gives you some K_{sp} values.

TABLE 16.2

Solubility Products of Some Slightly Soluble Ionic Compounds at 25°C

Compound	K_{sp}	Compound	K_{sp}
Aluminum hydroxide [Al(OH)$_3$]	1.8×10^{-33}	Lead(II) chromate (PbCrO$_4$)	2.0×10^{-14}
Barium carbonate (BaCO$_3$)	8.1×10^{-9}	Lead(II) fluoride (PbF$_2$)	4.1×10^{-8}
Barium fluoride (BaF$_2$)	1.7×10^{-6}	Lead(II) iodide (PbI$_2$)	1.4×10^{-8}
Barium sulfate (BaSO$_4$)	1.1×10^{-10}	Lead(II) sulfide (PbS)	3.4×10^{-28}
Bismuth sulfide (Bi$_2$S$_3$)	1.6×10^{-72}	Magnesium carbonate (MgCO$_3$)	4.0×10^{-5}
Cadmium sulfide (CdS)	8.0×10^{-28}	Magnesium hydroxide [Mg(OH)$_2$]	1.2×10^{-11}
Calcium carbonate (CaCO$_3$)	8.7×10^{-9}	Manganese(II) sulfide (MnS)	3.0×10^{-14}
Calcium fluoride (CaF$_2$)	4.0×10^{-11}	Mercury(I) chloride (Hg$_2$Cl$_2$)	3.5×10^{-18}
Calcium hydroxide [Ca(OH)$_2$]	8.0×10^{-6}	Mercury(II) sulfide (HgS)	4.0×10^{-54}
Chromium(III) hydroxide [Cr(OH)$_3$]	3.0×10^{-29}	Silver bromide (AgBr)	7.7×10^{-13}
Cobalt(II) sulfide (CoS)	4.0×10^{-21}	Silver carbonate (Ag$_2$CO$_3$)	8.1×10^{-12}
Copper(I) bromide (CuBr)	4.2×10^{-8}	Silver chloride (AgCl)	1.6×10^{-10}
Copper(I) iodide (CuI)	5.1×10^{-12}	Silver iodide (AgI)	8.3×10^{-17}
Copper(II) hydroxide [Cu(OH)$_2$]	2.2×10^{-20}	Silver sulfate (Ag$_2$SO$_4$)	1.4×10^{-5}
Copper(II) sulfide (CuS)	6.0×10^{-37}	Silver sulfide (Ag$_2$S)	6.0×10^{-51}
Iron(II) hydroxide [Fe(OH)$_2$]	1.6×10^{-14}	Strontium carbonate (SrCO$_3$)	1.6×10^{-9}
Iron(III) hydroxide [Fe(OH)$_3$]	1.1×10^{-36}	Strontium sulfate (SrSO$_4$)	3.8×10^{-7}
Iron(II) sulfide (FeS)	6.0×10^{-19}	Tin(II) sulfide (SnS)	1.0×10^{-26}
Lead(II) carbonate (PbCO$_3$)	3.3×10^{-14}	Zinc hydroxide [Zn(OH)$_2$]	1.8×10^{-14}
Lead(II) chloride (PbCl$_2$)	2.4×10^{-4}	Zinc sulfide (ZnS)	3.0×10^{-23}

Example 6. Calculating molar solubility of a precipitate under varying conditions.

The K_{sp} of zinc hydroxide, Zn(OH)$_2$, at 25 °C is 1.8×10^{-14}.

    a. Determine the molar solubility of zinc hydroxide in pure water at 25 °C.
    b. Determine the molar solubility of zinc hydroxide if 0.10 mol of Zn(NO$_3$)$_2$ are added to 1.0 L of a saturated solution of zinc hydroxide at 25 °C (assume no volume change).
    c. Determine the molar solubility of zinc hydroxide if the solution has an adjusted pH of 2.0 at 25 °C.

a. Molar solubility of Zn(OH)$_2$ in pure water.

General Strategy	Solution to Example 6a
Write the dissociation equation and K_{sp} expression. Solids do not appear in a K expression.	$\text{Zn(OH)}_2\,(s) \Leftrightarrow \text{Zn}^{2+}\,(aq) + 2\text{OH}^-\,(aq)$ $K_{sp} = [\text{Zn}^{2+}][\text{OH}^-]^2$

Place the data underneath the equation. The solid dissociates in a 1 to 2 ratio based on the coefficients of the equation. In solubility problems $-x$ may be substituted with $-s$ to refer to the amount that is soluble.	$\quad\quad\quad Zn(OH)_2\,(s) \Leftrightarrow Zn^{2+}\,(aq) + 2OH^-\,(aq)$ *initial* $\quad\quad$ solid $\quad\quad\quad$ 0 $\quad\quad\quad$ 0 *change* $\quad\quad -s \quad\quad\quad\quad +s \quad\quad +2s$ *equilibrium* $\quad$ solid $- s \quad\quad s \quad\quad\quad 2s$
Substitute into the K_{sp} expression and solve.	$1.8 \times 10^{-14} = (s)(2s)^2 = 4s^3$ $\quad [Zn^{2+}] = s = \;1.6 \times 10^{-5} = molar\;solubility\;of\;Zn(OH)_2$

b.   Molar solubility of $Zn(OH)_2$ in the presence of a common ion, Zn^{2+}, provided by the 0.10 M $Zn(NO_3)_2$ solution.

General Strategy	Solution to Example 6b
Write the dissociation equation and K_{sp} expression.	$Zn(OH)_2\,(s) \Leftrightarrow Zn^{2+}\,(aq) + 2OH^-\,(aq)$ $K_{sp} = [Zn^{2+}][OH^-]^2$
Place the data underneath the equation. The solid dissociates in a 1 : 2 ratio. The amount of Zn^{2+} that results from the dissociation of $Zn(OH)_2$ is insignificant when compared to the 0.10 M concentration.	$\quad\quad\quad\quad Zn(OH)_2\,(s) \Leftrightarrow Zn^{2+}\,(aq) \;+\; 2OH^-(aq)$ *initial* $\quad\quad\quad$ solid $\quad\quad\quad$ 0.10 mol $\quad\quad\quad$ 0 $\quad\quad\quad\quad\quad\quad\quad\quad\quad\quad$ 1.0 L *equilibrium* $\quad$ solid $- s \quad\quad$ 0.10 M $+ s \quad\quad 2s$ assume $+s$ is negligible and $[Zn^{2+}] = 0.10$ M.
Substitute into the K_{sp} expression and solve.	$1.8 \times 10^{-14} = (0.10)(2s)^2$ $\quad\quad\quad\quad = s = \;2.1 \times 10^{-7} = molar\;solubility\;of\;Zn(OH)_2$ Validate assumption $0.10\;M + (2.1 \times 10^{-7}) = 0.10\;M$

c.   Solubility of $Zn(OH)_2$ when the pH is adjusted to 2.0.

General Strategy	Solution to Example 6c
Write the dissociation equation and K_{sp} expression.	$Zn(OH)_2\,(s) \Leftrightarrow Zn^{2+}\,(aq) + 2OH^-\,(aq)$ $K_{sp} = [Zn^{2+}][OH^-]^2$
The pH is *defined* for this solution and thus the OH^- value is known. pH = 2.0 so pOH = 12.0 Place the data underneath the equation.	pH = 2.00 pH + pOH = 14.00 and pOH = 14.00 − 2.00 = 12.00 $[OH^-] = 10^{-pOH} = 1 \times 10^{-12}$ $\quad\quad\quad\quad Zn(OH)_2\,(s) \Leftrightarrow Zn^{2+}\,(aq) \;+\; 2OH^-\,(aq)$ *initial* $\quad\quad\quad$ solid $\quad\quad\quad$ 0 $\quad\quad\quad 1 \times 10^{-12}$ *equilibrium* $\quad$ solid $-s \quad\quad s \quad\quad\quad 1 \times 10^{-12}$

Substitute into the K_{sp} expression and solve.	$1.8 \times 10^{-14} = (s)(1 \times 10^{-12})^2$ $= s = 1.8 \times 10^{10} = molar\ solubility\ of\ Zn(OH)_2$

Note the decrease of solubility of $Zn(OH)_2$ (s) as you go from dissolving it in water (part a of Example 6) to dissolving it in the presence of the common ion, 0.10 M $Zn(NO_3)_2$ (part b in Example 6). Note the tremendous increase in solubility of $Zn(OH)_2$ in the presence of acid (part c in Example 6); this precipitate dissolves in strong acidic solution.

Predicting formation of a precipitate and ion concentration calculations

When mixing two solutions of soluble salts, a precipitate may occur, depending on the identity of the ions and their concentrations.

> **Take Note:** *You need to know the solubility rules in Chapter 4 of this review book and be familiar with the precipitates in Table 16.2 for the AP exam. You do not need to know the actual K_{sp} values, but you do need to have a sense of which precipitates are very insoluble and which ones are only moderately insoluble.*

You can predict whether a precipitate occurs by substituting the *initial concentrations of the ions* into the K_{sp} expression and calculating the reaction quotient Q (see Chapter 14 of this review book).

> If $Q = K_{sp}$, the precipitate is at equilibrium with the saturated solution and **no net change** occurs.

> If $Q > K_{sp}$, a **precipitate forms**, lowering the ion concentration until equilibrium is established and $Q = K$.

> If $Q < K_{sp}$, **no precipitate forms**. The solution is unsaturated; the ion concentrations are not yet at saturation levels.

Example 7. Predicting whether a precipitation reaction will occur.

> 10.0 mL of 0.010 M $Pb(NO_3)_2$ are mixed with 10.0 mL of 0.01 M NaI solution. Determine whether a precipitate will occur and justify your answer.

General Strategy	**Solution to Example 7**
Identify the precipitate, write the dissociation equation and the K_{sp} expression. Obtain the K_{sp} value from a solubility chart.	The precipitate is PbI_2. The dissociation equation is: $PbI_2 \Leftrightarrow Pb^{2+} + 2I^-$. $K_{sp} = [Pb^{2+}][I^-]^2 = 1.4 \times 10^{-8}$

To determine whether a precipitate occurs, calculate Q and compare Q to the K_{sp} value.	$Q = [Pb^{2+}]_{initial}[I^-]^2_{initial}$ The initial concentration of $[Pb^{2+}] = \dfrac{0.010\text{ M} \times 0.010\text{ L}}{0.020\text{ L}}$ $\qquad\qquad\qquad\qquad\qquad = 0.0050\text{ M }[Pb^{2+}]$ The initial concentration of $[I^-] = \dfrac{0.010\text{ M} \times 0.010\text{ L}}{0.020\text{ L}}$ $\qquad\qquad\qquad\qquad\qquad = 0.0050\text{ M }[I^-]$ $Q = [0.0050][0.0050]^2 = 1.2 \times 10^{-7}$ $Q > K$ since $1.2 \times 10^{-7} > 1.4 \times 10^{-8}$. *A precipitate will form* until the ion concentrations are lowered to the point that $Q = K$.

Example 8. Calculating ion concentrations in a precipitation reaction.

Determine all ion concentrations at equilibrium when 10.0 mL of 0.050 M $Pb(NO_3)_2$ are mixed with 20.0 mL of 0.010 M Na_2CrO_4.

General Strategy	Solution to Example 8
Identify the precipitate.	The precipitate is $PbCrO_4$.
Write the dissociation equation.	The dissociation equation is: $PbCrO_4 \Leftrightarrow Pb^{2+} + CrO_4^{2-}$.
Write the K_{sp} expression.	$K_{sp} = [Pb^{2+}][CrO_4^{2-}]$
Obtain the K_{sp} value from a solubility chart.	$K_{sp} = [Pb^{2+}][CrO_4^{2-}] = 2.0 \times 10^{-14}$
To determine the ion concentrations, the first step is to determine if a precipitate forms. In other words, you need to calculate Q and compare Q to K_{sp}.	$Q = [Pb^{2+}]_{initial}[CrO_4^{2-}]_{initial}$ The initial concentration of $[Pb^{2+}] = \dfrac{0.050\text{ M} \times 0.010\text{ L}}{0.030\text{ L}}$ $\qquad\qquad\qquad\qquad\qquad = 0.017\text{ M }[Pb^{2+}]$ The initial concentration of $[CrO_4^{2-}] = \dfrac{0.010\text{ M} \times 0.020\text{ L}}{0.030\text{ L}}$ $\qquad\qquad\qquad\qquad\qquad = 0.0067\text{ M }[CrO_4^{2-}]$ $Q = [0.017][0.0067] = 1.1 \times 10^{-4}$ $Q > K$ since $1.1 \times 10^{-4} > 2.0 \times 10^{-14}$. A precipitate will continue to form until $Q = K$.

For the spectator ion concentrations, such as NO_3^- and Na^+, calculate the moles of ion present and divide by the total volume of solution.	The NO_3^- concentration $$= \frac{(0.050\,mol/L\,Pb(NO_3)_2) \times 0.010\,L}{0.030\,L} \times \frac{2\,mol\,NO_3^-}{1\,mol\,Pb(NO3)2}$$ $= 0.033\ M\ NO_3^-$

The Na^+ concentration

$$[Na^+] = \frac{0.010\,mol/L\,Na_2CrO_4 \times 0.020\,L}{0.030\,L} \times \frac{2\,mol\,Na^+}{1\,mol\,Na_2CrO_4}$$

For the ions involved in precipitation, assume that the maximum amount of precipitate forms for an instant.

$= 0.013\ M\ Na^+$

$$Pb_2 + CrO_4^{2-} \Leftrightarrow PbCrO_4\,(s)$$

initially (mol)	5.0×10^{-4}	2.0×10^{-4}	0
assume maximum precipitate forms (mol).	-2.0×10^{-4}	-2.0×10^{-4}	$+\,2.0 \times 10^{-4}$

Assume all of the CrO_4^{2-} reacts and its concentration is 0 for an instant.

Determine how much solid redissolves using $-s$ to represent the change. The ions that form are $+s$ and follow the coefficient ratios of the balanced equation. But 's' is usually very small and negligible.

Amount (mol) left	3.0×10^{-4}	0	2.0×10^{-4}
Amount of solid that redissolves	$+s$	$+s$	$-s$
concentrations at equilibrium(M)	$\dfrac{3.0 \times 10^{-4}}{0.030\,L}+s$	s	solid

Substitute into the K_{sp} expression and solve.

assume +s negligible	$1.0 \times 10^{-2}\,M$	s	solid

Validate assumption.

$$2.0 \times 10^{-14} = [1.0 \times 10^{-2}][s]^2$$
$$s = 1.4 \times 10^{-6} = [\,CrO_4^{2-}\,]$$

$$1.0 \times 10^{-2}\,M + (1.4 \times 10^{-6}) = 1.0 \times 10^{-2}\,M = [Pb^{2+}]$$

Complex ion formation

A **complex ion** is made from a *metal ion bonded to one or more molecules or ions*. The production of a complex ion is an equilibrium reaction and has a **K_f, formation constant,** associated with it. K_f values are usually large, indicating that the reaction proceeds far to the right. The formation of a complex ion can cause a precipitate of low solubility to dissolve as the complex ion is formed. For example, AgCl has a low solubility in water but will dissolve in the presence of concentrated ammonia solution.

$$AgCl\ (s) \Leftrightarrow Ag^+\ (aq) + Cl^-\ (aq) \qquad\qquad K_{sp} = 1.6 \times 10^{-10}$$
$$Ag^+\ (aq) + 2NH_3\ (aq) \Leftrightarrow Ag(NH_3)_2^+\ (aq) \qquad\qquad K_f = 1.5 \times 10^7$$

Take Note: *The complex ions that appear most often on the AP exam are:*
$Ag(NH_3)_2^+$, $Ag(CN)_2^-$, $Al(OH)_4^-$, AlF_6^{3-}, $Cu(NH_3)_4^{2+}$, $Fe(SCN)_2^+$, $Fe(SCN)_6^{3-}$,

$Ni(NH_3)_x^{2-}$, $Ni(OH)_4^{2-}$, $Zn(OH)_4^{2-}$, $Zn(NH_3)_6^{2+}$, $Zn(NH_3)_4^{2+}$, $CoCl_4^{2-}$.

SAMPLE MULTIPLE CHOICE QUESTIONS

1. What is the pH of a solution made from adding 0.010 mol of solid sodium butanate, $NaC_4H_7O_2$, to 100.0 mL of 1.0 M butanoic acid? Assume no volume changes upon addition of the solid. (The K_a for $HC_4H_7O_2 = 1 \times 10^{-5}$.)

   A. The pH is 3
   B. The pH is 4
   C. The pH is 5
   D. The pH is 6
   E. The pH is 7

2. Which of the following combinations would result in the formation of a basic buffer?

   A. 10.0 mL of 0.10 M NH_4Cl added to 10.0 mL of 0.10 M HCl
   B. 10.0 mL of 0.10 M HCl added to 5.0 mL of 0.10 M NaCl
   C. 25.0 mL of 1.0 M NaF added to 20.0 mL of 1.0 M HF
   D. 10.0 mL of 0.10 M NH_4Cl added to 10.0 mL of 0.10 M NaOH
   E. 10.0 mL of 0.10 M NH_4Cl added to 5.0 mL of 0.10 M NaOH

3. Which of the following indicator(s) is/are appropriate to use when titrating a solution of hydrofluoric acid with sodium hydroxide?

Indicator	Endpoint pH range
Thymol blue	1.2 – 2.8
Bromophenol blue	3.0 – 4.6
Methyl red	4.2 – 6.3
Phenolphthalein	8.3 – 10

A. Phenolphthalein
B. Phenolphthalein and methyl red
C. Bromophenol blue, methyl red
D. Methyl red
E. Methyl red, bromophenol blue, and thymol blue

4. Which of the following compounds has the lowest molar solubility?

Compound	K_{sp}
$Co(OH)_2$	1.3×10^{-15}
$PbCO_3$	3.3×10^{-14}
Ag_2SO_4	1.4×10^{-5}
$BaCO_3$	8.1×10^{-9}
Hg_2Cl_2	1.1×10^{-18}

A. $Co(OH)_2$
B. $PbCO_3$
C. Ag_2SO_4
D. $BaCO_3$
E. Hg_2Cl_2

5. What is the approximate pH when 0.40 g of NaOH is added to 50.0 mL of 0.10 M HCOOH? Assume no volume change. (The K_a for HCOOH = 1.8×10^{-4}.)

A. pH 11
B. pH 5
C. pH 8
D. pH 9
E. pH 13

6. What is the approximate pOH when 5 mL of 0.10 M HCl are added to 10 mL of 0.10 M NH_3? (The K_b for NH_3 is 1.8×10^{-5}.)

A. pOH is between 2 and 3
B. pOH is between 3 and 4
C. pOH is between 4 and 5
D. pOH is between 5 and 6
E. pOH is between 8 and 9

7. What will happen when concentrated lye, NaOH solution, is added to a beaker containing a saturated solution of $Zn(OH)_2$ and a small amount of solid $Zn(OH)_2$?

A. More $Zn(OH)_2$ forms.
B. The pH of the solution decreases.
C. The solution gets cold.

D. The solid $Zn(OH)_2$ dissolves.

E. Nothing happens.

8. Using burets, a student titrated a solution of chlorous acid with a standardized solution of NaOH and obtained the following data:

	Trial 1	Trial 2	Trial 3
Initial reading of base (mL)	5.03	6.50	10.01
Final reading of base (mL)	26.03	26.00	29.61
Initial reading of acid (mL)	10.10	5.12	8.76
Final reading of acid (mL)	30.10	25.12	28.76

The discrepancy in the results of the first trial may be accounted for by which of the following reason(s)?

I. Using too little indicator in trial 1.

II. Adding distilled water to the beaker containing the chlorous acid during trial 1.

III. Not rinsing the insides of the buret with NaOH solution before filling the buret for trial 1.

IV. Forgetting to fill the tip of the buret before starting trial 1.

A. Statement I only

B. Statement II only

C. Statements II and III only

D. Statements III and IV only

E. All four statements

9. Which of the following is a buffer solution with an approximate pH of 3? (The K_a for HF $= 1 \times 10^{-3}$ and K_a for $CH_3COOH = 2 \times 10^{-5}$.)

A. 50.0 mL of 1.0 M HF and 25.0 mL of 1.0 M NaOH

B. 20.0 mL of 1.0 M CH_3COOH and 10.0 mL of 1.0 M NaOH

C. 10.0 mL of 0.001 M HCl and 10.0 mL of 0.001 NaCl

D. 50.0 mL of 1.0 M HF and 50.0 mL of 1.0 M NaOH

E. 20.0 ml of 3.0 M HF

10. What is the molar solubility of silver sulfide, Ag_2S? (The K_{sp} for Ag_2S is 6.0×10^{-51}.)

A. 1.1×10^{-17}

B. 3.0×10^{-17}
C. 8.5×10^{-17}
D. 2.4×10^{-25}
E. 1.5×10^{-25}

11.   Consider a solution made by combining 0.50 mol of CH_3COOH and 0.50 mol of CH_3COONa in enough water to make 100.0 mL of solution. Which of the following statements are true?

         I.   It is an acidic buffer.
         II.   Adding CH_3COOK solution will increase the pH of the solution.
         III.   The Brønsted bases are H_2O and CH_3COO^-.
         IV.   When 0.60 mol of NaOH are added, it becomes a basic buffer.

      A.   Statements I and II
      B.   Statements I and III
      C.   Statements I, II, and III
      D.   Statements I and IV
      E.   All the statements are true.

12.   What is the chromate ion concentration, $[CrO_4^{2-}]$, when 20.0 mL of 0.100 M $AgNO_3$ are mixed with 80.0 mL of 0.0100 M Na_2CrO_4? (The K_{sp} for $Ag_2CrO_4 = 9.0 \times 10^{-12}$.)

      A. 5.6×10^{-5}
      B. 5.6×10^{-7}
      C. 3.0×10^{-6}
      D. 2.6×10^{-8}
      E. 1.0×10^{-9}

13.   A solution *may* contain the ions Ba^{2+}, Ag^+, and Cu^{2+}. When sodium chloride solution is added, a white precipitate forms. The supernatant is decanted, potassium sulfate solution is added to the supernatant and no reaction is observed. When sodium hydroxide solution is added, a blue solid forms. Which ion(s) is/are present in the solution?

      A. Ag^+ and Cu^{2+}
      B. Ag^+
      C. Ba^{2+} and Ag^+
      D. Cu^{2+}
      E. Ba^{2+} and Cu^{2+}

14.   20.0 mL of weak acid are titrated with 0.10 M NaOH solution. The volume versus pH titration data are provided below.

Volume of base (mL)	pH
0.0	1.9
10.0	3.8
13	4.0
20	4.1
22	4.2
23	4.3
26	8.0
30	12.0

What is the approximate value for the K_a for the weak acid?

A. 1×10^{-2}
B. 1×10^{-4}
C. 1×10^{-8}
D. 1×10^{-12}
E.   Cannot be determined from these data

15.   Which of the following will increase the molar solubility of Ag_2CO_3? (The K_{sp} for Ag_2CO_3 is 8.1×10^{-12}.)

A.   Add HCl solution
B.   Add AgNO3 solution
C.   Add NaCN solution
D.   A and C
E.   A, B, and C

16.   What is the hydrogen ion concentration, $[H^+]$, when 20.0 mL of 0.10 M NaOH are added to 20.0 mL of 0.10 M hypobromous acid, HOBr? (The K_a for HOBr is 2×10^{-9}.)

A. 1×10^{-5}
B. 1×10^{-7}
C. 2×10^{-8}
D. 2×10^{-11}
E. 5×10^{-4}

17.   What is the $[H_3O^+]$ concentration when 40.0 mL of 0.10 M nitrous acid, HNO_2, are added to 10.0 mL of 0.10 M HCl? (The K_a for HNO_2 is 4.0×10^{-4}.)

A. 2×10^{-2}
B. 2×10^{-1}
C. 1×10^{-3}
D. 9×10^{-3}
E. 4.5×10^{-3}

18.   Which of the following titration(s) has/ have a neutral equivalence point where the pH = 7.0?

    A.   Titrating HNO_2 with NaOH solution
    B.   Titrating HNO_3 with NaOH solution
    C.   Titrating NH_3 with HCl solution
    D.   Titrating HF with KOH solution
    E.   The equivalence points in titrations A to D all have a pH = 7.0.

19.   Which of the following precipitates have their molar solubility affected by pH changes?

    A. $AgCl$, $BaSO_3$, $BaSO_4$, $CaCO_3$, Ag_2S, $Zn(OH)_2$
    B. $BaSO_3$, $BaSO_4$, $CaCO_3$, Ag_2S, $Zn(OH)_2$
    C. $AgCl$, $BaSO_3$, $CaCO_3$, Ag_2S, $Zn(OH)_2$
    D. $BaSO_3$, $CaCO_3$, Ag_2S, $Zn(OH)_2$
    E. $AgCl$, $BaSO_3$, $BaSO_4$, $CaCO_3$, $Zn(OH)_2$

20.   What is the molar solubility of CaF_2 if 0.10 mol of NaF are added to 1.0 L of a saturated solution of CaF_2? (The K_{sp} for CaF_2 is 4.0×10^{-11}.)

    A. 1.0×10^{-3}
    B. 4.0×10^{-10}
    C. 4.0×10^{-9}
    D.   0.10
    E. 1.0×10^{-11}

Comprehension Questions

1)   Dichloroacetic acid, CCl_2HCO_2H, is a weak monoprotic acid that reacts with water according to the following equation:

$$CCl_2HCO_2H\ (aq) + H_2O\ (l) \leftrightarrow H_3O^+\ (aq) + CCl_2HCO_2^-\ (aq) \qquad K_a = 4.47 \times 10^{-2}$$

    a)   Write an equilibrium constant expression for the dissociation of dichloroacetic acid in water.

    b)   Determine the initial concentration, expressed in moles per liter, of a solution of this acid that would have a pH of 3.80.

    c)   For the solution mentioned in part b, calculate the $[H^+]$, $[OH^-]$, and pOH.

d)   Consider the titration of a dichloroacetic acid solution with 0.1105 M sodium hydroxide.

      i)   What is the concentration of the dichloroacetic acid if 75.00 mL of its solution can be titrated to the equivalence point by 45.35 mL of the sodium hydroxide solution?

      ii)   Will the pH of the reaction mixture in i) its equivalence point be greater than, less than, or equal to 7? Explain your answer.

e)   The closely related trichloroacetic, CCl_3CO_2H, and chloroacetic, $CClH_2CO_2H$, acids are stronger and weaker acids, respectively, than dichloroacetic acid. Account for these differences based on molecular structure.

2)   The following questions concern the solubilities of various copper and silver salts:

a)   Silver sulfide dissolves in water according to the following equation:

$$Ag_2S\ (s) + H_2O\ (l) \leftrightarrow 2Ag^+\ (aq) + S^{2-}\ (aq) \qquad\qquad Ksp = 6.0 \times 10^{-51}$$

      i)   Write an equilibrium constant expression for the dissolution of silver sulfide in water.

      ii)   What would the molar concentration of silver ions be in a saturated solution of silver sulfide? (Assume that the solution is formed by dissolving excess silver sulfide in water.)

      iii)   What is the molar solubility of silver sulfide?

b)   How many grams of silver sulfide would need to be dissolved in a swimming pool (10. m long by 5.0 m wide by 2.0 m deep) to create a saturated solution?

c)   Copper(II) sulfide dissolves in water according to the following equation:

$$CuS\ (s) + H_2O\ (l) \leftrightarrow Cu^{+2}\ (aq) + S^{2-}\ (aq) \qquad\qquad Ksp = 6.0 \times 10^{-37}$$

      i)   Write an equilibrium constant expression for the reaction.

      ii)   What is the molar solubility of copper(II) sulfide in pure water?

iii)   What is the molar solubility of copper(II) sulfide in a 1.0×10^{-3} M solution of sodium sulfide?

3)   Diethylamine is a weak base that undergoes base hydrolysis according to this equation:

$$(C_2H_5)_2NH \ (aq) + H_2O \ (l) \leftrightarrow OH^- \ (aq) + (C_2H_5)_2NH_2^+ \ (aq)$$

a)   Write an equilibrium constant expression for the reaction.

b)   A 0.150 M solution of this base has a pH of 11.99. What is the K_b of diethylamine?

c)   Consider the titration of 25.00 mL of the above solution with 0.100 M hydrobromic acid, HBr.

i)   What is the pH of the neutralization reaction at its equivalence point?

ii)   What is the pH of the reaction mixture after 15.00 mL of acid has been added?

d)   Diethylamine, as a pure liquid or aqueous solution, has a distinct and disagreeable odor. Formation of its hydrochloride or hydrobromide salt leads to an odorless compound or solution. Explain this phenomenon.

ANSWERS TO SAMPLE MULTIPLE CHOICE QUESTIONS

1.   B

	HA	$\Leftrightarrow$	H^+ +	A^-
initial	1 mol/L			0.010 mol/0.10 L
changes	$-x$		$+x$	$+x$
equilibrium(M)	$1-x$		x	$0.1 + x$

Assume $1 - x = 1$ and $0.1 + x = 0.1$

$$1 \times 10^{-5} = \frac{[H^+][0.1]}{[1]}$$

$[H^+] = 1 \times 10^{-4}$ and $pH = -log \ (1 \times 10^{-4}) = 4.0$

Validate assumptions: $1 - (1 \times 10^{-4}) = 1$ and
$0.1 + (1 \times 10^{-4}) = 0.1$

2.   E

A basic buffer is the combination of a weak base and the conjugate acid of that weak base. In choice E, half of the original NH_4Cl (0.01 L × 0.10 M NH_4Cl = 0.001 mol NH_4Cl initially) has been neutralized by NaOH so there are 5×10^{-4} mol of NH_3 and 5×10^{-4} mol of NH_4Cl present. The amounts of base and conjugate acid are very small, so the buffering capacity of this buffer is also small.

3.   A

At the equivalence point, all the HF has been neutralized by the NaOH and is in the form of F^-. F^- is a base and undergoes a hydrolysis reaction with water: $F^- + H_2O \Leftrightarrow HF + OH^-$. The equivalence point is basic. The pH curve rises steeply in the pH range of 7 to 11. Any indicator that has an endpoint that falls in this range would work, such as *phenolphthalein*.

4.   B

$PbCO_3$ has the lowest molar solubility of 1.8×10^{-7}.

$$PbCO_3 \Leftrightarrow Pb^{2+} + CO_3^{2-}$$

Equilibrium conc.(M)   solid s s

$$K_{sp} = 3.3 \times 10^{-14} = [Pb^{2+}][CO_3^{2-}] = s^2$$
$$s = [Pb^2] = molar\ solubility\ of\ PbCO_3 = 1.8 \times 10^{-7}$$

The value for the K_{sp} for $Co(OH)_2$ is smaller than K_{sp} for $PbCO_3$ but to calculate the molar solubility of $Co(OH)_2$ you have to take the cube root of K_{sp}, thus the resulting molar solubility of $Co(OH)_2$ is greater than for $PbCO_3$.

$$Co(OH)_2 \Leftrightarrow Co^{2+} + 2OH^-$$

Equilibrium conc.(M)   solid s $2s$

$$K_{sp} = 1.3 \times 10^{-15} = [Co^{2+}][OH^-]^2 = 4s^3$$
$$s = molar\ solubility = 1 \times 10^{-5}$$

5.   E

$$0.40\,g\,NaOH \times \frac{1\,mol\,NaOH}{40.0\,g\,NaOH} = 0.010\,mol\,of\,NaOH$$

$$0.10\,\frac{mol}{L}\,HCOOH \times 0.050\,L = 0.005\,mol\,HCOOH$$

$$NaOH + HCOOH \rightarrow H_2O + HCOONa$$

Initially (mol)   0.010   0.0050
Change   − 0.005   − 0.005   +0.005   + 0.005

After the reaction has taken place, there are two bases: a strong base, NaOH, and a weak base, HCOONa. The strong base determines the pH of the solution.
$[OH^-] = 0.005$ mol/0.050 L $= 0.10$ M OH^-. pOH $= 1$ and *pH = 13*.

6.   C

A strong acid is added to a weak base. The acid neutralizes the base, forming the conjugate acid NH_4^+. Determine the concentrations of NH_3 and NH_4^+ that result after neutralization has taken place and substitute these concentrations into the K_b expression. Solve for pOH.

0.10 $\underline{mol}$ HCl $\times 0.005$ L $= 0.0005$ mol HCl added $= 0.0005$ mol H^+
            L

0.10 $\underline{mol}$ $NH_3 \times 0.01$ L $= 0.001$ mol NH_3 initially
            L

Neutralization reaction: NH_3 + H^+ → NH_4^+

Initial (mol)	0.001	0.0005	0
Change (mol)	−0.0005	− 0.0005	+0.0005
Final (mol)	0.0005	0	0.0005

Equilibrium reaction that determines pOH of solution:

                        NH_3 + H_2O → NH_4^+ + OH^-

Equilibrium (M) $\dfrac{0.0005 - x}{0.012 \text{ L}}$ $\dfrac{0.0005 + x}{0.012 \text{ L}}$ x

Assume $- x$ and $+ x$ negligible

$$1.8 \times 10^{-5} = \frac{(0.0005 \,/\, 0.012)(x)}{(0.0005 \,/\, 0.012)}$$

$x = [OH^-] = 1.8 \times 10^{-5}$, $pOH = - \log (1.8 \times 10^{-5}) = 5 - \log 1.8$
                                        $= pOH$ *between 4 and 5* (i.e., *4.74*)

A shortcut in the problem is to recognize that when moles of NH_3 = moles of NH_4^+, you are at the midpoint of the titration and $K_b = [OH^-]$.

7.   D

The *precipitate $Zn(OH)_2$ dissolves in the presence of strong base* to form the soluble complex $Zn(OH)_4^{2-}$ according to the equation:

$$Zn(OH)_2\ (s) + 2OH^-\ (aq) \rightarrow Zn(OH)_4^{2-}\ (aq)$$

8.   D

In trial 1, 21 mL of NaOH are used to titrate 20 mL of chlorous acid. In trials 2 and 3, 19.5 and 19.6 mL of NaOH are used to titrate the 20 mL of acid. More NaOH was required in trial 1. If the buret is not rinsed with NaOH, the first trial will have diluted NaOH in the buret and more NaOH will be required for the titration. If the buret tip is not filled initially, but is filled at the final volume reading, then more NaOH is used in that trial.

9.   A

When *25.0 mL of 1 M NaOH have been added to 50.0 mL of 1.0 M HF*, half of the HF has been neutralized, so moles of HF = moles of F^-. This is the half-way point of the titration, at which point the HA and F^- terms cancel out of the K_a expression and $K_a = [H^+] = 1 \times 10^{-3}$. Thus the pH of the buffer is 3.

10. A

The precipitate, Ag_2S, establishes an equilibrium in water. The silver and sulfide ions are present in a 2 to 1 ratio. To solve the problem, you write the dissociation equation, express the ion concentrations in terms of $2s$ and s, and then substitute into the K_{sp} expresssion and solve.

$$Ag_2S\ (s) \Leftrightarrow 2Ag^+\ (aq) + S^{2-}\ (aq)$$

Equilibrium   solid $2s$ s

$$K_{sp} = [Ag^+]^2[S^{2-}] = (2s)^2(s) = 6.0 \times 10^{-51}$$
$$s = [Ag^+] = molar\ solubility\ of\ Ag_2S = 1.1 \times 10^{-17}$$

11. C

It is an acidic buffer since a weak acid and its conjugate base are present in significant amounts (statement I). CH_3COOK is a soluble salt; the CH_3COO^- ion is a common ion and will drive the reaction to the left in accord with Le Chatelier's principle. A net reaction favoring the reactants reduces the H_3O^+ concentration and therefore pH increases (statement II). A Brønsted base accepts a proton. In the forward reaction H_2O accepts a proton; in the reverse reaction CH_3COO^- accepts a proton (statement III). Statement IV is false. You have a basic solution but it is not a buffer.

12. B

You calculate the moles of silver and chromate ion present in solution. Determine Q. If $Q > K_{sp}$, a precipitate occurs. Assume a maximum of precipitate forms (i.e., one ion is completely consumed). Substitute ion concentrations into the K_{sp} expression and solve.

$$0.100 \; \underline{mol} \; AgNO_3 \times 0.020 \; L = 0.0020 \; mol \; AgNO_3 = 0.0020 \; mol \; Ag^+$$
$$L$$
$$0.0100 \; mol \; Na_2CrO_4 \times 0.080 \; L = 0.00080 \; mol \; Na_2CrO_4 = 0.00080 \; mol \; CrO_4^{2-}$$

$Q = [Ag^+]^2[CrO_4^{2-}] = (0.002 / 0.010 \; L)^2 \; (0.00080 / 0.010 \; L)$. $Q > K_{sp}$, so a precipitation occurs.

	$2Ag^+ \; (aq) +$	$CrO_4^{2-} \; (aq) \;$ →	$Ag_2CrO_4 \; (s)$
Initially (mol)	0.0020	0.00080	0
Change (mol)	−0.0016	−0.00080	+0.00080 solid mol
Final (mol)	0.0004	0	0.00080 solid mol

Take these *ion amounts*, convert them to molarity, and substitute into the precipitation equilibrium reaction. Then use the K_{sp} expression to solve for the ion concentrations.

	$Ag_2CrO_4 \; (s) \quad \Leftrightarrow$	$2Ag^+ \quad +$	CrO_4^{2-}
Initially (M)		0.0004 mol / 0.1 L	0
Change (M)		+ 2s	+ s
Final (M)		$4.0 \times 10^{-3} + 2s$	s
Assume +2s negligible			

$$K_{sp} = 9.0 \times 10^{-12} = (4.0 \times 10^{-3})^2 \; (s)$$
$$s = \; 5.6 \times 10^{-7} \; = [CrO_4^{2-}]$$
Validate assumption: $4.0 \times 10^{-3} + 2 \; (5.6 \times 10^{-7}) = 4.0 \times 10^{-3}$

13. A

When chloride was added, the white precipitate AgCl formed. That means the Ag^+ *ion is present*. The addition of sulfate ion did not result in a reaction, so no barium ion is present (i.e., no BaSO₄ formed). The addition of a hydroxide ion resulted in the formation of the blue precipitate Cu(OH)₂. *The Cu^{2+} ion is present*.

14. B

When 26 mL of base have been added there is a sharp rise in the pH, indicating that the endpoint has been reached. Therefore, when 13.0 mL of base have been added, the halfway point has been reached. At this point half of the weak acid, HA, has been neutralized and converted to A^-. The HA and A^- cancel out of the K_a expression and $K_a = [H^+]$. When 13 mL of base have been added, the pH is 4.0 and thus $[H^+] = 1 \times 10^{-4} = K_a$.

15. D

The Ag_2CO_3 has an equilibrium reaction in water:

$$Ag_2CO_3 (s) \Leftrightarrow 2Ag^+ (aq) + CO_3^{2-} (aq)$$

When *HCl* is added it reacts with the CO_3^{2-} to form H_2O and CO_2. As the CO_3^{2-} is used, more Ag_2CO_3 dissolves to replace the CO_3^{2-} in accordance with Le Chatelier's principle. Along the same lines of reasoning, the *NaCN* reacts with the Ag^+ ion to form $Ag(CN)_2^{2-}$, a soluble complex, so that more Ag_2CO_3 dissolves.

16. D

When solving titration problems, you need to establish where along the titration curve you are to determine the appropriate equation that determines the pH. In this case, you are at the equivalence point of the titration; the moles of acid equals the moles of base added. All of the HOBr has been converted to OBr^-. The hydrolysis reaction of OBr^- in water determines the pH. OBr^- is a base so you need to use the K_b expression.

O.10 M NaOH $\times$ 0.020 L = 0.0020 mol NaOH
0.10 M HOBr $\times$ 0.020 L = 0.0020 mol HOBr

	OBr^-	$+ H_2O$	$\Leftrightarrow$	HOBr	$+$	OH^-
Initially (M)	0.0020 mol/0.040 L			0		0
Change (M)	$-x$			$+x$		$+x$
Equilibrium (M)	$0.050 - x$			x		x

Assume $-x$ negligible

$$K_b = \frac{K_w}{K_a} = \frac{1 \times 10^{-14}}{2 \times 10^{-9}} = 5 \times 10^{-6} = \frac{x^2}{(0.050)}$$

$$x = [OH^-] = 5 \times 10^{-4}$$

$$[H^+][OH^-] = 1 \times 10^{-14}$$

$$[H^+] = \frac{1 \times 10^{-14}}{5 \times 10^{-4}} = 2 \times 10^{-11}$$

17. A

You have a strong acid, HCl, and a weak acid, HNO_2, in the same beaker. The strong acid determines the H^+ concentration. The contribution of H^+ from HNO_2 is negligible.

$$0.10 \text{ M HCl} \times 0.010 \text{ L} = 0.001 \text{ mol HCl} = 0.0010 \text{ mole } H^+$$
$$[H^+] = 0.0010 \text{ mol} / 0.050 \text{ L} = 0.020 \text{ M}$$

18. B

Titrating a strong base with a strong acid results in an equivalence point pH of 7. $HNO_3 + NaOH \rightarrow H_2O + NaNO_3$. The salt is neutral in such cases.

19. D

All the compounds in (D) have product ions that react with H^+ or OH^- and are thus sensitive to pH. If a product ion is removed during a reaction, the precipitate dissolves to replace the ion as predicted by Le Chatelier's principle.

Precipitate equilibrium reactions

$BaSO_3 \Leftrightarrow Ba^{2+} + SO_3^{2-}$

$CaCO_3 \Leftrightarrow Ca^{2+} + CO_3^{2-}$

$Ag_2S \Leftrightarrow 2 Ag^+ + S^{2-}$

$Zn(OH)_2 \Leftrightarrow Zn^{2+} + 2OH^-$

pH sensitive reactions

$SO_3^{2-} + 2 H^+ \Leftrightarrow H_2SO_3$

$CO_3^{2-} + 2H^+ \rightarrow H_2O + CO_2$

$S^{2-} + 2H^+ \Leftrightarrow H_2S$

$Zn(OH)_2 + 2OH^- \rightarrow Zn(OH)_4^{2-}$

20. C

The common ion F^- from NaF will suppress the molar solubility of CaF_2.

$$CaF_2 \text{ (s)} \Leftrightarrow Ca^{2+} + 2F^-$$

Equilibrium (M)   solid s $0.10 + 2s$

Assume $+ 2s$ is negligible

$$K_{sp} = 4.0 \times 10^{-11} = [Ca^{2+}] [F^-]^2 = s (0.10 \text{ M})^2$$
$$s = 4.0 \times 10^{-9} = [Ca^{2+}] = molar \ solubility \ of \ CaF_2$$

Validate assumption: $[F^-] = 0.10 \text{ M} + 2(4.0 \times 10^{-9}) = 0.10 \text{ M}$

Answers to Comprehension Questions

1)   a) $K_a = \dfrac{[H_3O^+][CCl_2HCO_2^-]}{[CCl_2HCO_2H]} = 4.47 \times 10^{-2}$

b)   A good method for solving this kind of problem is to use the "ICE-box." In this case the equilibrium values for both products are known, and the initial value for the reactant must be determined. The pH gives us the equilibrium hydronium ion concentration ($[H_3O^+] = 10^{-pH}$), which will be equal to the equilibrium concentration of the conjugate base owing to the stoichiometry of the reaction and the initial concentrations.

$$CCl_2HCO_2H \ (aq) \quad + \quad H_2O \ (l) \quad \leftrightarrow \quad H_3O^+ \ (aq) \quad +$$
$$CCl_2HCO_2^- \ (aq)$$

Initial	y M	NA	0 M	0 M
Change	$-x$	NA	$+x$	$+x$
Equilibrium	$y-x$ M	NA	x M	x M

In this case the value of x is known ($x = 10^{-3.80} = 1.58 \times 10^{-4}$). The value of the initial concentration of acid can then be solved for by substituting the equilibrium values in the chart into the equilibrium constant expression. The "ICE-box" can be rewritten as follows:

$$CCl_2HCO_2H \ (aq) \quad + \quad H_2O \ (l) \quad \leftrightarrow \quad H_3O^+ \ (aq) \quad + \quad CCl_2HCO_2^- \ (aq)$$

Initial	y M	NA	0 M	0 M
Change	-1.58×10^{-4}	NA	$+1.58 \times 10^{-4}$	$+1.58 \times 10^{-4}$
Equilibrium	$y - 1.58 \times 10^{-4}$ M	NA	1.58×10^{-4} M	1.58×10^{-4} M

Substituting into the equilibrium constant expression:

$$K_a = 4.47 \times 10^{-2} = \frac{[H_3O^+][CCl_2HCO^-]}{[CCl_2HCO_2H]} = \frac{(1.58 \times 10^{-4})^2}{y - 1.58 \times 10^{-4}}$$

$$(4.47 \times 10^{-2}) \times (y - 1.58 \times 10^{-4}) = (1.58 \times 10^{-4})^2$$

$$4.47 \times 10^{-2}y - 7.06 \times 10^{-6} = 2.50 \times 10^{-8}$$

$y = (2.50 \times 10^{-8} + 7.06 \times 10^{-6}) / (4.47 \times 10^{-2}) = \mathbf{1.59 \times 10^{-4}}$ **M= initial concentration of acid.**

c)   To solve part b), the hydronium, $\mathbf{H_3O^+}$, ion concentration was already determined, $\mathbf{1.58 \times 10^{-4}}$ **M**. The hydroxide ion concentration can then be determined by dividing the value of K_w by the hydronium ion concentration ($[H_3O^+] \times [OH^-] = K_w = 1 \times 10^{-14}$).

$(1.58 \times 10^{-4}) \times [OH^-] = 1 \times 10^{-14}$

$[OH^-] = (1 \times 10^{-14}) / (1.58 \times 10^{-4}) = 6.39 \times 10^{-11}$

The pOH can be simply determined by using the relationship pH + pOH = 14. Therefore, **pOH = 14 – pH = 14 – 3.80 = 10.20**

       d)   i)   Dichloroacetic acid and sodium hydroxide react in a 1:1 ratio as indicated in the balanced equation below:

$$CCl_2HCO_2H \ (aq) \quad + \quad NaOH \ (aq) \quad \leftrightarrow \quad H_2O \ (l) \quad + \quad NaCCl_2HCO_2 \ (aq)$$

Since this is true, the number of moles of acid and base added will be identical at the equivalence point. Moles can be determined by multiplying volume times molarity for each solution. Therefore, at the equivalence point:

volume of acid × molarity of acid = volume of base × molarity of base

$0.07500 \text{ L} \times x \text{ M} = 0.04535 \text{ L} \times 0.1105 \text{ M}$

$x \text{ M} = (0.04535 \text{ L} \times 0.1105 \text{ M}) / 0.07500 \text{ L}$

$x = 0.06682$ M

       ii)   **The pH at the equivalence point will be greater than 7.** This can be explained or predicted in a variety of ways. You can simply state that this neutralization reaction is occurring between a strong base and a weak acid, and therefore the solution will be slightly basic at the equivalence point. This is due to the fact that at the equivalence point the solution contains a salt made up from the conjugate base of a weak acid, which is itself a weak base, and the conjugate acid of a strong base, which is a neutral cation or spectator. This combination of weakly basic anion and neutral cation leads to a slightly basic solution overall.

e)   Lewis structures for the three acids mentioned are shown.

chloroacetic acid   dichloroacetic acid   trichloroacetic acid

The fundamental reason for the increased the acidity of trichloroacetic acid versus dichloroacetic and chloroacetic acids is the increased electronegativity of chlorine versus hydrogen. Recall that electronegativity is the tendency of an element to draw shared electrons in a bond toward itself. As chlorine atoms are substituted for hydrogen in the above structures, electron density is redistributed towards the chlorine atoms and away from the carboxyl group. This inductive electron withdrawing has the ultimate effect of weakening the $O - H$ bond in the molecule or, alternatively, stabilizing the anion formed as each of these acids liberates a proton in aqueous solution. The net effect is shifting the equilibrium mix toward the ionized species and away from the nonionized acid (i.e., greater K_a as you go from monochloro-to trichloro).

2)   a)   i)   Equilibrium constant expressions for solubility equilibria follow the same rules that apply to all other equilibria. They are ultimately a *product* of ion concentrations (not a *ratio* of concentrations) in these cases because there is nothing on the reactant side of the equation that is included in the expression (i.e., the reactants are always a pure liquid and a pure solid and therefore are not included in the equilibrium expression). $K_{sp} = [Ag^+]^2[S^{2-}] = 6.0 \times 10^{-51}$

          ii)   If a solution is created by dissolving an ionic compound in water, you can predict the relative amounts of product ions that are formed using the balanced equation. A simplified "ICE-box" can be used to illustrate this process (with practice, the simplicity of the situation may make it unnecessary).

$$Ag_2S\ (s)\quad +\quad H_2O\ (l)\quad \leftrightarrow\quad 2Ag^+\ (aq)\quad +\quad S^{2-}\ (aq)$$

Initial	NA	NA	0 M	0 M
Change	NA	NA	+ 2x	+ x
Equilibrium	NA	NA	2x M	x M

The equilibrium values from the "ICE-box" can be substituted into the equilibrium constant expression to solve for the value of x.

$$K_{sp} = [Ag^+]^2[S^{2-}] = [2x]^2[x] = 4x^3 = 6.0 \times 10^{-51}$$

$$x = (6.0 \times 10^{-51} / 4)^{1/3} = 1.14 \times 10^{-17}$$

The molar concentration of silver ions will be equal to $2x$ or $\textbf{2.29} \times \textbf{10}^{\textbf{-17}} \textbf{ M}$

iii)   Molar solubility is the maximum amount of a substance that will dissolve at a given temperature expressed in moles of solute per liter of solution. In this case, 1 mol of sulfide ions will be produced for every 1 mol of silver sulfide that dissolves. Therefore the equilibrium molar concentration of sulfide ions, or x from part ii), will be numerically equal to the molar solubility of the salt. **The molar solubility of silver sulfide is 1.14×10^{-17} mol L.**

b)   The molar solubility calculated in a) iii) along with the volume of water contained in the swimming pool and the molar mass of silver sulfide will allow us to calculate the amount needed to create a saturated solution. The volume can be calculated by multiplying length times width times depth and then converting this value to liters.

volume of pool = 10. m $\times$ 5.0 m $\times$ 2.0 m = 100 m^3 $\times$ (1000 L / 1 m^3) = 1.0×10^5 L of water

1.0×10^5 L of water $\times$ 1.14×10^{-17} mol Ag$_2$S per liter of water $\times$ 247.80 g/mol = $\textbf{2.82} \times$ $\textbf{10}^{\textbf{-10}}$ **g.** We can see that it takes two tenths of a nanogram to saturate a solution in a swimming pool. In other words, this stuff is really insoluble!

c)   i) $K_{sp} = [Cu^{+2}][S^{2-}] = 6.0 \times 10^{-37}$

ii)   See the explanation for part a) i). In this case the ions are produced in a 1:1 ratio to each other and in a 1:1 ratio to the starting salt. Therefore the variable x can be used to represent the equilibrium molar concentration of each ion in solution as well as the molar solubility of this salt.

$$K_{sp} = [Cu^{+2}][S^{2-}] = [x][x] = 6.0 \times 10^{-37}$$

molar solubility of copper(II) sulfide = $x = \textbf{7.75} \times \textbf{10}^{\textbf{-19}} \textbf{ M}$

iii)   This part of the problem is a so-called common ion situation. The poorly soluble copper(II) sulfide is trying to dissolve into a solution that already contains sulfide ions. The presence of these ions will inhibit dissolution of the solid salt. The extent of dissolution or molar solubility can be determined once again by using the "ICE-box" in a manner similar to 2 a). The only difference this time will be that the initial concentration of sulfide ions will be 1.0×10^{-3} M instead of 0 M.

$$\text{CuS } (s) \quad + \quad \text{H}_2\text{O } (l) \quad \leftrightarrow \quad \text{Cu}^{2+} (aq) \quad + \quad \text{S}^{2-} (aq)$$

Initial	NA	NA	0 M	1.0×10^{-3} M
Change	NA	NA	$+x$	$+x$
Equilibrium	NA	NA	x M	$1.0 \times 10^{-3} + x$ M

Substituting these equilibrium values, in terms of x, into the equilibrium constant expression will allow for calculation of the molar solubility of copper(II) sulfide in the sodium sulfide solution. In this case, the molar solubility will be equal to the equilibrium copper ion concentration or x. This is due to the fact that the only source of copper ions in solution will come from dissolution of the copper(II) sulfide, and one copper ion will form for every one formula unit of copper(II) sulfide that dissolves, that is, a 1:1 ratio.

Also, because of the poor solubility of this copper salt, we can predict, without solving a quadratic equation, that the size of x will be negligible compared to 1.0×10^{-3}. Therefore we can replace the equilibrium concentration value for sulfide ions with the constant term 1.0×10^{-3} M.

$$K_{sp} = [Cu^{2+}][\, S^{2-}] = [x][\, 1.0 \times 10^{-3} + x \,] \approx [x][1.0 \times 10^{-3}] = 6.0 \times 10^{-37}$$

Therefore the molar solubility or $x = 6.0 \times 10^{-37} / 1.0 \times 10^{-3} = \mathbf{6.0 \times 10^{-34}}$ **M**

3)   a) $$K_b = \frac{[OH^-][(C_2H_5)_2\,NH_2^+]}{[(C_2H_5)_2\,NH]}$$

b)   To solve this problem, we will again turn to the "ICE-box." In previous problems the initial concentration and equilibrium constants were given and the equilibrium concentrations were sought. In this particular case the problem gives the initial and equilibrium concentrations (in a round-about way) and requests that we solve for the equilibrium constant itself.

$$(C_2H_5)_2NH\ (aq)\ +\ H_2O\ (l)\ \leftrightarrow\ OH^-\ (aq)\ +\ (C_2H_5)_2\,NH_2^+\ (aq)$$

Initial	0.150 M	NA	0 M	0 M
Change	$-x$	NA	$+x$	$+x$
Equilibrium	$0.150 - x$ M	NA	x M	x M

Given only the information in the chart, it would be impossible to solve for the equilibrium constant by plugging the values into the equilibrium constant expression (there would be two variables). However, the problem has given us information, pH in this case, from which we may calculate the equilibrium values. If we subtract the stated pH from 14 ($14 - 11.99 = 2.01$) we arrive at the pOH. Raising 10 to the power of $-$ pOH will give the equilibrium hydroxide ion concentration (i.e., $10^{-2.01} = 9.77 \times 10^{-3}$). The variable x then has a value of 9.77×10^{-3}. Substituting this value into the "ICE-box" yields the following result:

$$(C_2H_5)_2NH \ (aq) \quad + \quad H_2O \ (l) \quad \leftrightarrow \quad OH^- \ (aq) \quad + \quad (C_2H_5)_2 NH_2^+ \ (aq)$$

Initial	0.150 M	NA	0 M	0 M
Change	-9.77×10^{-3}	NA	$+9.77 \times 10^{-3}$	$+9.77 \times 10^{-3}$
Equilibrium	$0.150 - 9.77 \times 10^{-3}$ M	NA	9.77×10^{-3} M	9.77×10^{-3} M

These values may then be substituted into the equilibrium constant expression in part a) to determine the value of the equilibrium constant.

$$K_b = \frac{[OH^-][(C_2H_5)_2 NH_2^+]}{[(C_2H_5)_2 NH]} = \frac{[9.77 \times 10^{-3}][9.77 \times 10^{-3}]}{[0.150 - 9.77 \times 10^{-3}]} = \mathbf{6.81 \times 10^{-4}}$$

c)   i)   At the equivalence point, all of the acid and base will be used up and we are left with a solution of the salt diethylammonium hydrobromide. Diethylammonium is the weak conjugate acid of diethylamine and the bromide ions are the conjugate base (neutral or spectator in this case) of the hydrobromic acid (strong acid). To solve this part of the problem, you must first determine the amount of acid that needs to be added to reach the equivalence point and then the total volume of the solution (to calculate the concentrations of the species present at the equivalence point). Once these concentrations are known, they may be thought of and used as initial concentrations in a new "ICE-box" calculation concerning the weakly acidic cation diethylammonium.

HBr and diethylamine react in a 1:1 ratio according to the following equation:

$$(C_2H_5)_2NH \ (aq) + HBr \ (aq) \leftrightarrow Br^- \ (aq) + (C_2H_5)_2 NH_2^+ \ (aq)$$

We are given the concentrations of both acid and base and the volume of base added. The volume of acid added can be determined in a similar manner to 1c).

volume of base × concentration of base = volume of acid × concentration of acid

$$0.02500 \text{ L} \times 0.150 \text{ M} = x \text{ L} \times 0.100 \text{ M}$$

$$x \text{ L} = (0.02500 \text{ L} \times 0.150 \text{ M}) / 0.100 \text{ M} = 0.0375 \text{ L or } 37.5 \text{ mL}$$

The concentration of the diethylammonium ion formed will then be the number of moles of diethylamine added initially divided by the total volume of both solutions added.

moles of base added = $0.02500 \text{ L} \times 0.150 \text{ M} = 0.00375$ mol

total volume of solution = $0.02500 \text{ L} + 0.0375 \text{ L} = 0.0625$ L

concentration of diethylammonium ion = moles of ion / liters of solution = 0.00375 mol/ 0.0625 L = 6.00×10^{-2} M

This concentration will now be used as the initial concentration of a weak acid solution in an "ICE-box" calculation. The equilibrium hydronium ion concentration will then be used to calculate pH.

	$(C_2H_5)_2 NH_2^+ \ (aq)$ +	$H_2O \ (l)$	$\leftrightarrow$	$H_3O^+ \ (aq)$ +	$(C_2H_5)_2NH \ (aq)$
Initial	6.00×10^{-2} M	NA		0 M	0 M
Change	$-x$	NA		$+x$	$+x$
Equilibrium	6.00×10^{-2} M $- x$	NA		x M	x M

The equilibrium constant used for this part of the problem can be calculated using the relationship $K_a \times K_b = K_w$ or 1×10^{-14}. This relationship holds true for acid-base conjugate pairs, so the K_a of diethylammonium can be determined from the K_b of diethylamine.

$$K_a \text{ (diethylammonium)} = 1 \times 10^{-14} / 6.81 \times 10^{-4} = 1.47 \times 10^{-11}$$

Substituting . . .

$$K_a = \frac{[H_3O^+][(C_2H_5)_2 NH]}{[(C_2H_5)_2 NH_2^+]} = \frac{[x][x]}{[6.00 \times 10^{-2} \text{ M} - x]} = 1.47 \times 10^{-11}$$

$$x^2 = 8.81 \times 10^{-13} - 1.47 \times 10^{-11} x$$

$$x^2 + 1.47 \times 10^{-11} x - 8.81 \times 10^{-13} = 0$$

Solving the quadratic yields a value of 9.39×10^{-7}, which will equal the equilibrium hydronium ion concentration.

pH $= -\log [H_3O^+] = -\log 9.39 \times 10^{-7} = $ **6.03**

c)   ii)   To solve for the pH part-way to the equivalence point you need to take a two-step approach. First, the stoichiometry of the acid-base neutralization reaction must be calculated. In this case, the number of moles of diethylammonium ion produced must be calculated along with the number of moles of unreacted diethylamine. Their concentrations must be determined by dividing by the total volume of solution (25.00 mL + 15.00 mL = 40.00 mL or 0.04000 L in this case). The second step to solving this type of problem involves another calculation of equilibrium concentrations from initial concentrations ("ICE-box"). The initial concentrations used will be the final result from the stoichiometry calculation done above.

This reaction begins with 0.00375 mol of base (calculated in i)) to which is added 0.00150 mol of HBr (0.100 M × 0.0150 L = 0.00150 mol). They react in a 1:1 ratio, as noted, so all of the acid will be used up and 0.00150 mol of the base will be consumed as well. This will leave 0.00375 mol – 0.00150 mol or 0.00225 mol of base left over. At the same time, 0.00150 mol of diethylammonium ion will be produced. The concentrations of the reacting species in solution are calculated as shown (bromide is omitted because it is a spectator and the concentrations of hydroxide and hydronium will be determined with the "ICE-box").

$[(C_2H_5)_2NH] = 0.00225$ mol / 0.0400 L = 0.0563 M

$[(C_2H_5)_2NH_2{}^+] = 0.00150$ mol / 0.0400 L = 0.0375 M

These new "initial" concentrations will be used as follows:

	$(C_2H_5)_2NH_{(aq)}$ +	$H_2O_{(l)}$	$\leftrightarrow$	$OH^-{}_{(aq)}$ +	$(C_2H_5)_2NH_2{}^+{}_{(aq)}$
Initial	0.0563 M	NA		0 M	0.0375 M
Change	– x	NA		+ x	+ x
Equilibrium	0.0563 – x M	NA		x M	0.0375 + x M

$$K_b = \frac{[OH^-][(C_2H_5)_2NH_2{}^+]}{[(C_2H_5)_2NH]} = \frac{[x][0.0375+x]}{[0.0563-x]} = 6.81 \times 10^{-4}$$

$0.0375\, x + x^2 = 3.83 \times 10^{-5} + 6.81 \times 10^{-4}\, x$

$x^2 + 3.68 \times 10^{-2}\, x - 3.83 \times 10^{-5} = 0$

$x = [OH^-] = 1.01 \times 10^{-3}$ M

$pOH = -\log[OH^-] = -\log 1.01 \times 10^{-3} = 3.00$

$\mathbf{pH} = 14 - pOH = 14 - 3.00 = \mathbf{11.00}$

As we can see, the solution is still fairly basic. In this case you will also notice that the extent of base hydrolysis that occurs has diminished relative to the situation in which there was no diethylammonium present. This is just another example of the common ion effect.

        d)   Diethylamine, which is a liquid at room temperature, is a molecule and its solutions are therefore those of a molecule dissolved in water. The covalent bonds that hold together the individual atoms making up one of these molecules are very strong (on par with the strength of ionic bonds). However, the intermolecular attractions holding one diethylamine molecule to another or a diethylamine molecule to a water molecule are significantly weaker. This leads to the significant and measureable vapor pressure of small molecular species. Even at room temperature, well below the boiling point of diethylamine, many of the molecules in a sample of this substance will have sufficient energy to transfer from the liquid or aqueous phase into the gas phase. Once in the gas phase they can bind to the appropriate receptors in your nose or mouth, and be perceived by the sense of smell or taste.

Upon reaction with an acid, this base will form a salt. The diethylamine will receive a proton from the acid and therefore become positively charged diethylammonium. The acid (now in the form of a conjugate base), which lost a proton, is now negatively charged. These charges increase the intermolecular or interparticle forces holding every bit of the sample together. Essentially all ionic compounds are solids (most with negligible vapor pressures) at room temperature. The intermolecular forces holding ions to water molecules will similarly inhibit escape to the vapor phase from an aqueous solution. Since almost none of the substance will make it into the vapor phase, it cannot be perceived by the sense of smell.

CHAPTER 17
THE CHEMISTRY IN THE ATMOSPHERE

This chapter reviews the chemistry of the Earth's atmosphere, including:

- The composition of the atmosphere
- Depletion of the ozone in the stratosphere
- The greenhouse effect
- Acid rain
- Photochemical smog

> **Take Note**: *The Earth's atmosphere is not a required topic on the AP exam. You may find an occasional Multiple Choice question about the chemical basis of acid rain.*

The composition of the atmosphere

The composition of the Earth's atmosphere is 78.03% nitrogen, 20.99% oxygen, 0.033% carbon dioxide, and trace amounts of other gases. The chemical processes that occur in the atmosphere are quite complex. The atmosphere is divided into four layers in the following order from the surface of the Earth: the troposphere, stratosphere, mesosphere, and thermosphere (also called the ionosphere).

Depletion of the ozone in the stratosphere

Solar radiation that bombards the stratosphere promotes the photodissociation reaction of the oxygen molecules: $O_2 \rightarrow O + O$. These reactive O atoms combine with molecular oxygen, O_2, to form ozone, O_3, which helps to protect living organisms from UV radiation.

Chlorofluorcarbons (*CFCs*), *halogen substituted hydrocarbons*, were used as refrigerants and aerosol propellants in the past. They are relatively inert chemically and very volatile. CFCs slowly diffuse into the stratosphere where UV radiation promotes their decomposition. Reactive chlorine atoms are released. These very reactive chlorine atoms then attack the ozone molecules, decomposing them into diatomic oxygen molecules.

Nitric oxides also destroy stratospheric ozone. Their reactions are similar to the reactions of CFCs.

The greenhouse effect

The **greenhouse effect** describes *the trapping of heat near the Earth's surface by gases in the atmosphere, particularly carbon dioxide.* Methane, CFCs, and nitrogen oxides also contribute to the greenhouse effect. These gases help maintain the Earth's temperature. However, excessive amounts of these gases produced by human activity are contributing to global warming, which poses the threat of climatic change on the planet.

Acid rain

Nitrogen oxides and sulfur oxides are the two major components of acid rain. Nitrogen oxides are produced primarily by car exhaust. Sulfur oxides are produced from the burning of fossil fuels. These nonmetal oxides combine with water in the atmosphere to produce acids (see Chapter 4 of this review book):

$$2NO_2 \ (g) \quad + \quad H_2O \ (g) \quad \rightarrow \quad HNO_2 \ (aq) + H^+ \ (aq) + \quad NO_3^- \ (aq)$$

$$SO_3 \ (g) \quad + \quad H_2O \ (g) \quad \rightarrow \quad 2H^+ \ (aq) \quad + \quad SO_4^{2-} \ (aq)$$

Acid rain can corrode limestone and marble ($CaCO_3$), which are used as building materials and for sculptures. A typical reaction is:

$$CaCO_3 \ (s) \quad + \quad H_2SO_4 \ (aq) \quad \rightarrow \quad CaSO_4 \ (s) \quad + \quad H_2O \ (l) \quad + \quad CO_2 \ (g)$$

Photochemical smog

Photochemical smog is *formed by the reactions of automobile exhaust in the presence of sunlight.* The primary pollutants from automobile exhaust include NO, CO, and unburned hydrocarbons. These primary pollutants set in motion a complex set of photochemical reactions that produce secondary pollutants, such as NO_2 and O_3, that are responsible for the buildup of brown smog.

CHAPTER 18
ENTROPY, FREE ENERGY, AND EQUILIBRIUM

This chapter reviews the topics associated with thermodynamics, including

- The three laws of thermodynamics
- Qualitative assessment of spontaneous processes
- Calculating entropy changes for a reaction
- Calculating Gibbs free energy changes for a reaction
- Calculating free energy and the equilibrium constant

The three laws of thermodynamics

Thermodynamics, *the study of the interconversion of heat and other forms of of energy*, is based on three axioms:

First Law of Thermodynamics: *energy can be converted from one form to another, but cannot be created or destroyed.* The first law can be expressed in mathematical terms as $\Delta E = q + w$ (see Chapter 6 of this review book).

Second Law of Thermodynamics: *In any spontaneous process, there is an increase in the entropy of the universe.*

Third Law of Thermodynamics: *The entropy of a perfect crystalline substance is zero at absolute zero, 0 K.*

Take Note: *The interrelationships of enthalpy, H, entropy, S, and Gibbs free energy, G, are the primary concepts in thermodynamics. You must be able to apply these concepts both qualitatively and quantitatively on the AP exam.*

Qualitative assessment of spontaneous processes

A **spontaneous process** *occurs when reactants are brought together under a particular set of conditions and a reaction occurs.* The *two factors* that determine whether a reaction is spontaneous or not are the *changes in enthalpy and the changes in entropy.*

Enthalpy, *H*, is defined as $H = E + (PV)$ (see Chapter 6 of this review book). The change in enthalpy for a system is the heat gained or lost by the system at constant pressure, $\Delta H = \Delta E + \Delta(PV) = q_p$. *A decrease in enthalpy favors a spontaneous reaction.*

Entropy, *S*, is *a measure of the randomness or disorder of a system.* Solids are the most ordered state of matter and gases are the most disordered state of matter. *An increase in entropy favors a spontaneous reaction.*

If the enthalpy for a reaction decreases (ΔH is negative) and the entropy increases (ΔS is positive), the forward reaction will be spontaneous at all temperatures. If the enthalpy for a reaction increases (ΔH is positive) and the entropy decreases (ΔS is negative), the forward reaction is never spontaneous. If one factor favors spontaneity and the other does not, then the temperature of the reaction needs to be considered and Gibbs free energy, ΔG, must be calculated to determine whether the reaction is spontaneous. *A negative ΔG value indicates a spontaneous reaction.*

Gibbs free energy (*G*) (or simply free energy), is defined as:

$$\Delta G^0 = \Delta H^0 - T\Delta S^0 \text{ where}$$

ΔG^0 is Gibbs free energy (at standard conditions) in kJ /mol
ΔH^0 is enthalpy (at standard conditions) in kJ /mol
ΔS^0 is entropy (at standard conditions) in J /mol
T is temperature in Kelvin

Take Note: *When you perform thermodynamic calculations on the AP exam, be sure that your units for ΔG⁰ and ΔH⁰, which are in kilojoules (kJ/mol), agree with your units for ΔS⁰, which are usually given in joules (J/mol). The failure to make this conversion is a common error.*

The relationships between enthalpy, entropy, and free energy and the spontaneity of a reaction are summarized in the following table:

Gibbs Free Energy	Spontaneity of Reaction
$\Delta G^0 = \Delta H^0 - T\Delta S^0$ $(-)$ $(+)$	The sign of ΔG^0 is (–) and the reaction is spontaneous at all temperatures. The reaction is exothermic and the entropy is increasing. Both enthalpy and entropy favor spontaneity. The reaction is spontaneous at all temperatures.
$\Delta G^0 = \Delta H^0 - T\Delta S^0$ $(+)$ $(-)$	The sign of ΔG^0 is (+) and the reaction is not spontaneous in the forward direction. The reaction is endothermic and the entropy is decreasing. Neither enthalpy nor entropy favors spontaneity in the forward reaction.

$\Delta G^0 = \Delta H^0 - T\Delta S^0$ $\quad\quad\;\;(-)\quad\quad\;(-)$	The sign of ΔG^0 could be (+) or (–) depending on the temperature. This reaction is spontaneous at low temperatures where the unfavorable entropy term is not so significant.
$\Delta G^0 = \Delta H^0 - T\Delta S^0$ $\quad\quad\;\;(+)\quad\quad\;(+)$	The sign of ΔG^0 could be (+) or (–) depending on the temperature. This reaction is spontaneous at high temperatures where the favorable entropy term is significant.

Example 1. Qualitative evaluation of spontaneity of a reaction based on ΔH and ΔS.

Consider the reaction:

$$NH_4NO_3\,(s) \;\rightarrow\; NH_4^+\,(aq) \;+\; NO_3^-\,(aq) \quad\quad \Delta H^0 = 25\text{ kJ/mol}$$

(a) Predict the change in entropy when solid ammonium nitrate dissolves in water. Does ΔS favor the spontaneity of the reaction? Explain.
(b) Does the enthalpy factor favor the spontaneity of the reaction? Explain.

Solution to Example 1.
(a) *Entropy increases* as the solid dissolves and thus the *change in entropy favors spontaneity*. The solid state is more ordered than the liquid state and therefore $\Delta S > 0$.
(b) *ΔH does not favor spontaneity* of the reaction. A positive ΔH value means it is an endothermic reaction in the forward direction. A decrease in ΔH favors spontaneity.

Example 2. Qualitative evaluation of spontaneity of a reaction based on ΔH and ΔS.

Consider the the acid-base neutralization reaction:

$$H^+\,(aq) \;+\; OH^-\,(aq) \;\rightarrow\; H_2O\,(l) \quad\quad \Delta H^0 = -56.2\text{ kJ/mol}$$

(a) Predict the change in entropy for the reaction. Does ΔS favor the spontaneity of the reaction? Explain.
(b) Does the enthalpy change favor the spontaneity of the reaction? Explain.
(c) What is the driving force for this reaction?

Solution to Example 2.

(a) *ΔS is negative and does not favor spontaneity.* Two ions have more disorder than one molecule and thus $\Delta S < 0$.

(b) *ΔH favors spontaneity.* ΔH is negative, indicating that the forward reaction is exothermic and thus ΔH favors spontaneity.

(c) *ΔH is the driving force for this reaction.* Neutralization is a spontaneous reaction. Therefore, the ΔH change, which favors spontaneity, is dominant over the ΔS change, which does not.

Example 3. Predicting entropy changes.

Predict whether the entropy change is greater or less than zero for each of the following processes and explain:

(a) freezing hexane
(b) dissolving sodium chloride in water
(c) heating nitrogen from 10 °C to 80 °C
(d) N_2 (g) + 3H_2 (g) $\Leftrightarrow$ 2NH_3 (g)

Solution to Example 3.

(a) *ΔS decreases; $\Delta S < 0$.* Upon freezing, hexane molecules are held in a rigid arrangement that has less disorder than the liquid state.

(b) *ΔS increases; $\Delta S > 0$.* Sodium chloride dissolving in water allows the sodium and chloride ions to leave their fixed positions in the crystal lattice and enter the more random liquid state.

(c) *ΔS increases; $\Delta S > 0$.* The heating process increases molecular motions.

(d) *ΔS decreases; $\Delta S < 0$.* Four moles of gaseous reactants are converted to two moles of gaseous product. When the number of moles of gas change in a reaction, that change usually determines the sign of ΔS.

Calculating entropy changes for a reaction

Standard entropy, S^0, is the *absolute entropy of a substance at standard conditions, 1 atm and 25 °C* (see Table 18.1) and given in the units: $\dfrac{J}{mol \cdot K}$.

Standard Entropy Values ($S°$) for Some Substances at 25°C

TABLE 18.1

Substance	$S°$ (J/K · mol)
$H_2O(l)$	69.9
$H_2O(g)$	188.7
$Br_2(l)$	152.3
$Br_2(g)$	245.3
$I_2(s)$	116.7
$I_2(g)$	260.6
C(diamond)	2.4
C(graphite)	5.69
CH_4 (methane)	186.2
C_2H_6 (ethane)	229.5
He(g)	126.1
Ne(g)	146.2

Using standard entropy values for reactant and product molecules, the entropy change for a reaction is calculated using the equation:

$$\Delta S^0_{rxn} = \Sigma S^0_{(products)} - \Sigma S^0_{(reactants)} \quad \text{where}$$

ΔS^0_{rxn} is the entropy change for a reaction at standard conditions in J/mol·K

$\Sigma S^0_{(products)}$ is the sum of standard entropies of the product molecules in J/mol·K

$\Sigma S^0_{(products)}$ is the sum of the standard entropies of the reactant molecules in J/mol·K

Take Note: *AP exam questions involving entropy will supply the standard entropy values to you in the problem or will provide a means by which you can calculate them.*

Example 4. Calculating the entropy for a reaction.

Calculate the standard entropy change for the oxidation of sulfur dioxide at standard conditions:

$$2SO_2\,(g) \quad + \quad O_2\,(g) \quad \Leftrightarrow \quad 2SO_3\,(g)$$

given these data:

$$S^0_{SO_2} = 248 \ \frac{J}{mol \cdot K}$$

$$S^0_{O_2} = 205 \ \frac{J}{mol \cdot K}$$

$$S^0_{SO_3} = 256 \ \frac{J}{mol \cdot K}$$

General Strategy	Solution to Example 4
Write the appropriate equation for the data provided.	$\Delta S^0_{rxn} = \Sigma S^0_{(products)} - \Sigma S^0_{(reactants)}$
Substitute into the equation. Use the coefficients from the balanced equation. Be sure to be careful about the sign (+ or –) when doing the calculations.	$\Delta S^0_{rxn} = [2\,S^0_{SO_3}] - [S^0_{O_2} + 2\,S^0_{SO_2}]$ $= [2(256\,\frac{J}{mol \cdot K})] - [(205\,\frac{J}{mol \cdot K}) + 2(248\,\frac{J}{mol \cdot K})]$ $= -189\,\frac{J}{mol \cdot K}$ Note the negative sign. A decrease in entropy is expected since 3 mol of reactant gas form 2 mol of product gas.

Calculating Gibbs free energy changes for a reaction

The **standard Gibbs free energy change for a reaction** is *the free energy change for a reaction when it occurs under standard-state conditions, 25 °C, 1 atm pressure.* Using standard free energy values for reactant and product molecules, the free energy change for a reaction is calculated using the equation:

$$\Delta G^0_{rxn} = \Sigma G^0_{(products)} - \Sigma G^0_{(reactants)} \quad \text{where}$$

ΔG^0_{rxn} is the free energy change for a reaction at standard conditions in
    kJ/mol·K

$\Sigma G^0_{(products)}$ is the sum of standard free energies of the product molecules in
    kJ/mol·K

$\Sigma G^0_{(reactants)}$ is the sum of the standard free energies of the reactant molecules in
    kJ/mol·K

Note that the G^0_f values for elements at standard conditions are defined as 0.

The results of calculations for the free energy changes of a reaction yield three general conditions for ΔG_{rxn}, as summarized below:

Value ΔG_{rxn}	Implication for the Reaction
$\Delta G_{rxn} < 0$	The reaction is **spontaneous** in the forward direction.
$\Delta G_{rxn} > 0$	The reaction is **not spontaneous** in the forward direction.
$\Delta G_{rxn} = 0$	The reaction is **at equilibrium.**

When a *phase transition* occurs (such as the melting or boiling point), the system is at equilibrium and $\Delta G = 0$. Consequently,

$$\Delta G = \Delta H - T\Delta S$$
$$0 = \Delta H - T\Delta S$$

$$\text{for boiling}: \Delta H_{vap} = T_b\Delta S_{vap}$$
$$\text{for melting}: \Delta H_{fus} = T_m\Delta S_{fus}$$

Example 5. Calculating the Gibbs free energy for a reaction under standard conditions.

Calculate the standard free energy change for the reaction

$$C_3H_8\,(g) + 5O_2\,(g) \rightarrow 3CO_2(g) + 4H_2O\,(g)$$

given these data:

$$\Delta G^\circ_f\ C_3H_8 = -23\ \frac{kJ}{mol}$$

$$\Delta G^\circ_f\ CO_2 = -394.4\ \frac{kJ}{mol}$$

$$\Delta G^\circ_f\ H_2O = -237.2\ \frac{kJ}{mol}$$

General Strategy	Solution to Example 5
Write the appropriate equation for the data provided.	$\Delta G^0_{rxn} = \Sigma\Delta G^0_{(products)} - \Sigma\Delta G^0_{f\,(reactants)}$
Substitute into the equation. Use the coefficients from the balanced equation when substituting. Be sure to be careful about the sign (+ or −) when	$\Delta G^0_{rxn} = [3\Delta G^0_f(CO_2) + 4\Delta G^0_f(H_2O)] -$ $\qquad\qquad [1\Delta G^0_f(C_3H_6) + 5\Delta G^0_f(O_2)]$ $= [3(-394.4\ \frac{kJ}{mol}) + 4(-237.2\ \frac{kJ}{mol})] -$

doing the calculations.	$[(-23\frac{kJ}{mol}) + 5(0\frac{kJ}{mol})]$
	$= -2111 \text{ kJ}$
	$= -2100\ kJ$ (2 significant figures)
	The negative ΔG value indicates the reaction is spontaneous.

Example 6. Calculating ΔS_{vap}.

The molar heat of vaporization of water is 40.6 $\frac{kJ}{mol}$ at 100. °C and 1 atm.

Determine the entropy of vaporization of water at this temperature and pressure.

General Strategy	Solution to Example 6
Write the appropriate equation for the data provided.	$\Delta G = \Delta H - T\Delta S$ At a phase transition $\Delta G = 0$ and $0 = \Delta H - T\Delta S$ For boiling: $\Delta H_{vap} = T_b \Delta S_{vap}$ and $\Delta S_{vap} = \dfrac{\Delta H_{vap}}{T_b}$
Substitute into the equation: $\Delta S_{vap} = \dfrac{\Delta H_{vap}}{T_b}$ Convert temperature to Kelvin. Watch the energy units. Entropy is reported as $\dfrac{J}{mol \cdot K}$.	$40.6 \text{ kJ} \times \dfrac{1000 \text{ J}}{1 \text{ kJ}} = 40600 \text{ J/mol·K}$ $T_b = 100\ °C + 273 \text{ K} = 373 \text{ K}$ $\Delta S_{vap} = \dfrac{\Delta H_{vap}}{T_b} = \dfrac{40,600 J/mol}{373K} = 109\ \dfrac{J}{mol \cdot K}$ Note the positive sign of the entropy change. This is expected from a phase change of liquid to gas.

Calculating free energy and the equilibrium constant

During the course of a chemical reaction, not all reactants and products will necessarily be in their standard states. If they are not, ΔG at standard conditions can be calculated according to the following equation:

$$\Delta G = \Delta G^0 + RT\ln Q \quad \text{where}$$

ΔG is Gibbs free energy at nonstandard conditions
ΔG^0 is Gibbs free energy at standard conditions

$$R \text{ is } 8.31 \ \frac{J}{mol \cdot K}$$

T is the temperature in Kelvin

Q is the reaction quotient

When equilibrium is reached $Q = K$ and $\Delta G = 0$ and thus $\boldsymbol{\Delta G^0 = - RT\ln K}$.
Table 18.4 summarizes the ΔG^0 and K relationship:

TABLE 18.4

Relation Between ΔG° and K as Predicted by the Equation $\Delta G^\circ = -RT \ln K$

K	ln K	ΔG°	Comments
> 1	Positive	Negative	Products are favored over reactants at equilibrium.
= 1	0	0	Products and reactants are equally favored at equilibrium.
< 1	Negative	Positive	Reactants are favored over products at equilibrium.

Example 7. Calculating the equilibrium constant given ΔG^0_{rxn}.

Consider the following equilibrium reaction at standard conditions:

$$BaSO_4(s) \Leftrightarrow Ba^{2+}(aq) + SO_4^{2-}(aq)$$

The ΔG^0_{rxn} for this equilibrium is 56.8 kJ/mol. Determine the equilibrium constant for this reaction.

General Strategy	Solution to Example 7
Write the appropriate equation for the data provided.	$\Delta G^0_{rxn} = -RT\ln K$
Rewrite the equation in terms of the *desired* variable, substitute into the equation and solve.	$\Delta G^0_{rxn} = -RT \ln K$ $\ln K = -\dfrac{\Delta G^\circ_{rxn}}{RT}$
*Be sure your units agree. ΔG^0 is given in kJ/mol while the R value has J/mol·K units.	$^*\Delta G^0_{rxn} = 56.8 \text{ kJ} \times \dfrac{1000 \text{ J}}{1 \text{ kJ}} = 56{,}800 \text{ J}$

$$\ln K = - \frac{(56{,}800J)}{\left(8.31 \frac{J}{mol \cdot K}\right)(298K)}$$

$$K = 1.09 \times 10^{-10}$$

The positive value for ΔG^0_{rxn} indicates that this reaction is not spontaneous in the forward direction. It follows that the value of K is less than 1.

Take Note: *The Free Response section of the AP exam contains a list of formulas and many definitions and constants as reference material. This list, under the heading thermodynamics, contains many of the thermodynamic equations you will need. However, you will have to know how to apply each equation.*

SAMPLE MULTIPLE CHOICE QUESTIONS

Questions 1 – 3 refer to the following reaction and thermodynamic data.

$$PCl_5 \,(g) \quad \Leftrightarrow \quad PCl_3 \,(g) \quad + \quad Cl_2 \,(g) \qquad \Delta H^0 \;=\; 90 \text{ kJ}$$

$$S^0{}_{PCl_5} \quad = \quad 360 \;\frac{J}{mol \cdot K}$$

$$S^0{}_{PCl_3} \quad = \quad 310 \;\frac{J}{mol \cdot K}$$

$$S^0{}_{Cl_2} \quad = \quad 220 \;\frac{J}{mol \cdot K}$$

1. Calculate the ΔS^0 for the reaction.

   A. −50. J
   B. 0.0 J
   C. 170 J
   D. −440 J
   E. 530 J

2. Calculate the free energy change, ΔG^0, for the reaction.

   A. 39 kJ
   B. 51 kJ
   C. 150 kJ
   D. 130 kJ
   E. 80 kJ

3. Above what temperature would this reaction become spontaneous?

   A. 366 K
   B. 0 °C
   C. 598 °C
   D. 256 °C
   E. 700 K

4. At 25.0 °C, a 0.10 M solution of a weak base has a pH of 9. Which of the following expressions is correct for the calculation of the free energy for the ionization of this weak base?

   A. $-(8.31)(298)\ln 1 \times 10^{-11}$
   B. $-(0.0821)(298)\ln 1 \times 10^{-11}$
   C. $-(8.31)(298)\ln 1 \times 10^{-9}$
   D. $-(0.0821)(298)\ln 1 \times 10^{-9}$
   E. None of the above

5. A reaction is spontaneous only at high temperatures. Which of the following is most likely true for these conditions?

   A. $\Delta G < 0, \ \Delta H < 0, \ \Delta S < 0$
   B. $\Delta G > 0, \ \Delta H < 0, \ \Delta S > 0$
   C. $\Delta G < 0, \ \Delta H > 0, \ \Delta S < 0$
   D. $\Delta G > 0, \ \Delta H < 0, \ \Delta S > 0$
   E. $\Delta G < 0, \ \Delta H > 0, \ \Delta S > 0$

6. Which of the following correctly represents the relationship between free energy and the equilibrium constant for a reaction that proceeds spontaneously in the forward direction at 25 °C and 1 atm?

   A. $\Delta G > 0, \ K > 1$
   B. $\Delta G < 0, \ K > 1$
   C. $\Delta G > 0, \ K = 1$
   D. $\Delta G < 0, \ K = 1$
   E. $\Delta G < 0, \ K < 1$

7. Consider the following reactions. Which one(s) will most likely have a positive entropy value?

   I. $H_2 (g) \ + \ CuO (s) \ \rightarrow \ H_2O (g) \ + \ Cu (s)$

   II. $2C_2H_6 (g) \ + \ 7O_2 (g) \ \rightarrow \ 4CO_2 (g) \ + \ 6H_2O (g)$

   III. $COCl_2 (g) \ \rightarrow \ CO (g) \ + \ Cl_2 (g)$

   A. I only
   B. II only
   C. III only
   D. II and III
   E. I and II

8.   A 10.0-g sample of naphthalene is placed in a beaker. Heat is applied and the naphthalene melts at 80.1 °C. Which of the following is true at that temperature?

   A. $\Delta G < 0$, $\Delta H > 0$, $\Delta S > 0$
   B. $\Delta G = 0$, $\Delta H > 0$, $\Delta S > 0$
   C. $\Delta G > 0$, $\Delta H > 0$, $\Delta S > 0$
   D. $\Delta G < 0$, $\Delta H < 0$, $\Delta S > 0$
   E. $\Delta G = 0$, $\Delta H < 0$, $\Delta S > 0$

9.   An ideal gas occupies a volume of 10.0 L. It is allowed to expand to 14.0 L against a constant external pressure of 7.0 atm, while the temperature remains constant. Which of the following statements correctly describes the system?

   A.   The heat, q, equals zero.
   B.   The change in the internal energy, ΔE, of the system is zero.
   C.   The work done by the system is +28.0 L·atm.
   D.   The work done on the system is +98 L·atm.
   E.   All of the statements above are false.

10.   One mole of each of NH_3 (s), NH_3 (l), and NH_3 (g) are all at the same temperature. Which of the following statements is correct?

   A. NH_3 (s) has the highest entropy.
   B. NH_3 (l) has the highest entropy.
   C. NH_3 (g) has the highest entropy.
   D. NH_3 (s) and NH_3 (l) have the same entropy.
   E.   They all have the same entropy since the temperature is the same.

Comprehension Questions

1)   Consider the following spontaneous reaction occurring at 298 K:

$$2C_2H_2 \ (g) + 5O_2 \ (g) \rightarrow 4CO_2 \ (g) + 2H_2O \ (l)$$

   a)   Calculate ΔH^0_{rxn} for the above process given the following data:

Species	ΔH^0_f kJ / mol
C_2H_2 (g)	226.6
CO_2 (g)	–393.5
$H_2O(l)$	–285.8

b)   Predict the sign of ΔS^0_{rxn} and justify your answer using thermodynamic principles.

c)   What effect would lowering the temperature have on ΔG^0_{rxn}? Would there ever be a temperature at which the process would become nonspontaneous? Explain.

d)   Which quantity, ΔH^0_{rxn} or ΔS^0_{rxn}, promotes spontaneity in this reaction? Explain.

2)   a)   Calculate ΔG^0_{rxn} for the following process at 25.0 °C using the data provided:

$$2SO_2 (g) + O_2 (g) \rightarrow 2\,SO_3 (g)$$

Species	ΔG^0_f (kJ/mol at 298 K)
$SO_2 (g)$	−300.13
$SO_3 (g)$	−371.04

b) ΔH^0_{rxn} for the above process is − 197.9 kJ at 25.0 °C. Calculate the standard entropy change associated with this reaction.

c)   Temperature will have an effect on the spontaneity of this reaction. Calculate the temperature at which the reaction will become nonspontaneous in the forward direction.

3)   Consider the following reaction:

$$2NO (g) + O_2 (g) \rightarrow 2NO_2 (g) \qquad \Delta S^0 = -146.5 \text{ J/K at } 25.0 \text{ °C}$$

a)   Does the sign of ΔS^0 make sense for the above reaction? Give a justification based on thermodynamic principles.

b)   i)   Calculate ΔG^0_{rxn} for the above reaction at 25.0 °C given the following enthaply data collected at 25.0 °C:

$$\tfrac{1}{2}N_2 (g) + O_2 (g) \rightarrow NO_2 (g) \qquad \Delta H^0 = 33.1 \text{ kJ/mol}$$

$$\frac{1}{2}N_2\ (g) + \frac{1}{2}O_2\ (g) \rightarrow NO\ (g) \qquad \Delta H^0 = 90.29\ kJ/mol$$

      ii)   Is this reaction spontaneous under standard conditions? Explain.

c)   Calculate K_{eq} for the reaction under consideration at 25.0 °C.

d)   Suppose the reaction were set up at a temperature of 300 Kelvin and with the following initial gas pressures: $P_{NO} = 0.254$ atm, $P_{O_2} = 0.871$ atm, and $P_{NO_2} = 1.20$ atm. What would the value of ΔG be under these conditions?

ANSWERS TO SAMPLE MULTIPLE CHOICE QUESTIONS

1.   C

The standard absolute entropy values for each substance are provided. Therefore,

$$\Delta S^0_{rxn} = \Sigma S^0_{(products)} - \Sigma S^0_{(reactants)}$$

$$= [S^0_{PCl_3} + S^0_{Cl_2}] - [S^0_{PCl_5}]$$

$$= [1\ mol\ (310\frac{J}{mol \cdot K}) + 1\ mol\ (220\frac{J}{mol \cdot K})] - [1\ mol\ (360\frac{J}{mol \cdot K})]$$

$$= 170\ \frac{J}{K}$$

2.   A

Since the enthalpy change (ΔH^0) is provided and the entropy change has been calculated in Question 1, the free energy can be calculated using the Gibbs free energy equation:

$$\Delta G^0 = \Delta H^0 - T\Delta S^0$$

$$= (90\ kJ) - (298\ K)\ (0.170\ \frac{kJ}{K})$$

$$= 39\ kJ$$

3. D

At equilibrium neither the forward nor the reverse reaction is favored and $\Delta G = 0$. Therefore,

$$\Delta G^0 = \Delta H^0 - T\Delta S^0$$

$$0 = \Delta H^0 - T\Delta S^0 \quad \text{or} \quad \Delta H^0 = T\Delta S^0$$

$$90 \text{ kJ} = T(0.170 \text{ kJ})$$

$$T = 529 \text{ K} \quad (\text{or since } K = C + 273, \ 529 \text{ K} - 273 = 256 \text{ °C})$$

The reaction is nonspontaneous ($\Delta G^0 > 0$) at standard conditions. The entropy value favors spontaneity ($\Delta S^0 > 0$). If the temperature is above 529 K (256 °C) the reaction becomes spontaneous.

4. C

$$\Delta G^0 = -RT \ln K$$

This is a weak base. Since pH = 9, then the pOH = 5 and $[OH^-] = 1 \times 10^{-5}$ M

	B	+	H₂O	⇔	BH⁺	+	OH⁻
Initial	0.100				0		0
Change	$-x$		×		$+x$		$+x$
Equilibrium	0.100 $-$ ~~10⁻⁵~~				1×10^{-5}		1×10^{-5}

Therefore,

$$= -(8.31 \frac{J}{mol \cdot K})(298 \text{ K}) \ln \frac{[BH^+][OH^-]}{[B]}$$

$$= -(8.31)(298) \ln \frac{[1 \times 10^{-5}][1 \times 10^{-5}]}{0.100}$$

$$= -(8.31)(298) \ln 1 \times 10^{-9}$$

5. E

$$\Delta G^0 = \Delta H^0 - T\Delta S^0$$

ΔG^0 is negative because the reaction is spontaneous: $\Delta G^0 < 0$.
Since the reaction is spontaneous only at high temperatures, ΔH^0 is positive, $\Delta H^0 > 0$. ΔS^0 must be positive ($\Delta S^0 > 0$), but at high temperatures,

$-T \Delta S^0$ or $-(+)$ yields a negative quantity, resulting in a $\Delta G^0 < 0$.

6.   B

For a reaction to proceed spontaneously, ΔG^0 must be negative, $\Delta G^0 < 0$. Since the reaction proceeds spontaneously in the forward direction, product formation is favored and $K > 1$.

7.   D

A positive entropy value indicates that there has been an increase in the randomness in the system. In both II and III there are a greater number of moles of gaseous product than were present in the reactants.

8.   A

In order to melt, a substance needs heat. Therefore, $\Delta H^0 > 0$. When the solid melts, the liquid state results which, has a higher degree of randomness. Therefore, $\Delta S^0 > 0$. Since the process *occurs* as heat is added, then $\Delta G^0 < 0$.

9.   B

At constant temperature there is no change in internal energy of the gas, $\Delta E = 0$.

10.   C

The gaseous state has the highest disorder or randomness.

Answers to Comprehension Questions

1)   a)   Given the standard heats of formation for the reactants and products, you need only plug values into this equation to calculate ΔH^0_{rxn}.

$$\Delta H^0_{rxn} = \Sigma \Delta H^0_{f(products)} - \Sigma \Delta H^0_{f(reactants)}$$

ΔH^0_{rxn} = [(2 mol H_2O × –285.8 kJ/mol) + (4 mol CO_2 × –393.5 kJ/mol)] – [(2 mol C_2H_2 × 226.6 kJ/mol) + (5 mol O_2 × 0 kJ/mol)] = **–2599 kJ**

Note that the standard heat of formation of oxygen, and any element in its standard state, is always zero.

b) The sign for ΔS^0_{rxn} should be negative in this case. Entropy is disorder, so a negative ΔS indicates that disorder is decreasing or order is increasing. As a general rule, gases are more disordered than liquids, and liquids more disordered than solids. And, 2 mol of a gas has more disorder than 1 mol. In this case, 7 mol of gaseous reactants are transformed into 4 mol of gaseous products and 2 mol of liquid products. So, we have an overall decrease in the number of moles and a decrease in the number of moles of gas. This should lead to an overall decrease in disorder or an overall increase in order (whichever you prefer).

c) The relationship between ΔG^0, ΔH^0, and ΔS^0 is:

$$\Delta G^0 = \Delta H^0 - T\Delta S^0 \quad \text{(where } T \text{ is the absolute temperature)}$$

For this reaction, both ΔH^0 and ΔS^0 are negative. This means that spontaneity of the reaction will be temperature dependent. As the temperature increases, the $T\Delta S^0$ term will become an increasingly large negative number. But, since this value is being subtracted from the total, it will lead to higher values for ΔG^0. ΔG^0 will eventually become positive, and the forward reaction will no longer be spontaneous (i.e., the reaction will be spontaneous in the reverse direction).

d) In this case ΔH^0, which is negative, exothermic, is the driving force behind spontaneity. Exothermic processes that increase in disorder tend to favor spontaneity. In this case, the reaction becomes more ordered when proceeding from reactants to products.

2) a) ΔG^0 may be calculated in much the same way that ΔH^0 was in 1a).

$\Delta G^0_{rxn} = \Sigma \Delta G^0_{f\,(products)} - \Sigma \Delta G^0_{f\,(reactants)}$

ΔG^0_{rxn} = [(2 mol SO_3 × –371.04 kJ/mol)] – [(2 mol SO_2 × –300.13 kJ/mol) + (1 mol O_2 × 0 kJ/mol)] = **–141.8 kJ**

  b) Using the relationship $\Delta G^0 = \Delta H^0 - T\Delta S^0$ and the data provided, you can easily substitute and calculate the desired quantity.

b)   Using the relationship $\Delta G^0 = \Delta H^0 - T\Delta S^0$ and the data provided, you can easily substitute and calculate the desired quantity.

$$\Delta S^0 = (\Delta H^0 - \Delta G^0) / T$$

$$\Delta S^0 = (-197.9 \text{ kJ} - (-141.8 \text{ kJ})) / 298.15 \text{ K}$$

$$\Delta S^0 = -1.88 \times 10^{-1} \text{ kJ/K} = -188.2 \text{ J/K}$$

c)   Using the same equation as in b), set $\Delta G^0 = 0$ and solve for T. Above this temperature the reaction will be spontaneous in the reverse direction.

$$\Delta G^0 = \Delta H^0 - T\Delta S^0 = 0$$

$$\Delta H^0 = T\Delta S^0$$

$$T = \Delta H^0 / \Delta S^0 = -197.9 \times 10^3 \text{ J} / -188 \text{ JK}^{-1} = \textbf{1053 K}$$

At temperatures slightly above 1053 Kelvin, the reverse reaction should become spontaneous.

3)   a)   In this reaction, 3 mol of gaseous reactants become 2 mol of gaseous products. This corresponds with a decrease in disorder, and that would be represented by a negative ΔS^0. The magnitude of the number also makes sense as entropy changes often can be expressed in joules, as opposed to kilojoules.

b)   i) ΔH^0 will be calculated as in 1a). The reactions given are formation reactions for the nonelemental reactant and product. Once ΔH^0 is calculated, ΔG^0 can be calculated using the relationship $\Delta G^0 = \Delta H^0 - T\Delta S^0$.

$$\Delta H^0_{rxn} = \Sigma \Delta H^{f0}_{(products)} - \Sigma \Delta H^0_{f\,(reactants)}$$

$\Delta H^0_{rxn} = [(2 \text{ mol NO}_2 \times 33.1 \text{ kJ/mol})] - [(2 \text{ mol NO} \times 90.29 \text{ kJ/mol}) + (1 \text{ mol O}_2 \times 0 \text{ kJ/mol})] = -114 \text{ kJ}$

$\Delta G^0 = \Delta H^0 - T\Delta S^0 = -114 \text{ kJ} - (298.15 \text{ K} \times (-146.5 \times 10^{-3} \text{ kJ/K})) = \textbf{-70.3 kJ}$

ii)   The value of ΔG^0 is negative, indicating a spontaneous reaction in the forward direction.

c) K_{eq} can be calculated using the relationship $\Delta G^0 = -RT\ln K$. It's important to remember to express both ΔG^0 and the gas constant, R, in terms of joules.

$$\ln K = - \Delta G^0 / (R \times T)$$

$$K = e^{-[\Delta G_0 / (R \times T)]} = e^{-[(-70300\,\text{J}) / (8.314\,\text{J·mol}^{-1}\text{·K}^{-1} \times 298.2\,\text{K})]} = e^{+28.4} = \mathbf{2.07 \times 10^{12}}$$

d) In this situation we need to use the following relationship that includes ΔG^0 and a term to modify the free energy change by a value related to the absolute temperature and the reaction quotient, Q. Remember that the expression defining Q is the same as that for K_{eq} except that nonequilibrium values will be plugged into Q.

$$\Delta G = \Delta G^0 + RT\ln Q$$

Values for reactants and products in the gas phase should be expressed in terms of partial pressures instead of concentrations when calculating Q.

$$Q = \frac{(P_{NO_2})^2}{(P_{NO^2})(P_{O_2})} = \frac{(1.20)^2}{(0.254)^2(0.871)} = 25.6$$

$\Delta \boldsymbol{G} = -70300\,\text{J} + (8.314\,\text{J·mol}^{-1}\text{·K}^{-1} \times 300\,\text{K} \times \ln 25.6) = -114000\,\text{J} + 8090\,\text{J} = -62200\,\text{J} = \mathbf{-62.2\ kJ}$

CHAPTER 19
ELECTROCHEMISTRY

This chapter reviews **electrochemistry**, *the branch of chemistry that deals with the interconversion of chemical energy and electrical energy*, which is of tremendous practical significance. Important topics associated with electrochemistry include:

- Oxidation-reduction (redox) reactions
- Galvanic cells
- Standard reduction potentials and electromotive force
- The spontaneity of redox reactions
- The Nernst equation
- Electrolysis
- Quantitative aspects of electrolysis

The concepts in this chapter connect many of the topics you have studied previously, especially equilibrium and thermodynamics, to practical applications in chemistry. As you review this material, it is important to pay particular attention to these connections.

> **Take Note:** *The questions in the Free Response section of the AP exam will frequently require you to integrate multiple concepts. In this chapter, you will see examples of the type and depth of content synthesis that is required on the AP exam.*

Oxidation-reduction (redox) reactions

Redox reactions involve the transfer of electrons from one substance to another. In **oxidation** *an element loses electrons, which results in an increase in the element's oxidation number*. In **reduction** *an element gains electrons, which results in a decrease in the element's oxidation number*. When you check the oxidation numbers in the following single replacement reaction, you will see that zinc is oxidized and copper is reduced (see Chapter 4 of this review book).

$$\underset{Zn\,(s)}{\overset{0}{}} \; + \; \underset{Cu(NO_3)_2\,(aq)}{\overset{+2 \;\; +5\,-2}{}} \; \rightarrow \; \underset{Zn(NO_3)_2\,(aq)}{\overset{+2 \;\; +5\,-2}{}} \; + \; \underset{Cu\,(s)}{\overset{0}{}} \quad \text{(oxidation numbers)}$$

Many redox reactions are more complex than the simple example above. They often involve oxoanions such as chromate, CrO_4^{2-}, permanganate, MnO_4^-, and sulfate, SO_4^{2-}, in acidic or basic media. To *balance complex redox reactions*, you need to follow six simple steps:

1. Write the unbalanced equation for the reaction in ionic form.
2. Separate the oxidation and reduction half-reactions.

3. Balance <u>each</u> half-reaction first to conserve atoms and second to conserve charge by adding electrons. In an acidic medium, add H_2O to balance O atoms and then H^+ to balance H atoms.
4. Multiply one or both half-reactions to equalize the number of electrons, that is, the number of electrons lost in oxidation must equal the number of electrons gained in reduction. Now add the two half-reactions together so that the electrons on both sides cancel, and obtain the final equation.
5. Inspect the equation. If a species such as H^+ or H_2O is common to both sides of the equation, complete the balancing by subtracting the one with the smaller coefficient from both sides of the equation.
6. Check the final equation to confirm that atoms and charge are conserved.

Example 1. Balancing a redox reaction.

Balance the following redox reaction that occurs in an acidic solution:

$$Cr_2O_7^{2-} + C_2O_4^{2-} \rightarrow Cr^{3+} + CO_2$$

General Strategy	Solution to Example 1
Write the unbalanced ionic equation.	$Cr_2O_7^{2-} + C_2O_4^{2-} \rightarrow Cr^{3+} + CO_2$
Separate the oxidation and reduction half-reactions.	$\overset{+3\ -2}{C_2O_4^{2-}} \rightarrow \overset{+4\ -2}{CO_2} \qquad \text{oxidation}$ $\overset{+6\ -2}{Cr_2O_7^{2-}} \rightarrow \overset{+3}{Cr^{3+}} \qquad \text{reduction}$
Balance each half-reaction for atoms and charge.	$C_2O_4^{2-} \rightarrow 2CO_2 + 2e^-$ $6e^- + 14H^+ + Cr_2O_7^{2-} \rightarrow 2Cr^{3+} + 7H_2O$
Multiply the half-reactions to equalize the electrons. Then <u>add</u> the two half-reactions to obtain the net equation.	$[C_2O_4^{2-} \rightarrow 2CO_2 + 2e^-]3$ $\underline{6e^- + 14H^+ + Cr_2O_7^{2-} \rightarrow 2Cr^{3+} + 7H_2O}$ $3C_2O_4^{2-} + 14H^+ + Cr_2O_7^{2-} \rightarrow 6CO_2 + 2Cr^{3+} + 7H_2O$

To balance reactions that occur in a basic medium:

1. Balance the equation as if it were in an acidic medium first.

2. Then add enough OH⁻ ions to equal the number of H⁺ ions. Add this number of OH⁻ ions to both sides of the equation. H⁺ and OH⁻ make water.
3. Inspect the equation. If a species such as H⁺ or H_2O is common to both sides of the equation, complete the balancing by subtracting the one with the smaller coefficient from both sides of the equation.
4. Check the final equation to confirm that atoms and charge are conserved.

Galvanic cells

A spontaneous redox reaction can be used to generate electricity. The apparatus in which the reaction is carried out is called a **galvanic or voltaic** cell (see Figure 19.1). There are *two major compartments, each referred to as a* **half-cell**. The oxidation reaction takes place in one compartment and the reduction in the other.

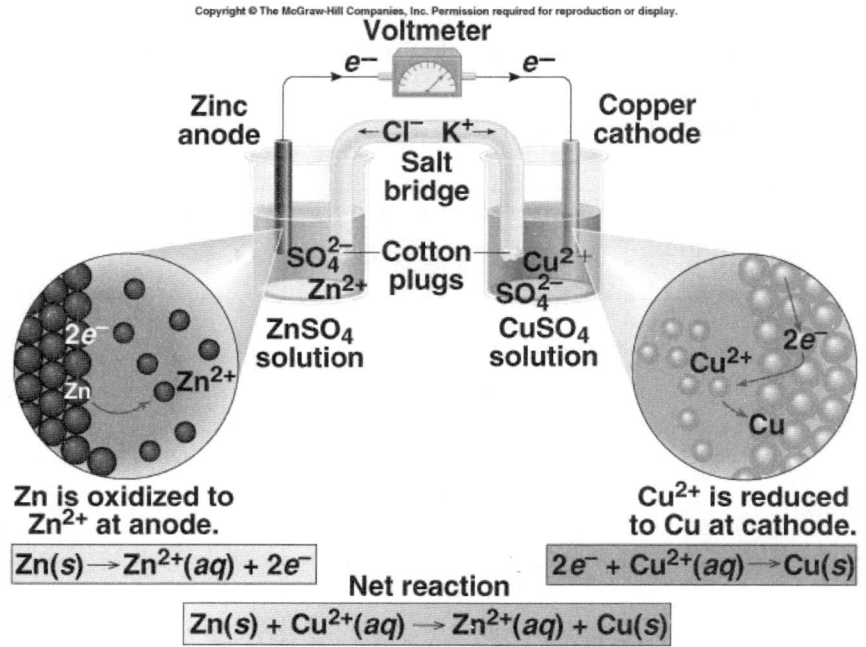

Figure 19.1

Oxidation, *the loss of electrons,* occurs in one compartment at the *electrode* called the **anode,** whose charge is *negative*. **Reduction,** *the gain of electrons,* occurs in the other compartment at the *electrode* called the **cathode,** whose charge is *positive*. The **salt bridge** completes the circuit by allowing ions to migrate and maintain electrical neutrality in each half-cell. Cations migrate to the cathode and anions migrate to the anode.

In a galvanic cell, electrons flow from the anode to the cathode because of the difference in electrical potential between the electrodes. The *difference in electrical potential between the anode and the cathode* is measured by a voltmeter in **volts** and is called **cell voltage.** Two other terms, the **electromotive force** (emf, *E*) and **cell potential**, are also used to *denote the cell voltage.*

Standard reduction potentials and electromotive force

Standard reduction potentials, E^0, are based on the reduction potential of hydrogen being assigned a value of 0.00 V. These **standard reduction potentials** given in Table 19.1 *are associated with reduction half-reactions under standard conditions, 1 atm pressure, 25 °C, and 1.00 M solutions.* Remember that the more positive the reduction potential, the more likely the reduction is to occur. As the reduction potential becomes negative, the substance is less likely to be reduced.

The standard reduction potentials can be used to determine the **cell potential** or *voltage* of an electrochemical cell. For example, consider the cell made by combining a zinc half-cell with a silver half-cell. The reduction potential for zinc, $Zn^{2+} + 2e^- \rightarrow Zn$, is –0.74 V. The reduction potential for silver, $Ag^+ + 1e^- \rightarrow Ag$, is +0.80 V. Therefore, silver will be reduced and zinc will be oxidized. The cell potential can be calculated:

$$
\begin{array}{lll}
Zn \rightarrow Zn^{2+} + 2e^- & +0.74\ V & \text{oxidation potential} \\
\underline{(Ag^+ + 1e^- \rightarrow Ag)2} & \underline{+0.80\ V} & \underline{\text{reduction potential}} \\
Zn + 2Ag^+ \rightarrow Zn^{2+} + 2Ag & +1.54\ V & \text{cell potential, } E^0_{cell}
\end{array}
$$

or

$$E^0_{cell} = E^0_{oxidation} + E^0_{reduction}$$

It is important to note that:

1. The half-cells are reversible. *When the half-cell is reversed, the sign of the potential changes.*
2. Changing the stoichiometric coefficients to balance the half-cell <u>DOES NOT</u> affect the E^0 because electrode potentials are intensive properties. Remember that intensive properties depend on the nature of the substance, not the amount.

TABLE 19.1

Standard Reduction Potentials at 25 °C*

Half-Reaction	E^0 (V)
$F_2(g) + 2e^- \longrightarrow 2F^-(aq)$	+2.87
$O_3(g) + 2H^+(aq) + 2e^- \longrightarrow O_2(g) + H_2O$	+2.07
$Co^{3+}(aq) + e^- \longrightarrow Co^{2+}(aq)$	+1.82
$H_2O_2(aq) + 2H^+(aq) + 2e^- \longrightarrow 2H_2O$	+1.77
$PbO_2(s) + 4H^+(aq) + SO_4^{2-}(aq) + 2e^- \longrightarrow PbSO_4(s) + 2H_2O$	+1.70
$Ce^{4+}(aq) + e^- \longrightarrow Ce^{3+}(aq)$	+1.61
$MnO_4^-(aq) + 8H^+(aq) + 5e^- \longrightarrow Mn^{2+}(aq) + 4H_2O$	+1.51
$Au^{3+}(aq) + 3e^- \longrightarrow Au(s)$	+1.50
$Cl_2(g) + 2e^- \longrightarrow 2Cl^-(aq)$	+1.36
$Cr_2O_7^{2-}(aq) + 14H^+(aq) + 6e^- \longrightarrow 2Cr^{3+}(aq) + 7H_2O$	+1.33
$MnO_2(s) + 4H^+(aq) + 2e^- \longrightarrow Mn^{2+}(aq) + 2H_2O$	+1.23
$O_2(g) + 4H^+(aq) + 4e^- \longrightarrow 2H_2O$	+1.23
$Br_2(l) + 2e^- \longrightarrow 2Br^-(aq)$	+1.07
$NO_3^-(aq) + 4H^+(aq) + 3e^- \longrightarrow NO(g) + 2H_2O$	+0.96
$2Hg^{2+}(aq) + 2e^- \longrightarrow Hg_2^{2+}(aq)$	+0.92
$Hg_2^{2+}(aq) + 2e^- \longrightarrow 2Hg(l)$	+0.85
$Ag^+(aq) + e^- \longrightarrow Ag(s)$	+0.80
$Fe^{3+}(aq) + e^- \longrightarrow Fe^{2+}(aq)$	+0.77
$O_2(g) + 2H^+(aq) + 2e^- \longrightarrow H_2O_2(aq)$	+0.68
$MnO_4^-(aq) + 2H_2O + 3e^- \longrightarrow MnO_2(s) + 4OH^-(aq)$	+0.59
$I_2(s) + 2e^- \longrightarrow 2I^-(aq)$	+0.53
$O_2(g) + 2H_2O + 4e^- \longrightarrow 4OH^-(aq)$	+0.40
$Cu^{2+}(aq) + 2e^- \longrightarrow Cu(s)$	+0.34
$AgCl(s) + e^- \longrightarrow Ag(s) + Cl^-(aq)$	+0.22
$SO_4^{2-}(aq) + 4H^+(aq) + 2e^- \longrightarrow SO_2(g) + 2H_2O$	+0.20
$Cu^{2+}(aq) + e^- \longrightarrow Cu^+(aq)$	+0.15
$Sn^{4+}(aq) + 2e^- \longrightarrow Sn^{2+}(aq)$	+0.13
$2H^+(aq) + 2e^- \longrightarrow H_2(g)$	0.00
$Pb^{2+}(aq) + 2e^- \longrightarrow Pb(s)$	−0.13
$Sn^{2+}(aq) + 2e^- \longrightarrow Sn(s)$	−0.14
$Ni^{2+}(aq) + 2e^- \longrightarrow Ni(s)$	−0.25
$Co^{2+}(aq) + 2e^- \longrightarrow Co(s)$	−0.28
$PbSO_4(s) + 2e^- \longrightarrow Pb(s) + SO_4^{2-}(aq)$	−0.31
$Cd^{2+}(aq) + 2e^- \longrightarrow Cd(s)$	−0.40
$Fe^{2+}(aq) + 2e^- \longrightarrow Fe(s)$	−0.44
$Cr^{3+}(aq) + 3e^- \longrightarrow Cr(s)$	−0.74
$Zn^{2+}(aq) + 2e^- \longrightarrow Zn(s)$	−0.76
$2H_2O + 2e^- \longrightarrow H_2(g) + 2OH^-(aq)$	−0.83
$Mn^{2+}(aq) + 2e^- \longrightarrow Mn(s)$	−1.18
$Al^{3+}(aq) + 3e^- \longrightarrow Al(s)$	−1.66
$Be^{2+}(aq) + 2e^- \longrightarrow Be(s)$	−1.85
$Mg^{2+}(aq) + 2e^- \longrightarrow Mg(s)$	−2.37
$Na^+(aq) + e^- \longrightarrow Na(s)$	−2.71
$Ca^{2+}(aq) + 2e^- \longrightarrow Ca(s)$	−2.87
$Sr^{2+}(aq) + 2e^- \longrightarrow Sr(s)$	−2.89
$Ba^{2+}(aq) + 2e^- \longrightarrow Ba(s)$	−2.90
$K^+(aq) + e^- \longrightarrow K(s)$	−2.93
$Li^+(aq) + e^- \longrightarrow Li(s)$	−3.05

Increasing strength as oxidizing agent

Increasing strength as reducing agent

*For all half-reactions the concentration is 1 M for dissolved species and the pressure is 1 atm for gases. These are the standard-state values.

The spontaneity of redox reactions

Since galvanic cells produce spontaneous redox reactions, the cell potential can be related to the thermodynamic quantities you learned in Chapter 18. The relationship between Gibbs free energy, ΔG^0, and voltage, E^0_{cell}, is given by the equation:

$$\Delta G^0 = -nFE^0_{cell} \quad \text{where}$$

n is the number moles of electrons exchanged in the balanced equation

F is one Faraday, 96,500 coulombs(C)/mol of electrons

E^0_{cell} is the cell voltage in volts (J/C)

Example 2. Calculating the free energy, ΔG^0, from the emf of an electrochemical cell.

Calculate the standard free energy for the zinc-silver galvanic cell.

General Strategy	Solution to Example 2
Arrange the half-reactions to obtain the most positive E^0_{cell}.	$\begin{array}{lll} Zn & \rightarrow Zn^{2+} + 2e^- & +0.74 \text{ V} \\ (Ag^+ + 1e^- \rightarrow Ag)2 & & +0.80 \text{ V} \\ Zn + 2Ag^+ \rightarrow Zn^{2+} + 2Ag & & +1.54 \text{ V} \quad E^0_{cell} \end{array}$
Balance the redox reaction to determine <u>n.</u> Remember, balancing does not change the E^0_{cell}.	$n = 2$ mol of electrons $\qquad E^0_{cell} = 1.54$ V
Calculate the free energy using the formula: $\Delta G^0 = -nFE^0_{cell}$ Be sure to watch your signs and units.	$\Delta G^0 = -nFE^0_{cell}$ $= -(2 \text{ mol}) (96,500 \dfrac{C}{mol \; electrons})(1.54 \dfrac{J}{C})$ $= -297220$ J or -297 kJ

You know from Chapter 18 of this review book that free energy, ΔG^0, is related to the equilibrium constant, K, by the expression:

$$\Delta G^0 = -RT\ln K \quad \text{where}$$

$$R \text{ is } 8.31 \; \dfrac{J}{mol \cdot K} \text{ and}$$

T is the temperature in Kelvin

Example 3. Calculation of the equilibrium constant based on emf data.

Calculate the equilibrium constant for the zinc-silver cell in Example 2.

General Strategy	Solution to Example 3
The Standard Reduction Potentials Table has already been used to calculate the free energy for the zinc-silver cell in Example 2.	$\Delta G^0 = -297$ kJ
Calculate the equilibrium constant for the cell using $\Delta G^0 = -RT\ln K$. Watch your units and signs.	$\Delta G^0 = -RT\ln K$ $-297 \text{ kJ} = -(8.31 \dfrac{J}{mol \cdot K})(298 \text{ K}) \ln K$ $-297,000 \text{ J} = -(8.31 \dfrac{J}{mol \cdot K})(298 \text{ K}) \ln K$ $\dfrac{297,000\, J}{(8.31)(298)} = \ln K$ $119 = \ln K$ $K = 4.79 \times 10^{51}$ Notice that K is large and this reaction proceeds far to the right. This confirms that this reaction is spontaneous in the forward direction.

The Nernst equation

As a voltaic cell runs, the reactants are converted to products. As the concentrations of reactants and products change, the system is no longer at standard conditions and the emf changes. The Nernst equation allows you to calculate the voltage, E_{cell}, as the cell composition changes.

$$E_{cell} = E_{cell}^0 - \frac{0.0592\,V}{n} \log Q \qquad \text{where}$$

E_{cell}^0 is the standard cell potential

n is moles of electrons in the balanced equation

Q is the reaction quotient

Example 4. Calculating the voltage of a cell not at standard conditions.

Determine the measured emf for the zinc-silver cell in Example 2 if the concentration of the zinc in the zinc half-cell is 1.10 M and the concentration of the silver in the silver half-cell is 0.90 M.

General Strategy	Solution to Example 4
Write and balance the reaction.	$Zn\,(s)\; +\; 2Ag^+\,(aq) \to\; Zn^{2+}\,(aq)\; +\; 2Ag\,(s)$
Write the reaction quotient. Remember that solids are not included in the reaction quotient.	$Q = \dfrac{\left[Zn^{2+}\right]}{\left[Ag^+\right]^2}$
Use the Nernst equation to determine the measured voltage. $E_{cell} = E_{cell}^0 - \dfrac{0.0592V}{n}\log Q$ Remember that <u>n</u> is the total number of moles of electrons exchanged in the balanced redox reaction.	$E_{cell} = E_{cell}^0 - \dfrac{0.0592V}{n}\log Q$ $= 1.54\text{ V} - \dfrac{0.0592V}{2}\log\dfrac{\left[Zn^{2+}\right]}{\left[Ag^+\right]^2}$ $= 1.54\text{ V} - 0.0296\text{ V}\log\dfrac{[1.10]}{[0.90]^2}$ $E_{cell} = 1.4\text{ V}$

Example 5. An application of the Nernst equation.

The Danielle cell composed of a zinc half-cell and a copper half-cell has a measured voltage of 1.25 V. The standard cell potential is 1.10 V. If the concentration of the copper half cell is 0.60 M, what is the concentration of the zinc half-cell?

The standard reduction potentials for zinc and copper are:

$$Zn^{2+} + 2e^- \to Zn \quad -0.76\text{ V}$$
$$Cu^{2+} + 2e^- \to Cu \quad\;\; 0.34\text{ V}$$

General Strategy	Solution to Example 5
Using the Standard Reduction Potential Table, calculate the E^0 for the Danielle cell.	Zn will be oxidized and Cu reduced to obtain the most positive E^0.

Remember to arrange the half-reactions to obtain the most positive E^0 for the cell.	$Zn \rightarrow Zn^{2+} + 2e^-$ $\quad$ 0.76 V $Cu^{2+} + 2e^- \rightarrow Cu$ $\quad$ 0.34 V $Zn + Cu^{2+} \rightarrow Zn^{2+} + Cu$ $\quad$ 1.10 V
Using the Nernst equation, calculate the concentration of the zinc half-cell.	$$E_{cell} = E^0_{cell} - \frac{0.0592\,V}{n} \log Q$$ $$1.25\ V = 1.10\ V - \frac{0.0592\,V}{2\,mol} \log \frac{\left[Zn^{2+}\right]}{\left[Cu^{2+}\right]}$$ $$1.25\ V = 1.10\ V - \frac{0.0592\,V}{2\,mol} \log \frac{\left[Zn^{2+}\right]}{\left[0.60\right]}$$ $$[Zn^{2+}] = 4.8 \times 10^{-6}\ \frac{mol}{L}$$

Electrolysis

In contrast to spontaneous redox reactions where chemical energy is changed into electrical energy, **electrolysis** *uses electrical energy to cause a nonspontaneous chemical reaction to occur.* An **electrolytic cell** is *an apparatus for carrying out electrolysis.*

Oxidation occurs at the anode and reduction occurs at the cathode just as they did in galvanic cells. In contrast to the galvanic cells, *the charge on the anode is positive and the charge on the cathode is negative in electrolysis.* The battery serves as the "electron pump" driving the electrons to the cathode. For example, in the electrolysis of *molten* sodium chloride (NaCl is melted, no water is present), the following reactions occur:

$$2Cl^- \ (l) \rightarrow Cl_2 \ (g) \ + \ 2e^- \qquad \text{anode (oxidation)}$$
$$\underline{2Na^+ \ (l) \ + \ 2e^- \rightarrow 2Na \ (l)} \qquad \text{cathode (reduction)}$$
$$2Na^+ \ (l) \ + \ 2Cl^- \ (l) \rightarrow 2Na \ (l) \ + \ Cl_2 \ (g) \qquad \text{overall reaction}$$

Most electrolysis reactions in the chemistry lab occur in water solution. When electrolysis is carried out in water solution, there are usually two potential oxidation reactions and two potential reduction reactions. If, for example, you attempt to electrolyze an aqueous solution of sodium bromide, there are two possible oxidation reactions at the anode: the oxidation of a bromide ion, $2Br^- \rightarrow Br_2 + 2e^-$, and the oxidation of water, $2H_2O \ (l) \rightarrow O_2 \ (g) + 4H^+ \ (aq) + 4e^-$. There are also two potential reduction reactions at the cathode: the reduction of a sodium ion, $Na^+(aq) + 1e^- \rightarrow Na \ (s)$, and the reduction of water, $2H_2O \ (l) + 2e^- \rightarrow H_2 \ (g) \ + \ 2OH^-(aq)$. In short, *water is always a potential reactant.*

If you examine the Standard Reduction Potentials Table (Table 19.1), it will help you determine which reactions are most likely to occur.

Potential oxidation reactions:

$$2Br^- \rightarrow Br_2 \ + \ 2e^- \qquad\qquad\qquad -1.07 \text{ V}$$

$$2H_2O \ (l) \rightarrow O_2 \ (g) + \ 4H^+ \ (aq) + \ 4e^- \qquad -1.23 \text{ V}$$

Bromide has the higher oxidation potential at the anode so Br_2 forms at the anode.

Potential reduction reactions:

$$Na^+ \ (aq) + \ 1e^- \ \rightarrow \ \ Na \ (s) \qquad\qquad -2.71 \text{ V}$$

$$2H_2O \ (l) + \ 2e^- \ \rightarrow \ H_2 \ (g) \ + \ 2OH^- (aq) \qquad -0.83 \text{ V}$$

Water has the higher reduction potential at the cathode so $H_2(g)$ forms at the cathode.

Therefore, the overall electrolysis reaction is:

$$\begin{aligned}
2Br^- \ &\rightarrow Br_2 \ + \ \cancel{2e^-} & \text{(anode)} \\
\underline{2H_2O \ (l) + \ \cancel{2e^-} \ \rightarrow \ H_2 \ (g) \ + \ 2OH^- (aq)} & & \text{(cathode)} \\
2Br^- \ + \ 2H_2O \ (l) \ \rightarrow \ H_2 \ (g) \ + \ Br_2 \ (l) \ + \ 2OH^- (aq) & & \text{(overall)}
\end{aligned}$$

Quantitative aspects of electrolysis

The *mass of reactant consumed or product formed in electrolysis is directly proportional to the amount of electricity passed through the solution*. In electrolysis, the current applied is measured in amperes, A. The longer the current flows, the more electrons are involved and the more reactant is consumed and product formed. The relationship between the charge, the current, and time is:

$$C = A \times s \quad \text{where}$$

C is the charge in coulombs
A is the current in amps
s is the time in seconds

Example 6. Calculating the amount of product produced in electrolysis.

Calculate the mass of bromine liquid produced in the electrolysis of an aqueous solution of sodium bromide if a current of 2.50 A is passed through the solution for 3.0 hr.

General Strategy	Solution to Example 6
Determine the correctly balanced equation for the electrolysis reaction.	$2Br^- \rightarrow Br_2 + 2e^-$ $2H_2O\ (l) + 2e^- \rightarrow H_2\ (g) + 2OH^-(aq)$ $2Br^- + 2H_2O\ (l) \rightarrow H_2\ (g) + Br_2\ (l) + 2OH^-(aq)$
Determine the number of coulombs. $C = A \times s$	$1\ C = 1\ A \times 1\ s$ $= 2.50\ A \times 3\ hr \times \dfrac{60\ min}{1\ hr} \times \dfrac{60\ sec}{1\ min}$ $=\ 27,000\ C$
Using Faradays, moles of electrons exchanged, and molar mass, determine the grams of the *desired* material produced. Remember that 1 mole of electrons = 96,500 C (1 Faraday).	$27,000\ C \times \dfrac{1\ mol\ e^-}{96,500\ C} \times \dfrac{1\ mol\ Br_2}{2\ mol\ electrons} \times \dfrac{160.\ g}{1\ mol\ Br_2}$ $=\ 22.3\ g\ Br_2$ $=\ 22\ g\ Br_2$

Example 7. Determining the time to electrolyze a solution.

A 1.00 M aqueous solution of chromium(III) bromide will be electrolyzed to "chrome plate" a faucet. How many hours will it take to deposit 75.0 g of chromium metal from the solution when a current of 3.00 A is running through the solution?

General Strategy	Solution to Example 7
Determine the number of moles of electrons needed to plate out 1 mol of metal. Write the half reaction.	$Cr^{3+}(aq) + 3e^- \rightarrow Cr\ (s)$ 3 mol of electrons

Using dimensional analysis, determine the number of coulombs required. Use the molar mass, the number of moles of electrons needed to plate out the metal, and 96,500 coulombs = 1 mole of electrons.	$75.0 \text{ g Cr} \times \dfrac{1\, mol\, Cr}{52.0\, g\, Cr} \times \dfrac{3\, mol\, electrons}{1\, mol\, Cr} \times$ $\dfrac{96,500\, coulombs}{1\, mol\, electrons}$ $= \quad 418,000 \text{ coulombs}$
Determine the time required. $C = A \times s$	$C = A \times s$ $418,000 \text{ C} = 3.00 \text{ A} \times s$ $s = 139,333 \text{ sec} = 38.7 \text{ hr}$

SAMPLE MULTIPLE CHOICE QUESTIONS

Questions 1 and 2 refer to the following redox equation.

$$\_Fe^{2+}(aq) + \_H^+(aq) + \_Cr_2O_7^{2-}(aq) \rightarrow \_Fe^{3+}(aq) + \_Cr^{3+}(aq) + \_H_2O\ (l)$$

1. What is the correct balancing coefficient for the Fe^{3+} ion?

   A. 7
   B. 6
   C. 2
   D. 14
   E. 10

2. What is the oxidizing agent in Question 1?

   A. Cr^{3+}
   B. H^+
   C. Fe^{3+}
   D. $Cr_2O_7^{2-}$
   E. H_2O

3. Consider the following oxidation-reduction reaction:

   $$Fe^{2+}(aq) + Ni\ (s) \rightarrow Ni^{2+}(aq) + Fe\ (s)$$

   Calculate the free energy, ΔG^0, for this reaction, given the standard reduction potentials of nickel and iron ($1\ F = 96,500\ C$):

   $$Ni^{2+} + 2e^- \rightarrow Ni\ (s) \qquad E^0 = -0.24\ V$$
   $$Fe^{2+} + 2e^- \rightarrow Fe\ (s) \qquad E^0 = -0.44\ V$$

   A. $+38\ kJ$
   B. $+18\ kJ$
   C. $-1.9 \times 10^2\ kJ$
   D. $+1.5 \times 10^2\ kJ$
   E. $-4.0 \times 10^{-2}\ kJ$

4. Consider the following reaction:

   $$Cl_2\ (g) + 2NaOH\ (aq) \rightarrow NaOCl\ (aq) + NaCl\ (aq) + H_2O\ (l)$$

   Which of the following is a true statement with respect to this reaction?

A. The oxidation number of Cl_2 does not change.
B. The oxidation number of oxygen changes.
C. The oxidation number of sodium changes.
D. This reaction is a disproportionation reaction.
E. This reaction requires a catalyst in order to happen.

5. A 0.60 M solution of $Al_2(SO_4)_3$ is electrolyzed. If a current of 2.0 A is passed through the solution for a period of 30 min., which expression correctly predicts the grams of aluminum metal that will be plated out?

A. $\dfrac{(2.0)(30)(60)(27)}{(96,500)(3)}$

B. $\dfrac{(2.0)(30)(60)(27)}{(96,500)}$

C. $\dfrac{(2.0)(30)(27)}{96,500}$

D. $\dfrac{(2.0)(30)(27)}{(96,500)(3)}$

E. $\dfrac{(2.0)(30)(27)}{96,500}$

6. Consider the redox equation for the Danielle cell:

$$Cu^{2+}(aq) \quad + \quad Zn\,(s) \rightarrow Zn^{2+}\,(aq) + Cu\,(s)$$

The value for the equilibrium constant for this reaction is 1.65×10^{37} at 25 °C. Therefore, the E^0 and the ΔG^0 for this reaction:

A. are both positive
B. are both negative
C. E^0 is positive and ΔG^0 is negative
D. $E^0 = 0$ and $\Delta G^0 = 0$, since the system is at equilibrium
E. cannot be determined from the information provided

7. Consider the following standard reduction potentials:

$$Sn^{2+} + 2e^- \rightarrow Sn \qquad E^0 = -0.14\text{ V}$$
$$Co^{2+} + 2e^- \rightarrow Co \qquad E^0 = -0.28\text{ V}$$
$$Cu^{2+} + 2e^- \rightarrow Cu \qquad E^0 = +0.34\text{ V}$$

Which pair will result in a spontaneous reaction?

A. $Cu + Sn^{2+}$
B. $Co + Cu^{2+}$
C. $Sn + Co^{2+}$
D. $Cu^{2+} + Co^{2+}$
E. $Cu + Sn^{2+}$

8. A current equivalent to 0.500 Faradays is passed through a solution containing copper(II) sulfate, $CuSO_4$. How much copper can be plated out?

A. 64 g
B. 32 g
C. 16 g
D. 7.0 g
E. 3.2 g

9. Consider the following oxidation-reduction reaction:

$$Zn\ (s) + 2Ag^+\ (aq) \rightarrow Zn^{2+}\ (aq) + 2Ag\ (s)$$

The Nernst equation is $E = E^0 - \dfrac{0.0592V}{n} \log Q$.

Which of the following changes will *increase* the measured emf?

    I.   Increase the amount of solid zinc
    II.   Increase the concentration of the zinc ion
    III.   Increase the concentration of the silver ion
    IV.   Decrease the concentration of the zinc ion
    V.   Decrease the concentration of the silver ion

A. I, II
B. II, III
C. I, II, III
D. III, IV, V
E. III, IV

10. When an aqueous solution of copper(II) bromide is electrolyzed, what color change will occur at the anode?

A. lighter blue color
B. darker blue color
C. no change
D. purple
E. red-brown

Comprehension Questions

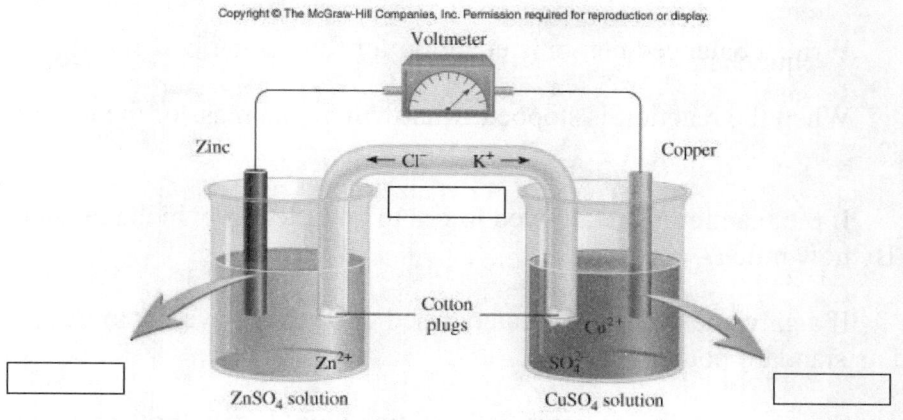

Net reaction

1)   Question 1 refers to the diagram.

        a)   Label the salt bridge, anode, and cathode in the diagram, and write a balanced net ionic equation for the galvanic cell in the appropriate box.

        b)   Calculate the standard potential for the cell depicted.

        c)   Label the direction of electron flow for the spontaneous reaction.

        d)   What purpose or function does the salt bridge serve, and how does it accomplish this task?

        e)   Calculate the standard free energy change for the above reaction, and describe how ΔG^0 and E^0 relate to spontaneity.

2)   A piece of copper wire with a mass of 6.50 g is submerged into 0.350 L of 0.500 M silver nitrate solution. The reaction is allowed to proceed until 1.00 g of silver metal has been removed from solution and deposited onto the surface of the copper metal.

        a)   Write a balanced net ionic equation for this scenario.

        b)   When the reaction is stopped, what will be the mass of the remaining copper wire?

        c)   If the reaction were allowed to run to completion, which reactant would be in excess? By how much?

        d)   If a galvanic cell were created based upon the above net ionic reaction, what would its standard potential be?

        e)   Suppose that the galvanic cell mentioned in d) were not set up under standard conditions, but with silver and copper ion concentrations of 0.500 M and 0.112 M, respectively. What would the new cell potential be if the reaction were run at 25 °C?

3)   Electrical contacts are often plated with a thin layer of gold because of gold's low susceptibility to corrosion and its excellent electrical conductivity. The gold layer can be applied through an electroplating process from a solution of gold cyanide, $Au(CN)_3$.

        a)   Suppose that an electrical timer switch (like the one found on many home clothes washing machines) has 45 contact points that need to be plated to a thickness of 0.0500 mm, and that each contact has a surface area of 4.00 mm^2. How many moles and what mass of gold would be needed to plate all of the contacts in the timer? (The density of gold is 19.30 g/cm^3.)

        b)   How long, in minutes, would it take to apply the layer of gold mentioned in a) in an electrolytic cell supplied with a current of 1.00 A?

        c)   In a certain manufacturing process, an electrolytic cell using 5.0 L of a gold cyanide solution having a concentration of 0.110 M was run at a current of 2.50 A for a total of 1.5 hr. What would the concentration of gold ions left in solution be at the end of this electroplating run?

4)   The table of reduction potentials shown can be used to answer the questions that follow.

Half-Reaction	E^0 (volts)
$Cl_2 + 2e^- \rightarrow 2Cl^-$	1.36
$O_2\,(g) + 4H^+ + 4e^- \rightarrow 2H_2O\,(l)$	1.23
$I_2 + 2e^- \rightarrow 2I^-$	0.53
$2H_2O\,(l) + 2e^- \rightarrow H_2\,(g) + 2OH^-$	−0.83
$Ba^{+2} + 2e^- \rightarrow Ba\,(s)$	−2.71

a)   A 2.0-V electric current was applied to a 1.0 M aqueous solution of barium chloride.

i)   Write a balanced oxidation half-reaction for the process occurring in the electrolytic cell in a).

ii)   Write a balanced reduction half-reaction for the process occurring in the electrolytic cell in a).

b)   The same 2.0-V electric current was applied to a 1.0 M aqueous solution of barium iodide.

i)   Write a balanced oxidation half-reaction for the process occurring in the electrolytic cell in b).

ii)   Write a balanced reduction half-reaction for the process occurring in the electrolytic cell in b).

c)   The above processes would not be useful for the production of chlorine gas. Suggest a method of electrolysis that would effectively produce chlorine gas.

d)   What is the basic difference between a galvanic cell and an electrolytic cell?

ANSWERS TO SAMPLE MULTIPLE CHOICE QUESTIONS

1.   B

Separate the oxidation half-reaction from the reduction half-reaction.

$$Fe^{2+} \rightarrow Fe^{3+} \qquad\qquad \text{oxidation}$$
$$Cr_2O_7^{2-} \rightarrow Cr^{3+} \qquad\qquad \text{reduction}$$

Balance each half-reaction for atoms and for charge:

$$Fe^{2+} \rightarrow Fe^{3+} + 1e^-$$
$$6e^- + 14H^+ + Cr_2O_7^{2-} \rightarrow 2Cr^{3+} + 7H_2O$$

Inspection shows that the oxidation half-reaction must be multiplied by six to have the electrons lost in the oxidation reaction equal the electrons gained in the reduction reaction. Therefore, the coefficient for the *Fe*$^{3+}$ *must be 6.*

2. D

The oxidizing agent is reduced. The oxidation number of the chromium is +6 in $Cr_2O_7^{2-}$ and +3 in Cr. Therefore, *the dichromate, $Cr_2O_7^{2-}$ is the oxidizing agent.*

3. A

The relationship between ΔG^0 and E^0 is given by the equation: $\Delta G^0 = -nFE^0_{cell}$. The desired reaction shows nickel being oxidized and it will have an E^0 of +0.24 V. The iron is reduced and has an E^0 of –0.44 V. This gives an emf for the cell, E^0_{cell}, of –0.20 V. The negative value for the E^0_{cell} is already an indication that the reaction is not spontaneous in the forward direction and the value of ΔG^0 should be positive.

$$\Delta G^0 = -nFE^0_{cell}$$
$$= -(2 \text{ mol } e^-)(\frac{96,500\,C}{1\,mol\,electrons})(-0.20\frac{J}{C})$$
$$= +36670 \text{ J}$$
$$= +38 \text{ kJ}$$

4. D

The reaction is a disproportionation. Chlorine, Cl_2, with an initial oxidation number of 0, has oxidation numbers of +1 and –1 in the products (see Chapter 4 of this review book).

5. A

$C = A \times s$

$= (2.0 \text{ A})(30 \text{ min})(\dfrac{60sec}{\text{min}}) \text{ C}$

$= (2.0 \text{ A})(30 \text{ min})(\dfrac{60sec}{\text{min}}) \text{ C} \times \dfrac{1\,mol\,electrons}{96,500\,C} \times \dfrac{1\,mol\,Al}{3\,mol\,electrons} \times \dfrac{27g}{1\,mol\,Al}$

$= \dfrac{(2.0)(30)(60)(27)}{(96,500)(3)}$

6. C

Since the equilibrium constant has a large value, the reaction proceeds far to the right. Product formation is heavily favored. Therefore, the ΔG^0 must be negative (spontaneous in the forward direction) and the E^0 must be positive.

7. B

The calculation of E^0_{cell} for each of the pairings gives a positive value only for B. Only this pairing will result in a spontaneous reaction.

$$\begin{array}{ll} Co\,(s) \;\rightarrow\; Co^{2+} + 2e^- & E^0 \;=\; +0.28 \text{ V} \\ \underline{Cu^{2+} + 2e^- \;\rightarrow\; Cu} & \underline{E^0 \;=\; +0.34 \text{ V}} \\ Co\,(s) + Cu^{2+} \rightarrow Co^{2+} + Cu\,(s) & E^0_{cell} = +0.62 \text{ V} \end{array}$$

8. C

Since 1 Faraday = 1 mol of e^-, 0.500 Faraday = 0.500 mol of e^-.
To plate out copper(II) requires 2 mol of electrons per mole of copper.
Therefore,

$0.50 \text{ mol } e^- \times \dfrac{1\,mol\,Cu}{2\,mol\,electrons} \times \dfrac{64\,g}{1\,mol\,Cu} = 16 \text{ g Cu}$

9. E

The Q in the Nernst equation, $E_{cell} = E^0_{cell} - \dfrac{0.0592 V}{n} \log Q$, for this reaction is

$\dfrac{[Zn^{2+}]}{[Ag^+]^2}$. Either an increase in the concentration of silver ion in the denominator or a decrease in the concentration of Zn^{2+} ion in the numerator will cause Q to be a smaller value. The smaller the quantity subtracted from E^0_{cell}, the larger E_{cell} will be.

10. E

Bromide ion is more easily oxidized than water at the anode. The element bromine will be produced and the color will be red-brown in water solution.

Answers to Comprehension Questions

1) a)

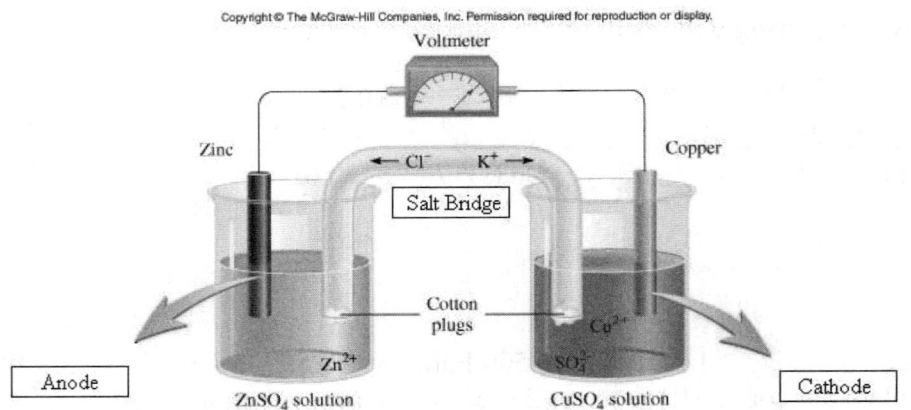

Net reaction

$$Zn(s) + Cu^{2+}(aq) \rightarrow Cu(s) + Zn^{2+}(aq)$$

b)   Standard cell potentials can be calculated in a number of ways. One simple way is to add the reduction potential of the substance being reduced to the oxidation potential of the substance being oxidized. In this case Zn metal is being oxidized and copper(II) ions are being reduced. To get an oxidation potential from a reduction potential table you need only change the sign in front of the voltage.

$$E^0_{cell} = E^0_{oxidation} + E^0_{reduction} = 0.76\,V + 0.34\,V = \mathbf{1.10\,V}$$

c)   Electrons are flowing from the zinc electrode (where they are being produced in an oxidation reaction) through the wire into the copper electrode (where they are being consumed in a reduction reaction).

d)   The salt bridge maintains charge neutrality in each of the half-cells. As zinc ions are being added to the solution on the left, anode, the zinc sulfate solution is becoming positively charged because of an imbalance between positive and negative ions. Similarly, the right compartment, cathode, is becoming more negatively charged as copper ions leave the solution and create a similar unbalance between cations and anions. The salt bridge compensates for this by allowing cations or anions to diffuse into these half-cell compartments.

e) $\Delta G^0 = -nFE^0$. The free energy change is related to the potential difference between the half-cells, the charge on electrons, and the number of electrons passing through this voltage difference. In this case $n = 2$ because 2 mol of electrons are changing ownership for every mole of substance oxidized or reduced. F is the Faraday constant and happens to equal the charge on a mole of electrons or 96,450 C.

$$\Delta G^0 = -nFE^0 = -2\ mol \times 96{,}500\ C/mol \times 1.10\ V = -2.12 \times 10^5\ V\ C = -2.12 \times 10^5\ J = \mathbf{-212\ kJ}$$

The spontaneity of a reaction cannot only be assessed by the sign of ΔG^0, which in this case is negative indicating a spontaneous reaction, but also by the sign of E^0_{cell}, which in this case is positive indicating a spontaneous reaction.

2)   a)   **Cu (s) + 2Ag$^+$ (aq) → 2Ag (s) + Cu^{2+} (aq)**

b)   The stoichiometry of the above reaction tells us that for every 2 mol of silver removed from solution, 1 mol of copper will become solvated ions. Calculation of the amount of copper consumed will begin with conversion of grams of silver to moles of silver and end with conversion of moles of copper to grams of copper.

1.00 g silver produced × (1 mol Ag / 107.87 g) × (1 mol Cu / 2 mol Ag) × (63.55 g Cu / 1 mol Cu) = 0.295 g Cu

6.50 g Cu (initial) – 0.295 g = **6.21 g Cu remaining**

        c)   A calculation of amount of product formed will be performed using each reactant as though there were an excess of the other. The calculation that yields the smallest amount of product started with the limiting reactant. The other reactant is in excess.

6.50 g Cu × (1 mol Cu/63.55 g Cu) × (2 mol Ag /1 mol Cu) = 0.205 mol Ag produced

0.350 L AgNO$_3$ × (0.500 mol AgNO$_3$ / L of solution) × (1 mol Ag$^+$ / 1 mol AgNO$_3$) × (1 mol Ag / 1 mol Ag$^+$) = 0.175 mol Ag produced.

Therefore, copper is in excess.

0.350 L AgNO$_3$ × (0.500 mol AgNO$_3$ / L of solution) × (1 mol Ag$^+$ / 1 mol AgNO$_3$) × (1 mol Cu / 2 mol Ag$^+$) × (63.55 g Cu / 1 mol Cu) = 5.56 g Cu used

Amount of Cu in excess = 6.50 g Cu (initial) – 5.56 g Cu (used) = **0.94 g Cu in excess**

        d)   Remember that $E^0_{cell} = E^0_{oxidation} + E^0_{reduction}$, and, in this case, copper is being oxidized and silver is being reduced.

$$E^0_{cell} = E^0_{oxidation} + E^0_{reduction} = -0.34\,\text{V} + 0.80\,\text{V} = \textbf{0.46 V}$$

        e)   In this case we must use the Nernst equation to calculate this nonstandard voltage.

$$E_{cell} = E^0_{cell} - (RT/nF)\,\ln Q$$

Q is the reaction quotient, which is equal to $[Cu^{2+}] / [Ag^+]^2$ (in this case the quantities within the brackets are nonequilibrium concentration values expressed in moles per liter).

E_{cell} = 0.46 V – [(8.314 J·mol^{-1}·K^{-1} × 298.15 K) / (2 mol × 96,500 C·mol^{-1})] ln (0.112 / 0.500^2) = 0.46 V – 0.0129 V ln (0.448) = **0.47 V**

3)   a)   This problem begins with the calculation of a volume of gold (surface area × thickness), which will be converted to a mass of gold, using the density, and then to a number of moles of gold through the use of gold's molar mass.

mass of gold per contact = density of gold × volume of gold = 19.30 g/cm^3 × (0.0500 mm × 4.00 mm^2) × (1cm^3 / 1000 mm^3) = 3.86 × 10^{-3} g Au per contact × 45 contacts = **1.74 × 10^{-1} g Au**

1.74 × 10^{-1} g Au × (1 mol Au / 196.97 g Au) = **8.82 × 10^{-4} mol Au**

b)   The unit of current or ampere is equal to 1 C of electrical charge per second (it's a rate of flow of electrical current). Faraday's constant relates an amount of charge to a number of electrons. Knowledge of the current flowing along with Faraday's constant and the fact that 3 mol of electrons are needed for every mole of gold ions reduced to gold atoms will allow for calculating the amount of time necessary to supply enough electrons to reduce the gold ions.

8.82 × 10^{-4} mol Au × (3 mol of electrons / 1 mol of Au) × (96,500 C / mol of electrons) × (1 sec / 1.00 C) × (1 min / 60 sec) = **4.25 min**

c)   In this instance we will calculate the number of moles of gold reduced during the 1.5 hr period and then subtract it from the total moles of gold at the start of the experiment. The new concentration can then be calculated, assuming no change in volume of the solution.

moles of Au removed from solution = 1.5 hours × (3600 sec / hr) × (2.5 C / sec) × (1 mol electrons / 96,500 C) × (1 mol Au / 3 mol of electrons) = 4.67 × 10^{-2} mol Au removed from solution.

Total moles initially = 5.0 L × 0.110 mol/L = 0.55 mol Au

Moles of Au final = 0.55 – 0.0467 = 0.50 mol Au final

Final concentration of Au = 0.50 mol / 5.0 L = **0.10 M**

4)   a)   i)   Whenever an aqueous solution is electrolyzed, there are several possibilities for the primary oxidation or reduction reaction. The substance in solution with the highest oxidation potential (lowest reduction potential) will be oxidized most readily. The substance with the highest reduction potential will be the substance most likely to be reduced. In this case, water is the substance present in a barium chloride solution (which contains barium ions, chloride ions, and water molecules) having the highest oxidation potential. Therefore,

$$\textbf{2H}_2\textbf{O } (\textit{l}) \; \rightarrow \textbf{O}_2\,(\textit{g}) + \textbf{4H}^+ + \textbf{4e}^-$$

ii)   The substance in this solution with the highest reduction potential is also water, therefore,

$$2H_2O \ (l) + 2e^- \rightarrow H_2 \ (g) + 2OH^-$$

The net result in this case would simply be electrolysis of water.

b)   i)   The same rules apply here, so . . .

$$2I^- \rightarrow I_2 + 2e^-$$

ii)   And . . . $2H_2O \ (l) + 2e^- \rightarrow H_2 \ (g) + 2OH^-$

c)   In this case water has a higher oxidation potential than chloride, and therefore no elemental chlorine is produced. Removal of the water is necessary so that the current can oxidize chloride ions, but solid NaCl doesn't conduct electricity. The solution is to electrolyze molten NaCl and collect the sodium metal and chlorine gas produced.

d)   The fundamental difference between these two types of cells is that one, the galvanic cell, contains a reaction that will spontaneously (without the application of external force or energy) form products, whereas the electrolytic cell requires the application of an electrical current (input of energy) to form products. All electrolytic cells have a positive ΔG^0. Galvanic cells with a positive cell potential will have a negative ΔG^0.

CHAPTER 20
METALLURGY AND THE CHEMISTRY OF METALS

This chapter reviews the following topics regarding metallic elements:

- Conductivity
- Periodic trends in metallic properties
- The alkali metals
- The alkaline earth metals
- The special case of aluminum

Most metals come from **minerals**, *naturally occurring substances with a wide range of chemical compositions* that are most commonly found in the Earth's crust, seawater, and the sea floor. A *mineral deposit that is concentrated enough to warrant an economical recovery of the desired metal* is called an **ore**.

Take Note: *The questions on the AP exam tend to focus on the trends within the periods and groups on the periodic table (see Chapter 8 of this review book) rather than detailed characteristics of a specific element. However, because of the unique chemistry of aluminum discussed in this chapter, aluminum is one element that does tend to appear on the AP exam.*

Conductivity

Metals conduct because of the nature of metallic bonding. Metals are visualized as having a fixed positive core and a **sea of valence electrons,** *valence electrons that are delocalized over the entire crystal.* The mobility of these electrons makes metals good conductors of heat and electricity.

Semiconductors are *elements that do not normally conduct electricity, but will conduct at elevated temperatures or when combined with a small amount of certain other elements*. The elements silicon and germanium are especially suited for this purpose and are used extensively in electronic components such as computer chips.

Periodic trends in metallic properties

Metals are generally shiny, solid at room temperature (with the exception of mercury), good conductors of heat and electricity, **malleable** (*can be hammered flat*), and **ductile** (*can be drawn into a wire*). On the periodic table, metallic character decreases from left to right in a period, but increases from top to bottom in a group. Metals tend to have low electronegativities, form cations, and have positive oxidation numbers.

The alkali metals

The alkali metals, Group 1A, are the most active metals on the periodic table. They have the smallest electronegativities on the periodic table and all form cations with a +1 oxidation number. They react easily and sometimes explosively with oxygen and water (see Chapters 4 and 8 of this review book).

The alkaline earth metals

The alkaline earth metals, Group 2A, have slightly higher electronegativities than the alkali metals. They are less reactive than the alkali metals but can still react with oxygen and water. Except for beryllium, the alkaline earth elements have similar properties, forming a +2 cation with the stable electron configuration of the preceding noble gas (see Chapters 4 and 8 of this review book).

The special case of aluminum

Aluminum is the third most plentiful element in the earth's crust. Aluminum is considered an active metal. Remember that there is a decrease in metallic properties as you go from left to right on the periodic table. Aluminum does not react with oxygen and water as do alkali and alkaline earth metals, but it does react with strong acids and strong bases:

$$2Al(s) + 6HCl(aq) \rightarrow 2AlCl_3 + 3H_2(g)$$

$$2Al(s) + 2NaOH(aq) \rightarrow 2NaAlO_2 + 3H_2(g)$$

Aluminum hydroxide is amphoteric (*can display properties of both acid and base*):

$$Al(OH)_3(s) + 3H^+(aq) \rightarrow Al^{3+}(aq) + 3H_2O(l)$$

$$Al(OH)_3(s) + OH^-(aq) \rightarrow Al(OH)_4^-(aq)$$

Take Note: *A common question on the AP exam, especially in question 4 in the Free Response section on reaction types, asks about the amphoteric properties of aluminum. The actual complex ion of aluminum with hydroxide ion is $Al(OH)_4^-$, but an answer of $Al(OH)_6^{3-}$ is also accepted.*

CHAPTER 21
NONMETALLIC ELEMENTS AND THEIR COMPOUNDS

This chapter reviews the properties of some of the principal nonmetallic elements. The topics covered include:

- General properties of nonmetals
- Properties of hydrogen, carbon, nitrogen, phosphorus, oxygen, and sulfur, and of the halogens

---

Take Note: *The questions on the AP exam tend to focus on the trends within the periods and groups on the periodic table rather than detailed characteristics of a specific element (see Chapter 8 of this review book).*

General properties of nonmetals

There are only 25 nonmetals on the periodic table. The properties of the nonmetals are more varied than those of metals. Some of the nonmetals are gases at room temperature: hydrogen, oxygen, nitrogen, fluorine, chlorine, and the noble gases. Only bromine is a liquid at room temperature. The remaining nonmetals are solids. Unlike metals, nonmetallic elements are poor conductors of heat and electricity. These elements also exhibit both positive and negative oxidation numbers.

The principal **metalloids**, which include *boron, silicon, germanium, and arsenic*, are semiconducting elements. These elements display properties characteristic of both metals and nonmetals.

Nonmetals, which are found on the upper right side of the periodic table, tend to have high electronegativities. The electronegativities tend to increase from left to right in a period and decrease from top to bottom in a family. Electronegativities do not apply to noble gases because these elements do not usually react.

Properties of hydrogen, carbon, nitrogen, phosphorus, oxygen, and sulfur, and of the halogens

Chapter 21 of Chang contains many details of the chemistry of these important nonmetals. This is superb reference material, but it is not frequently tested on the AP exam.

CHAPTER 22
TRANSITION METAL CHEMISTRY
AND COORDINATION COMPOUNDS

Chapter 22 reviews transition metals and their ability to form complex ions. The discussion in Chapter 22 is limited to the fourth period transition metals and includes:

- The general properties of transition metals
- Electron configurations and oxidation states
- Complex ion formation, ligands, and the coordination number
- Naming coordination compounds
- Formation and reactions of complex ions

Take Note: *The AP exam does not require an in depth knowledge of complex ions and coordination compounds. A* <u>basic</u> *knowledge of the elements that form complex ions, the coordination number for the ions, oxidation numbers of the metal ion in the complex ion, and basic reactions are the main ideas with which you need to be familiar. The examples included in the discussion that follows will provide you with sufficient knowledge for similar questions on the AP exam.*

The general properties of transition metals

Transition elements fill the *d subshell from d^1 to d^{10}*. Incompletely filled *d* subshells give rise to several notable properties, including distinctive coloring, the formation of paramagnetic compounds, catalytic activity, and the tendency to form complex ions.

Periodic trends in the transition metals are different from trends in the main group elements. Transition metals are also less reactive than the alkali and alkaline earth elements. In fact, many are unreactive or slow to react in acid. They generally have higher densities, higher melting points and boiling points, and higher heats of fusion and vaporization than the alkali and alkaline earth elements.

Electron configurations and oxidation states

The fourth period transition elements fill the $3d$ subshells. When these elements form cations, electrons are removed *first* from the $4s$ orbital and then from the $3d$ subshells. For example, the vanadium atom's electron configuration is $[Ar]4s^2 3d^4$. The vanadium ion, V^{2+}, has the electron configuration $[Ar]3d^4$.

Most of the transition metals exhibit multiple oxidation states. The most common oxidation states are +2 and +3, with +3 being more stable at the beginning of the series and +2 being more stable toward the end of the series. The highest oxidation state for a transition metal is +7 for manganese.

Complex ion formation, ligands, and the coordination number

Transition metals have the ability to form complex ions. The formation of a complex ion is a Lewis acid-base reaction. Complex ions form from a transition metal ion and **ligands**, *molecules or ions that have electron pairs to donate to the d subshells of the transition metal cation.* The empty *d* subshells of the metal cation can accept electron pairs; the metal acts as the Lewis acid. The ligands are the electron pair donors or Lewis bases. The bonds that hold the ligand to the metal are coordinate covalent bonds. The **coordination number** is *the number of ligands surrounding the central metal ion in a complex ion.* The most common coordination numbers are 4 and 6.

Some of the most common complex ions are:

Complex Ion	Name of Ion	Color of Ion
$Fe(CN)_6^{3-}$	hexacyanoferrate(III)	red
$Ag(NH_3)_2^+$	diamminesilver(I)	colorless
$Co(NH_3)_6^{3+}$	hexamminecobalt(III)	yellow
MnF_6^{4-}	hexafluoromanganate(II)	pink
$Cu(NH_3)_4^{2+}$	tetrammine copper(II)	blue
$Zn(OH)_4^{2-}$	tetrahydroxozincate(II)	colorless
$Al(H_2O)_6^{3+}$	hexaaquaaluminum(III)	colorless

> **Take Note:** *On the AP exam, you will occasionally be asked to determine the most likely coordination number for a complex ion. A good rule of thumb to determine the number of ligands is to double the oxidation number on the metal cation (see examples). This usually gives you the coordination number. Two notable exceptions are $FeSCN^{2+}$ and $Al(OH)_4^-$.*

Naming coordination compounds

Coordination compounds *are compounds that contain at least one complex ion.* The complex ion can be the cation or the anion in the compound. In order to name the coordination compound, the oxidation number of the transition metal in the complex ion must first be determined. Consider the complex ion CoF_6^{3-}. A fluoride ion has a –1 oxidation number. Since there are six fluoride ions, the total oxidation value for the six fluoride ions is –6. However, the overall charge on the complex ion is 3–. Therefore, the oxidation number on the cobalt metal must be +3.

The rules for naming coordination compounds are:

- The cation is always named before the anion.
- Within a complex ion, the ligands are named first in alphabetical order. The metal ion is named last.
- The names of anionic ligands end with *o*. Neutral ligands are simply the name of the molecule. The exceptions are H_2O (aquo), CO (carbonyl), and NH_3 (ammine).
- When several ligands of the same type are present, the Greek prefixes *di-, tri-, tetra-*, etc., precede the ligand name.
- The oxidation number of the metal is written in Roman numerals following the name of the metal.
- If the complex ion is an anion, the metal name ends in *-ate*.

Example 1. Naming coordination compounds with a complex cation.

Write the systemic name for the following coordination compound: $Co(NH_3)_4Cl_3$.

General Strategy	Solution to Example 1
Determine the oxidation number on the metal in the complex ion. The complex cation in this compound is: $Co(NH_3)_4^{3+}$	$Co(NH_3)_4Cl_3$ Since the ammonia is a neutral ligand, and Cl is –1, then –1 × 3 = –3, making Co = +3.
Name the cation first.	In this cobalt compound, the complex ion is first. It has the neutral ligand NH_3 that is named *ammine*. There are four of these ligands, hence tetraammine. tetraammine cobalt(III)
Name the anion.	Chloride
Name the coordination compound.	*tetraammine cobalt(III) chloride*

Example 2. Naming coordination compounds with a complex anion.

Name the following coordination compound: $K_3[Fe(CN)_6]$.

General Strategy	Solution to Example 2
Determine the oxidation number of the metal in the complex ion.	$K_3[Fe(CN)_6]$ Potassium is +1. +1 × 3 = +3. Therefore, the $[Fe(CN)_6]$ = –3 overall. A cyanide polyatomic ion,

	CN, has a charge of 1^-. $-1 \times 6 = -6$. Therefore, the Fe has an oxidation state of $+3$.
The anion is: $Fe(CN)_6^{3-}$	
Name the cation first.	The cation is K^+, potassium.
Name the anion.	The complex ion, $[Fe(CN)_6]$, has six cyanide ligands; hence, hexacyano. The metal, Fe, is in the complex anion and is thus named fer*rate*(III). The anion is hexacyanoferrate(III)
Name the coordination compound.	*potassium hexacyanoferrate(III)*

Formation and reactions of complex ions

The formation of complex ions usually occurs when a transition metal ion and a concentrated solution of a potential ligand like ammonia are mixed. For example, when concentrated ammonia is added to an aqueous solution of nickel nitrate, a complex ion forms:

$$Ni^{2+}(aq) + 4NH_3\ (aq) \rightarrow Ni(NH_3)_4^{2+}(aq)$$

A particular case of complex ion formation involves some of the insoluble transition metal hydroxides. For example, when insoluble $Zn(OH)_2$ has a concentrated solution of sodium hydroxide (NaOH) added to it, the soluble complex ion, $Zn(OH)_4^{2-}$ will form:

$$Zn(OH)_2(s) + OH^-(aq) \rightarrow Zn(OH)_4^{2-}(aq)$$

> **Take Note:** *While aluminum is not a transition metal, its hydroxide will undergo a complex ion formation reaction similar to zinc:*
> $Al(OH)_3(s) + OH^-(aq) \rightarrow Al(OH)_4^-(aq)$. $Al(OH)_6^{3-}$ *is also accepted as an answer as the product of this reaction. These complex ion formations with zinc and aluminum frequently appear on the AP exam.*

A *common reaction of complex ions* is ligand **exchange or substitution**. You can view this as a form of single replacement reaction (see Chapter 4 of this review book). An example of ligand exchange would be the displacement of a chloride ion from insoluble silver chloride by molecular ammonia, to produce the soluble diammine silver complex ion:

$$AgCl(s) + 2NH_3\ (aq) \rightarrow Ag(NH_3)_2^+(aq) + Cl^-(aq)$$

> **Take Note:** *On the AP exam, if you see a question that involves a reaction between a* <u>transition metal ion</u> *and a* <u>concentrated</u> *solution of a potential ligand like ammonia, then the product is most likely a complex ion.*

CHAPTER 23
NUCLEAR CHEMISTRY

This chapter reviews the basics of nuclear chemistry. The topics include:

- Natural radioactivity: alpha, beta, and gamma radiation
- Predicting alpha or beta decay
- The kinetics of radioactive decay
- Nuclear transmutation
- Nuclear fission
- Nuclear fusion
- Binding energy and mass defect

> **Take Note:** *Historically, the AP exam has included just a few questions on nuclear chemistry either as Multiple Choice or Free Response questions. The exam questions have tended to ask you to predict the products of nuclear decay, determine which elements are likely to undergo radioactive decay, and distinguish between fission and fusion. The examples included in the discussion that follows will provide you with sufficient knowledge for similar questions on the AP test.*

Natural radioactivity: alpha, beta, and gamma radiation

Nuclear reactions differ from ordinary chemical reactions because the nucleus of the atom, and hence the identity of the element, changes. Nuclear reactions are accompanied by the absorption or release of enormous amounts of energy and mass is not conserved (see mass defect discussion).

The spontaneous emission by unstable nuclei of particles or electromagnetic radiation, or both, is known as **radioactivity.** The three main types of naturally occurring radioactivity are summarized below:

Type of Radioactivity	Symbol	Particle	Penetrating Power
Alpha particle	$_2^4He^{2+}$ or α (alpha)	A helium nucleus	The radioactive particle with the least penetrating power
Beta particle	$_{-1}^{0}\beta$	A fast moving electron	More penetrating than alpha particles
Gamma radiation	γ	High energy electromagnetic radiation (not a particle)	The most penetrating radiation

Many elements have isotopes that are radioactive and all elements with an atomic number greater than 83 are radioactive.

The disintegration of a radioactive nucleus is often the beginning of a **radioactive decay series,** *a sequence of nuclear reactions that eventually results in a stable isotope.* It is important for you to be able to balance the nuclear reaction for each of the steps in a radioactive decay series. Consider, for example, the alpha decay of $^{226}_{88}Ra$:

$$^{226}_{88}Ra \ \rightarrow \ ^{222}_{86}Rn \ + \ ^{4}_{2}He^{2+}$$

or

$$^{226}_{88}Ra \ \rightarrow \ ^{222}_{86}Rn \ + \ \alpha$$

Notice that in alpha decay, the atomic number decreases by two and the mass number by four. In the beta decay below, notice that the atomic number increases by one but the mass number remains the same after the nuclear transformation:

$$^{214}_{82}Pb \ \rightarrow \ ^{214}_{83}Bi \ + \ ^{0}_{-1}\beta$$

The *beginning radioactive isotope* is referred to as the **parent isotope** and *the product* is referred to as the **daughter isotope.**

Example 1. Predicting the products of nuclear reactions.

Identify the "X" in each of the following nuclear equations:

a) $^{238}_{92}U \ + \ ^{1}_{0}n \ \rightarrow \ ^{94}_{36}Kr \ + \ \mathbf{X} \ + \ 3\,^{1}_{0}n$

b) $\mathbf{X} \rightarrow \ ^{14}_{7}N \ + \ ^{0}_{-1}\beta$

c) $^{232}_{90}Th \ \rightarrow \ ^{228}_{88}Rn \ + \ \mathbf{X}$

Solution to Example 1.

The general strategy is to make sure the sum of the atomic numbers and the sum of the mass numbers are conserved as you go from reactants to products in a nuclear equation.

a) $^{238}_{92}U \ + \ ^{1}_{0}n \ \rightarrow \ ^{94}_{36}Kr \ + \ \mathbf{X} \ + \ 3\,^{1}_{0}n$

The sum of atomic numbers of the reactants is 92. Since Kr has an atomic number of 36, X must have an atomic number of 56.

The total mass of the reactants is 239. The product masses include Kr with a mass of 94 and the three neutrons which contribute a mass of 3. Therefore, the mass of X must be 142 and $X = {}^{142}_{56}Ba$.

b) $X \rightarrow {}^{14}_{7}N + {}^{0}_{-1}\beta$

The sum of the atomic numbers of the products is 6. Therefore X has an atomic number of 6.
The total mass of the products is 14. Therefore, the mass of X must be 14 and $X = {}^{14}_{6}C$.

c) ${}^{232}_{90}Th \rightarrow {}^{228}_{88}Rn + X$

The atomic number of the reactant is 90. Since one product has an atomic number of 88, X has an atomic number of 2.

The total mass of the reactant is 232. Since one product has a mass of 228, the mass of X must be 4 and $X = {}^{4}_{2}He^{2+}$ (which can also be referred to as an alpha particle, α).

Take Note: *A common question in the Multiple Choice section of the AP exam is one that requires you to supply the missing particle in a radioactive decay process.*

Predicting alpha or beta decay

You might ask the question, what type of radioactive decay will an element undergo? Scientists have determined that the type of radioactive decay that an isotope will undergo depends on the *neutron-to-proton ratio* in the isotope. Elements that have a neutron-to-proton ratio falling on the **band of stability** (see Figure 23.1) are *stable isotopes, and do not undergo radioactive decay.* Notice that for the first 20 elements, the isotopes that have a 1:1 neutron-to-proton ratio are stable.

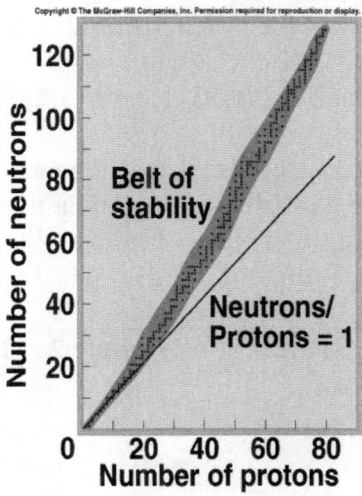

Figure 23.1

An element's neutron-to-proton ratio and the band of stability can be used to predict the type of radioactive decay an isotope will undergo. For example, an isotope with a neutron-to-proton ratio that places it to the left of the band of stability has too many neutrons. This element will undergo beta decay to create more protons. This will create a neutron to proton ratio that will place the element on the band of stability and result in a stable isotope.

The kinetics of radioactive decay

All radioactive decay reactions obey **first-order** kinetics (see Chapter 13 of this review book). The number of nuclei at time zero (N_0) and at time t (N_t) is given by the equation:

$$\ln \frac{N_t}{N_0} = -kt \text{ where}$$

k is the rate constant (the unit varies)
t is the time (the unit varies)

The **half-life** of a radioactive isotope is *the time it takes for half of the atoms in a radioactive sample to undergo radioactive decay*. The half-life is unique to each radioactive isotope. It can be calculated by the first-order half-life equation:

$$t_{\frac{1}{2}} = \frac{0.693}{k} \text{ where}$$

$t_{\frac{1}{2}}$ is the half-life of the radioisotope (unit varies)
k is the rate constant

Example 2. Calculations involving the kinetics of radioactive decay.

A 10.0-g sample of radioactive Np-93 decays to 2.80 g over 4 hr.

a) Determine the rate constant for the reaction.
b) Determine the half-life of this radioactive element.
c) If a 10.0-g sample of Np-93 is allowed to decay for 1.00 hr, how many grams of Np-93 remain?

General Strategy	Solution to Example 2a, 2b, 2c
Remember that radioactive decay obeys first-order kinetics. $$\text{Ln}\frac{N_t}{N_0} = -kt$$	a) $$\ln\frac{N_t}{N_0} = -kt$$ $$\ln\frac{2.80\,g}{10.0\,g} = -k\,(4\ \text{hr})$$ $$k = 0.318\ hr^{-1}$$
The half-life is given by the equation: $$t_{\frac{1}{2}} = \frac{0.693}{k}$$	b) $$t_{\frac{1}{2}} = \frac{0.693}{k}$$ $$= \frac{0.693}{0.318\,hr^{-1}}$$ $$= 2.17\ hr$$
The amount remaining is calculated based on first-order kinetics. $$\ln\frac{N_t}{N_0} = -kt$$	c) $$\ln\frac{N_t}{N_0} = -kt$$ $$\ln\frac{N_t}{10.0\,g} = -(0.318\ \text{hr}^{-1})(1\ \text{hr})$$ $$N_t = 7.27\ g$$

Nuclear transmutation

It is possible to produce radioactive elements artificially. This process is called artificial **transmutation**. Elements can be bombarded with alpha particles, neutrons, and small nuclei to synthesize larger nuclei. The **transuranium elements**, *those with atomic numbers greater than 92*, are all created by transmutation.

Nuclear fission

Bombardment of an element containing a large nucleus by one or more neutrons creates an unstable neutron-to-proton ratio. The large, unstable nucleus undergoes **nuclear fission** where *it divides to form smaller nuclei and in the process, releases one or more neutrons and a great deal of energy (see diagram).*

$$\ce{^{235}_{92}U} + \ce{^{1}_{0}n} \rightarrow \ce{^{90}_{38}Sr} + \ce{^{143}_{54}Xe} + 3\ce{^{1}_{0}n}$$

The neutrons released in the fission process bombard other nuclei close by. If there is enough fissionable material, it can result in a **nuclear chain reaction,** *a self-sustaining sequence of nuclear fission reactions.* Nuclear power plants are designed to harness the energy from a controlled chain reaction to generate electricity.

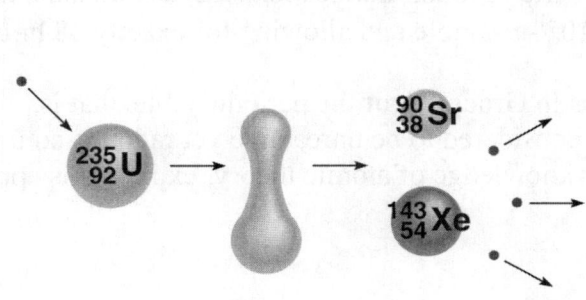

Nuclear fusion

Nuclear fusion results when *two smaller nuclei combine to form a larger one.* The most common fusion reaction, which occurs in the sun, fuses hydrogen nuclei to produce a helium nucleus. Because fusion reactions occur only at extremely high temperatures, they are often called **thermonuclear reactions.**

Fusion reactions are also accompanied by the release of an enormous amount of energy. Nuclear fusion could be a valuable energy source. However, the conditions required to create a controlled fusion reaction are hard to achieve.

Binding energy and mass defect

Einstein discovered the *interconvertability of mass and energy,* which is summarized in the equation $E = mc^2$. Therefore, the mass of a nucleus is a direct measure of its energy content. However, *the mass of the nucleus is always very slightly less than the sum of the separate masses of its constituent nucleons.* This small mass difference is called the **mass defect.** When the mass defect is expressed as energy applying Einstein's equation, it is called the **binding energy** of the nucleus. The binding energy is *the energy required to break up a nucleus into its component protons and neutrons.* The binding energy is a quantitative measure of nuclear stability.

Comprehension Questions

1)   Radon is a toxic gas that can accumulate in the basements of houses. It is a health hazard because, as a gas, it can be inhaled and, once inhaled, it decays to form alpha particles and a solid product that remains in the lungs.

   a)   Write a balanced nuclear equation indicating the primary decay product of radon.

   b)   The half-life of radon-222 is 3.824 days. Calculate the final mass of radon starting with a 1.50×10^{-4}-g sample and allowing for exactly 48 hr of decay.

   c)   Radon is in Group 8A of the periodic table, that is, it is a noble gas. These elements are normally considered to be unreactive yet radon is an unstable, reactive substance. Using your knowledge of atomic theory, explain this apparent paradox.

2)   a)   Differentiate between nuclear fission, nuclear fusion, and nuclear transmutation.

   b)   Describe what the term "belt of stability" means.

   c)   As atomic number increases, the neutron-to-proton ratio for known stable isotopes also increases. Provide a rationalization for this observed phenomenon based on your understanding of nuclear structure, stability, and forces.

3)   Technetium-95 is a short-lived isotope with an atomic mass of 94.90766 amu.

   a)   Calculate the nuclear binding energy in joules for a mole of technetium atoms.

   b)   Technetium-95 decays by electron capture. Write the nuclear symbol for its decay product.

   c)   A 10.0-g sample of technetium-95 was allowed to decay for 49.5 hr, at which time 1.8 g of the sample remained. Calculate the half-life of technetium-95.

Answers to Comprehension Questions

1)   a) $^{222}_{86}Rn \rightarrow \, ^{218}_{84}Po \, + \, ^4_2\alpha$

    b)   Nuclear decay is first order, and the half-life of a first-order reaction is equal to 0.693/k (k is the rate constant for the process). If this relationship is used to calculate the rate constant, then a concentration vs. time relationship can be used to calculate the final mass of the substance.

$t \, \frac{1}{2} = 3.824$ days $\times 24$ hr/day $= 91.78$ hr $= 0.693 \, / \, k$

$k = 7.55 \times 10^{-3} \, hr^{-1}$

ln (final amount) – ln (initial amount) $= -kt$

$\ln (X \, g \, / \, 1.50 \times 10^{-4} \, g) = - \, (7.55 \times 10^{-3} \, hr^{-1} \times 48 \, hr)$

$X \, / \, 1.50 \times 10^{-4} \, g = e^{\, - (7.55 \, \times \, 10^{-3} \, hr^{-1} \, \times \, 48 \, hr)}$

$X = \mathbf{1.04 \times 10^{-4} \, g}$

    c)   The simple answer here is that we are talking about two different types of reactivity. When it is stated that the noble gases are unreactive, people are referring to the formation and cleavage of chemical bonds; bonds created by sharing or trading electrons. This is extranuclear, electron chemistry, and, with respect to "electron chemistry," radon is rather unreactive. The reactivity that radon is known for is nuclear decay. Radioactive substances have unstable nuclei that spontaneously decay or decompose over time. When this happens, they emit radiation in various forms and produce a different element (in this case polonium).

2)   a)   Fission is the process by which a heavy or large nucleus decomposes into lighter particles and nuclei. Fusion is a joining together of smaller nuclei to form a larger nucleus. Both of these phenomena can be accompanied by the release of significant amounts of energy. Transmutation is the process by which one type of nucleus is bombarded by other nuclei with the hope that the two will partially stick together to form a new element.

    b)   Graphing the number of neutrons versus the number of protons for the first 83 elements on the periodic table yields a slightly curved plot with a positive slope of slightly greater than one. The best fit line through these points is the belt of stability. It lies just above the line that would indicate a 1:1 neutron-to-proton ratio.

c)   The neutron-to-proton ratio for the first 20 elements or so is approximately 1:1. Beyond this point, the ratio begins to exceed 1. The rationalization for this phenomenon is that as more positively charged protons are put into the nucleus, it becomes increasingly unstable due to the charge repulsions experienced between particles at such short distances. If neutral particles such as neutrons are added, they increase the attractive forces holding the nucleus together due to additional strong and weak nuclear forces, without adding to the repulsive Coulombic interactions experienced by the protons.

3)   a)   Technetium has atomic number 43 so it has 43 protons in its nucleus. Tc-95 has a total of 95 protons and neutrons, so it must have 52 neutrons. If you were to add up the total mass of all these nucleons, the mass would exceed the atomic mass. The difference in mass, or mass defect, has been converted into energy to hold the nucleus together. This energy can be calculated by plugging the mass difference into the famous $E = mc^2$.

total mass of nucleons = (43 protons $\times$ 1.007825 amu) + (52 neutrons $\times$ 1.008665 amu) = 95.787055 amu.

mass difference = nucleon mass – isotopic mass = 95.787055 amu – 94.90766 amu = 0.879395 amu $\times$ (1 kg / 6.022 $\times$ 10^{26} amu) = 1.460304 $\times$ 10^{-27} kg

$E = mc^2$ = 1.460304 $\times$ 10^{-27} kg $\times$ (3.00 $\times$ 10^8 m/s)^2 = 1.314 $\times$ 10^{-10} J per nucleus

1.314 $\times$ 10^{-10} J / nucleus $\times$ 6.022 $\times$ 10^{23} nuclei / mol = **7.91 $\times$ 10^{13} J mol**

b)   In electron capture an inner electron decays into the nucleus and combines with a proton to form a neutron. The atomic number of the element decreases by 1 and the mass number remains constant. So . . . .

the symbol is $^{95}_{42}$**Mo**

c)   Using the concentration versus time relationship for first-order reactions, ln ([A]$_t$ / [A]$_0$) = –kt, the rate constant can be determined for this process. The half-life of the reaction can then be determined by dividing 0.693 by k.

ln (1.8 / 10) = –k $\times$ 49.5 hr
k = 0.0346 hr^{-1}

t ½ = 0.693 / k = 0.693 / 0.0346 hr^{-1} = **20.0 hr**

CHAPTER 24
ORGANIC CHEMISTRY

This chapter reviews some basic concepts in organic chemistry, including:

- The classes, properties, and reactions of simple hydrocarbons
- Isomers
- The chemistry of functional groups

> **Take Note:** *You must have some rudimentary knowledge of simple organic nomenclature, basic functional groups, simple reactions, and isomers for the AP chemistry exam. You should note that many of the weak acids discussed in this book are, in fact, organic acids, the generic formula for which is written as R-**COOH**.*

Organic chemistry is *the branch of chemistry that deals exclusively with carbon-containing compounds.*

The classes, properties, and reactions of simple hydrocarbons

The **aliphatic** hydrocarbons *do not contain any benzene rings.* The aliphatic hydrocarbons are divided into alkanes, alkenes, and alkynes. If you need to review the names of these compounds, refer back to Chapter 2 of this review book.

Alkanes have the general formula C_nH_{2n+2}. They are **saturated hydrocarbons,** that is, *they contain the maximum number of hydrogen atoms that can bond with the number of carbon atoms present.* The only intermolecular forces at work among alkane molecules are dispersion forces. As the carbon chain becomes longer, there is an increase in the strength of the dispersion forces and a corresponding increase in melting and boiling points. Consequently, lighter alkanes are usually gases and heavier ones tend to be liquids and solids.

The alkanes are relatively unreactive. They are used for fuel (**combustion reactions**) and can undergo **substitution reactions** with the halogens.

$$CH_4\ (g)\ +\ 2O_2\ (g)\ \rightarrow\ CO_2\ (g)\ +\ 2H_2O\ (g) \qquad \text{combustion}$$

$$CH_4\ (g)\ +\ Cl_2\ (g)\ \rightarrow\ CH_3Cl\ (g)\ +\ HCl\ (g) \qquad \text{substitution}$$

Alkenes have the general formula C_nH_{2n}. They *contain at least one double bond.* **Alkynes** have the general formula C_nH_{2n-2}. They *contain at least one triple bond.* Alkenes and alkynes are referred to as **unsaturated hydrocarbons** because *the carbon atoms in these compounds have fewer than the maximum number of hydrogen atoms.* Like the alkanes, the alkenes and alkynes are subject to increasingly stronger dispersion forces as the chain lengthens. Lighter alkenes and alkynes are gases and heavier ones are liquids and solids.

The alkenes and alkynes are more reactive than the alkanes due to the presence of the double and triple bonds. The main reactions of alkenes and alkynes are **combustion and addition reactions.** The combustion reactions result in the formation of carbon dioxide and water. Halogens, hydrohalic acids, and hydrogen, in the presence of a catalyst, add easily to the double bond in alkenes and the triple bonds in alkynes.

$$C_2H_4 \ (g) \ + \ 3O_2 \ (g) \ \rightarrow 2CO_2 \ (g) \ + 2H_2O \ (g) \ \text{combustion}$$

$$C_2H_4 \ (g) \ + \ Br_2 \ (g) \ \rightarrow \ C_2H_4Br_2 \ (l) \qquad \text{halogen addition}$$

$$C_2H_4 \ (g) \ + \ HBr \ (g) \ \rightarrow \ C_2H_5Br \ (l) \qquad \text{hydrohalic acid addition}$$

$$C_2H_4 \ (g) \ + \ H_2 \ (g) \ \rightarrow \ C_2H_6 \ (g) \qquad \text{hydrogen addition}$$

Aromatic hydrocarbons *contain one or more benzene rings, C_6H_6, and are extremely* stable. The delocalized pi (π) electrons in a benzene system can be hydrogenated, but only with difficulty. Substitution reactions onto the benzene ring also are possible.

Isomers

As the alkane chain becomes longer, branching can occur.   The result is a molecule having the *same empirical formula but a different structure.* When different structures with the same empirical formula are possible, they are referred to as **isomers** (see the diagram).

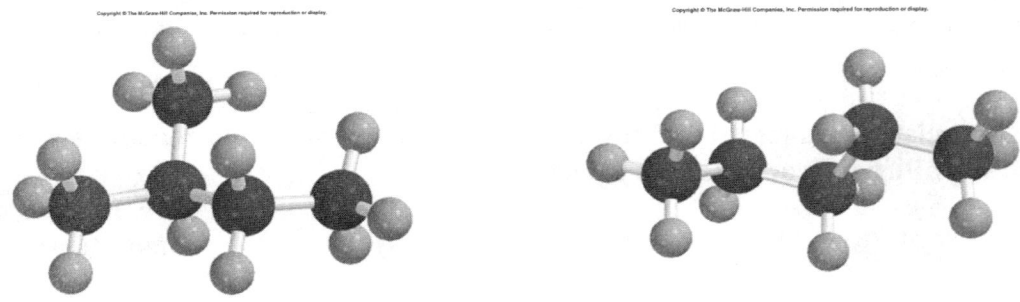

isopentane (branched chain)   *n*-pentane (straight chain)

In the case of alkenes, a **geometric isomer** can result. In the case of dichloroethylene, ClHC=CHCl, the chlorine atoms can be on the same side of the double bond (*cis*-isomer) or the chlorine atoms can be on the opposite sides of the double bond (***trans***-isomer) (see the diagram).

$$
\begin{array}{ccc}
\text{Cl} & & \text{Cl} \\
\diagdown & & \diagup \\
& \text{C} = \text{C} & \\
\diagup & & \diagdown \\
\text{H} & & \text{H}
\end{array}
\qquad\qquad
\begin{array}{ccc}
\text{Cl} & & \text{H} \\
\diagdown & & \diagup \\
& \text{C} = \text{C} & \\
\diagup & & \diagdown \\
\text{H} & & \text{Cl}
\end{array}
$$

cis-dichloroethylene   *trans*-dichloroethylene

cis- and *trans*- isomers tend to have distinctly different physical and chemical properties.

The chemistry of functional groups

You learned that inorganic chemistry has three basic classes of compounds: acids, bases, and salts. Organic chemistry has many more classes of compounds. After the primary hydrocarbons, the most common organic compounds are the oxygen- and nitrogen-containing compounds: alcohols, ethers, aldehydes and ketones, carboxylic acids, esters, and amines. Their properties and simple reactions are summarized in Table 24.4.

TABLE 24.4

Important Functional Groups and Their Reactions

Functional Group	Name	Typical Reactions
$\diagup$C=C$\diagdown$	Carbon-carbon double bond	Addition reactions with halogens, hydrogen halides, and water; hydrogenation to yield alkanes
—C≡C—	Carbon-carbon triple bond	Addition reactions with halogens, hydrogen halides; hydrogenation to yield alkenes and alkanes
—$\ddot{\text{X}}$: (X = F, Cl, Br, I)	Halogen	Exchange reactions: $CH_3CH_2Br + KI \longrightarrow CH_3CH_2I + KBr$
—$\ddot{\text{O}}$—H	Hydroxyl	Esterification (formation of an ester) with carboxylic acids; oxidation to aldehydes, ketones, and carboxylic acids

Structure	Name	Reactions
$\overset{..}{C}=\overset{..}{\underset{..}{O}}$	Carbonyl	Reduction to yield alcohols; oxidation of aldehydes to yield carboxylic acids
$-\overset{\displaystyle :O:}{\overset{\displaystyle \|}{C}}-\overset{..}{\underset{..}{O}}-H$	Carboxyl	Esterification with alcohols; reaction with phosphorus pentachloride to yield acid chlorides
$-\overset{\displaystyle :O:}{\overset{\displaystyle \|}{C}}-\overset{..}{\underset{..}{O}}-R$ (R = hydrocarbon)	Ester	Hydrolysis to yield acids and alcohols
$-\overset{..}{N}\overset{\displaystyle R}{\underset{\displaystyle R}{\big<}}$ (R = H or hydrocarbon)	Amine	Formation of ammonium salts with acids

CHAPTER 25
SYNTHETIC AND NATURAL ORGANIC POLYMERS

This chapter reviews synthetic and natural polymers. The discussion includes:

- The properties of polymers
- Synthetic organic polymers
- Proteins
- Nucleic acids

Take Note: *The AP exam has rarely included questions on synthetic or biological polymers.*

The properties of polymers

A **polymer** is *a molecular compound distinguished by a high molar mass, ranging up to millions of grams, and made up of many repeating units*. The properties of these macromolecules differ greatly from small, inorganic molecules. Many synthetic organic polymers, like nylon and plastics, are widely used in our technological society. Natural polymers, such as DNA, are the basis of all life processes.

Synthetic organic polymers

Synthetic organic polymers are made up of **monomers**, *simple repeating units*. These monomers form polymers in two ways: addition reactions and condensation reactions.

Addition reactions require an alkene monomer. The monomer's double bond is the source of a free radical that can easily add to free radicals of the same monomer. When a polymer is made up of *only one type of monomer*, the polymer is called a **homopolymer.** A **copolymer** is created when *two or more different monomers* react to create a polymer.

Condensation reactions result when two monomers condense out a water molecule between them and forge a polymer of alternate repeating monomers. The most famous of these is the condensation polymer created by adipic acid and hexamethylenediamine, which results in nylon-66.

Proteins

Amino acids are *compounds that contain at least one amino group ($-NH_2$) and at least one carboxyl group ($-COOH$).* There are 20 naturally occurring amino acids. Amino acids undergo condensation reactions to produce *long chains of amino acids* called **proteins.** Proteins play a key role in nearly every biological process. The human body contains an estimated 100,000 different kinds of proteins, each of which has a specific physiological function.

Nucleic acids

Nucleic acids are *high molar mass polymers that play an essential role in protein synthesis*. **Deoxyribonucleic acid (DNA)** and **ribonucleic acid (RNA)** are the two types of nucleic acids. DNA molecules are among the largest molecules known, having molar masses of up to tens of billions of grams. The RNA molecules vary greatly in size.

Advanced Placement Chemistry

Exam I

Note: For all questions, assume that the temperature is 298 K, the pressure is 1.00 atm, and solutions are aqueous unless otherwise noted.

Directions: Each set of lettered choices below refers to the numbered statements immediately following it. Select the one lettered choice that best answers each question or best fits each statement and then fill in the corresponding oval on the answer sheet. A choice may be used once, more than once, or not at all in each set.

Questions 1–4 refer to aqueous solutions containing 1:1 mole ratios of the following pairs of substances. Assume all concentrations are 0.10 M.

    A.   hydrochloric acid and sodium chloride
    B.   methylamine and methylammonium chloride
    C.   potassium hydroxide and ammonia
    D.   sodium hydroxide and hydrobromic acid
    E.   acetic acid and sodium acetate

1. A buffer with a pH greater than 7

2. A buffer with a pH less than 7

3. The solution with a pH of 7

4. The solution with the lowest pH

Use these answers for questions 5–8.

    A. dispersion forces
    B. VSEPR theory
    C. hydrogen bonding
    D. Valence Bond theory
    E. Lattice energy

5. Is used to explain why MgO has a much higher melting point than NaF.
6. Is used to explain why H_2O is a liquid at room temperature.

7. Is used to explain the fact that the six bonds in sulfur hexafluoride are equivalent.

8. Is used to explain the fact that the hydrogen-nitrogen-hydrogen bond angle in ammonia is approximately 107°.

Use these answers for questions 9–12.

A. $1s^2\, 2s^2\, 2p^6\, 3s^2\, 3p^5$
B. $1s^2\, 2s^2\, 2p^6\, 3s^2\, 3p^6$
C. $1s^2\, 2s^2\, 2p^6\, 2d^{10}\, 3s^2\, 3p^6$
D. $1s^2\, 2s^2\, 2p^6\, 3s^2\, 3p^6\, 3d^5$
E. $1s^2\, 2s^2\, 2p^6\, 3s^2\, 3p^6\, 3d^{10}$

9. The ground-state configuration of the iron(III) ion.

10. The ground-state configuration of the chloride ion.

11. The ground-state configuration of a common halogen atom.

12. The ground-state configuration of a common ion of an alkali metal.

Use these answers for questions 13–16.

A. nitric acid
B. nitrogen dioxide
C. hydrochloric acid
D. calcium carbonate
E. potassium permanganate

13. Is known as the oxidizing acid.
14. A strong oxidizing agent that changes color upon reduction.
15. Is known to be a natural neutralization agent.
16. Is known to contribute to photochemical smog.

Questions 17 – 20 refer to the diagram below:

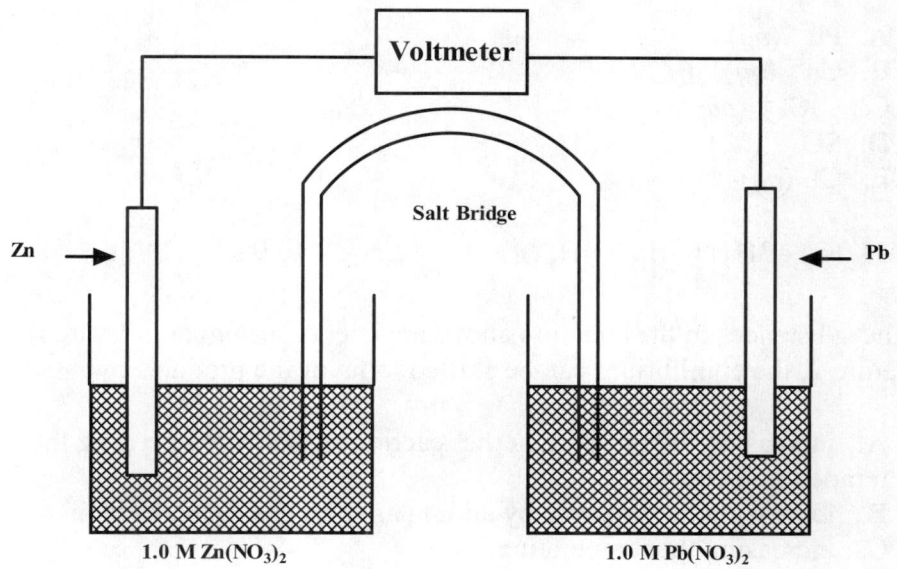

The spontaneous reaction that occurs when the cell above operates is:

$$Pb^{2+} + Zn(s) \rightarrow Pb(s) + Zn^{2+}$$

A. Voltage decreases.
B. Voltage increases.
C. Voltage becomes zero and remains at zero.
D. No change in voltage occurs.
E. Direction of voltage change cannot be predicted without additional information.

Which of the above occurs for each of the following circumstances?

17. A 20-mL sample of a 2.0M NaI solution is added to the beaker on the right.
18. The zinc electrode is replaced by a zinc electrode half the size of the original.
19. 100 mL of 2.0 M $Zn(NO_3)_2$ is added to the beaker on the left.
20. Current is allowed to flow for 15 min.

Directions: Each of the questions or incomplete statements below is followed by five suggested answers or completions. Select the one that is best in each case and then fill in the corresponding oval on the answer sheet.

21. A fluffy blue precipitate forms when 0.5 M NaOH (aq) is added to a 0.5 M solution of which of the following ions?

A. Pb^{2+} (aq)
B. Cu^{2+} (aq)
C. $C_2O_4^{2-}$ (aq)
D. SO_4^{2-} (aq)
E. Cl^- (aq)

22. $N_2(g) + 3H_2(g) \rightleftarrows 2NH_3(g)$ $\Delta E = -40.9$ kJ

When the substances in the equation above are at equilibrium at pressure P and temperature T, the equilibrium can be shifted to favor the products by

A. increasing the pressure in the reaction vessel while keeping the temperature constant
B. increasing the pressure by adding an inert gas such as argon
C. increasing the temperature
D. allowing some hydrogen gas to escape at constant P and T
E. adding a catalyst

23. Each of the following can act as both a Brønsted acid and a Brønsted base EXCEPT

A. H_2SO_3
B. $H_2PO_4^-$
C. HSO_4^-
D. H_2O
E. HCO_3^-

24. A student pipetted five 25.00-mL samples of acetic acid and transferred each sample to a separate beaker, diluted each sample with distilled water, and added a few drops of phenolphthalein to each sample. Each sample was then titrated with a sodium hydroxide solution to the appearance of the first permanent faint pink color. The following results were obtained.

Volumes of NaOH Solution

First sample..................22.25 mL
Second sample..............21.35 mL
Third sample.................21.37 mL
Fourth sample...............21.32 mL
Fifth sample..................21.38 mL

Which of the following is the most probable explanation for the variation in the student's results?

A. Varying amounts of water were added to the samples.
B. Less phenolphthalein was added to the first sample.
C. The student misread the buret.
D. The pipette was not rinsed with the acetic acid solution.
E. The buret was rinsed with distilled water but not rinsed with the NaOH solution.

25. Which of the following reactions has the largest positive value of ΔS?

A. $Na_2S\ (g) + 2HCl\ (g) \rightarrow 2NaCl\ (aq) + H_2S\ (g)$
B. $NH_4NO_3\ (s) \rightarrow N_2\ (g) + O_2\ (g) + 2H_2O\ (g)$
C. $2Na\ (s) + Cl_2\ (g) \rightarrow 2NaCl\ (s)$
D. $Cu_2O\ (s) + H_2\ (g) \rightarrow 2Cu\ (s) + H_2O\ (g)$
E. $H_2\ (g) \rightarrow 2H\ (g)$

26. In the periodic table, as the atomic number increases from 21 to 30, in general, what happens to the atomic radius?

A. It decreases only.
B. It decreases, then increases.
C. It remains fairly constant.
D. It increases only.
E. It increases, and then decreases.

27. Two flexible containers for gases are at the same temperature and pressure. One holds 16 g of oxygen and the other holds 22 g of carbon dioxide. Which of the following statements regarding these gas samples is TRUE?

A. The volume of the carbon dioxide container is the same as the volume of the oxygen container.
B. The number of molecules in the carbon dioxide container is greater than the number of molecules in the oxygen container.
C. The density of the carbon dioxide sample is the same as that of the oxygen sample.
D. The average kinetic energy of the carbon dioxide molecules is greater than the average kinetic energy of the oxygen molecules.
E. The average speed of the carbon dioxide molecules is greater than the average speed of the oxygen molecules.

28. What is the hybridization of Cl in the ClF_5 molecule?

A. sp^2
B. sp^3
C. dsp^2
D. dsp^3
E. d^2sp^3

Experiment	Initial [A] (mol L^{-1})	Initial [B] (mol L^{-1})	Initial Rate of Formation of product [C] (mol L^{-1} sec^{-1})
1	0.20	0.10	2.5×10^{-4}
2	0.20	0.20	2.5×10^{-4}
3	0.40	0.20	5.0×10^{-4}

29. The initial-rate data in the table above were obtained for the reaction represented below. What is the experimental rate law for the reaction?

$$A + B \rightleftharpoons 2C$$

A. rate = k[A]
B. rate = k[A] [B]
C. rate = k[A] [B]^2
D. rate = k[A]^2 [B]
E. rate = k[A] / [B]

30. In the titration of a weak base of unknown concentration with a solution of a strong acid, a pH meter was used to follow the progress of the titration. Which of the following represents the graph of the data collected for this experiment?

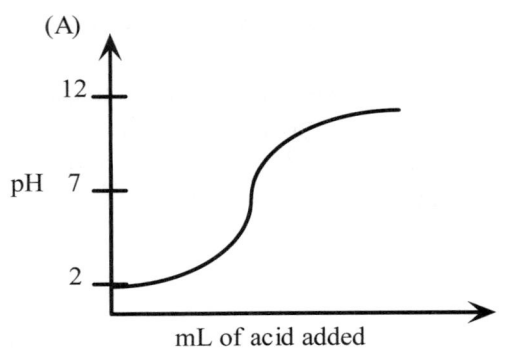

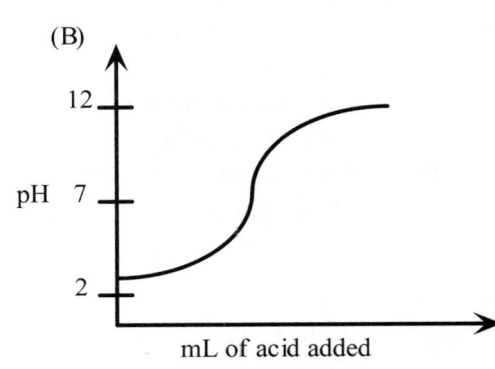

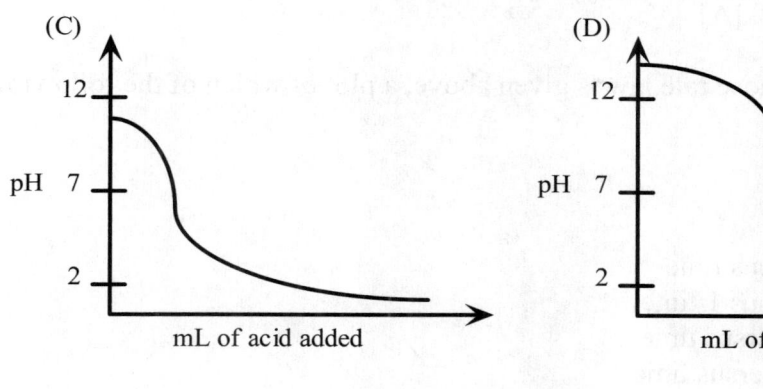

(C) — pH vs. mL of acid added

(D) — pH vs. mL of acid added

(E) — pH vs. mL of acid added

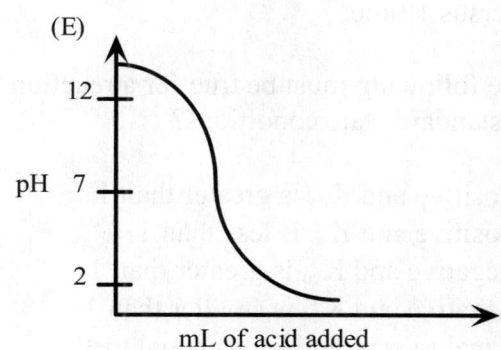

31. $^{109}_{47}Ag + ^{4}_{2}He \rightarrow 2^{1}_{0}n + \_?$

What is the missing product in the nuclear reaction represented above?

A. $^{111}_{49}In$

B. $^{112}_{49}In$

C. $^{111}_{47}Ag$

D. $^{112}_{47}Ag$

E. $^{105}_{45}Rh$

32. The Lewis dot structure of which of the following molecules would require resonance structures?

A. Br_2
B. O_2
C. NH_3
D. CH_4
E. N_2O

33.   rate $= k[A]$

For the reaction whose rate law is given above, a plot of which of the following is a straight line?

    A. [A] versus time
    B. [A] versus 1/time
    C. 1/[A] versus time
    D. ln [A] versus time
    E. ln [A] versus 1/time

34. Which of the following must be true for a reaction that proceeds spontaneously from initial standard state conditions?

    A. ΔG^0 is positive and K_{eq} is greater than 1
    B. ΔG^0 is positive and K_{eq} is less than 1
    C. ΔG^0 is negative and K_{eq} is greater than 1
    D. ΔG^0 is negative and K_{eq} is smaller than 1
    E. ΔG^0 is equal to zero and K_{eq} is equal to 1

35. Which of the following compounds contains the most pi bonds?

    A. $HC_2H_3O_2$
    B. $HClO_3$
    C. CO_2
    D. SO_2
    E. C_6H_6

Questions 36–37 refer to the following elements.

    A. sodium
    B. magnesium
    C. iodine
    D. mercury
    E. fluorine

36. Is a liquid in its standard state at 298 K.

37. Reacts violently with cold water to form a strong base.

Questions 38–39 refer to an electrolytic cell that involves the following half-reaction.

$$AgCl + e^- \rightarrow Ag + Cl^-$$

38. Which of the following occurs in the reaction?

    A. Ag is oxidized at the anode.
    B. AgCl is reduced at the cathode.
    C. Silver is converted from the −1 oxidation state to the 0 oxidation state.
    D. Cl^- acts as a reducing agent.
    E. Cl^- is oxidized at the anode.

39. A steady current of 3.0 A is passed though a copper-production (copper(II) ions) cell for 10 min. Which of the following is the correct expression for calculating the number of grams of copper produced? (1 Faraday = 96,500 C)

    A. $\dfrac{(2)(96,500)(10)}{(3)(63.5)}$

    B. $\dfrac{(3)(96,500)(10)}{(2)(63.5)(60)}$

    C. $\dfrac{(60)(63.5)(10)}{(3)(96,500)}$

    D. $\dfrac{(3)(60)(63.5)(10)}{(2)(96,500)}$

    E. $\dfrac{(3)(63.5)(10)}{(2)(96,500)}$

40. The half-life for radioactive element Z is 8.00 min. What mass of Z was originally present in a sample if 60.0 g is left after 40.0 min?

    A. 120. g
    B. 240. g
    C. 300. g
    D. 960. g
    E. 1920. g

41. Which of the following is a correct interpretation of the results of Thomson's experiments in which magnetic and electrical fields were applied to a cathode ray tube?

    A. Electrons in atoms are negative.
    B. The positive charge of an atom is found in a small dense region.
    C. Atoms consist of mostly empty space.
    D. Electrons' charge-to-mass ratio is a constant.
    E. Alpha particles are smaller than protons.

42. $2C_6H_6\,(l) + 15O_2\,(g) \rightarrow 12CO_2\,(g) + 6H_2O\,(l)$ $\Delta G^0 = -\,6400.\ \text{kJ/mol rxn}$

    $C\,(s) + O_2\,(g) \rightarrow CO_2\,(g)$ $\Delta G^0 = -\,400.\ \text{kJ/mol rxn}$

    $H_2\,(g) + \tfrac{1}{2}O_2\,(g) \rightarrow H_2O\,(l)$ $\Delta G^0 = -250.\ \text{kJ/mol rxn}$

What is the standard free energy change for the reaction below, as calculated from the data above?

$$6C\,(s) + 3H_2\,(g) \rightarrow C_6H_6\,(l)$$

    A. –250. kJ/mol rxn
    B. –100. kJ/mol rxn
    C. –50. kJ/mol rxn
    D. 50. kJ/mol rxn
    E. 100. kJ /mol rxn

43. $2X\,(g) + Y\,(g) \leftrightarrow 2Z\,(g)$

When the concentration of substance Y in the reaction above is tripled, all other factors being held constant, it is found that the rate of the reaction increases by a factor of nine. The most probable explanation for this observation is that

    A. The order of the reaction with respect to substance Y is three and the order with respect to X is six.
    B. Substance Y is involved in each of the steps in the mechanism of the reaction.
    C. Substance Y is involved in the rate-determining step of the mechanism, but is not involved in subsequent steps.
    D. The order of the reaction with respect to Y is two.
    E. The reactant with the smallest coefficient in the balanced equation generally has the greatest effect on the rate of the reaction.

44. Relatively fast rates of chemical reactions are associated with each of the following EXCEPT

    A.   the presence of a catalyst
    B.   high reaction temperature
    C.   high concentration of products
    D.   strong intramolecular forces in reactant molecules
    E.   low energy of activation

45. A 1.0-L sample of an aqueous solution contains 0.10 mol of $CaCl_2$ and 0.10 mol of NaCl. What is the minimum number of moles of $AgNO_3$ that must be added to the solution to precipitate all of the Cl^- as AgCl (s)? (Assume that AgCl is insoluble.)

    A.   0.10 mol
    B.   0.20 mol
    C.   0.30 mol
    D.   0.40 mol
    E.   0.60 mol

$$HCN\,(aq) + NH_3\,(aq) \Leftrightarrow NH_4^+\,(aq) + CN^-\,(aq)$$

46. The reaction represented above has an equilibrium constant equal to 1.12 at 500 K. Which of the following can be concluded from this information?

    A. $NH_3\,(aq)$ is a stronger base than $CN^-\,(aq)$.
    B. $NH_4^+\,(aq)$ is a stronger acid than HCN (aq).
    C.   The conjugate base of $NH_3\,(aq)$ is $CN^-\,(aq)$.
    D.   The equilibrium constant will increase with an increase in temperature.
    E.   The reaction is exothermic.

47.   A (g) + 2B $(g) \Leftrightarrow$ 2C (g)

When 0.50 mol of A and 1.25 mol of B are placed in an evacuated 1.00-L flask, the reaction represented above occurs. After the reactants and the product reach equilibrium and the initial temperature is restored, the flask is found to contain 0.25 mol of product C. Based on these results, the equilibrium constant, K_c for the reaction is:

A.   0.17
B.   0.63
C.   0.74
D.   1.1
E.   2.2

48. CO_2, $SiCl_4$, PCl_3, XeF_4, SF_6

Which of the following does not describe any of the molecules above?

A.   linear
B.   octahedral
C.   square planar
D.   tetrahedral
E.   trigonal bipyramidal

49. What is the pH of a 1.0×10^{-2} molar solution of a weak base with a K_b equal to 1.3×10^{-6} ?

A.   2
B.   between 4 and 7
C.   7
D.   between 7 and 11
E.   12

50. The volume of distilled water that should be **added** to 20.0 mL of 3.00 M NaOH (*aq*) to prepare a 0.200 M NaOH (*aq*) solution is approximately

A.   0.300 mL
B.   3.00 mL
C.   30.0 mL
D.   280. mL
E.   300. mL

$$N_2 \ (g) \ + \ 3H_2 \ (g) \ \Leftrightarrow 2NH_3 \ (g)$$

51. The reaction above takes place in a closed, rigid vessel. The initial pressure of $N_2 (g)$ is 1.0 atm and that of $H_2 (g)$ is 3.5 atm. There is no $NH_3 (g)$ initially present. The experiment is carried out at constant temperature. What is the total pressure in the container when the partial pressure of $NH_3 (g)$ reaches 0.90 atm?

    A.   0.10 atm
    B.   2.6 atm
    C.   3.1 atm
    D.   3.6 atm
    E.   5.4 atm

52. Appropriate uses of a visible-light spectrophotometer include which of the following?

I. Determining which ions are present in a solution that may contain Ni^{2+}, Co^{2+}, and Fe^{3+}

II. Measuring the conductivity of a solution of $CaCl_2$

III. Determining the concentration of a solution of $AgNO_3$

    A.   I only
    B.   II only
    C.   III only
    D.   I and II only
    E.   I and III only

53. $Cu^{2+} + 2Fe^{2+} \rightleftarrows Cu (s) + 2Fe^{3+}$

If the equilibrium constant for the reaction above is 2.5×10^{-4}, which of the following correctly describes the standard voltage, E^0, and the standard free energy change, ΔG^0, for this reaction?

    A. E^0 and ΔG^0 are both zero.
    B. E^0 and ΔG^0 are both positive.
    C. E^0 and ΔG^0 are both negative.
    D. E^0 is positive and ΔG^0 is negative.
    E. E^0 is negative and ΔG^0 is positive.

$$2X (s) + 3Ca^{2+} (aq) \rightarrow 3Ca (s) + 2X^{3+} (aq) \qquad E^0 = -1.21V$$
$$Ca^{2+} (aq) + 2e^- \rightarrow Ca (s) \qquad E^0 = -2.87 \text{ V}$$

54. According to the information above, what is the standard reduction potential for the half-reaction $X^{3+}(aq) + 3e^- \rightarrow X(s)$?

    A.   −4.08 V
    B.   −1.66 V
    C.   0.553 V
    D.   1.66 V
    E.   4.98 V

55. Which of the following techniques is most appropriate for the recovery of solid KCl from an aqueous solution of KCl?

    A.   Electrolysis
    B.   Filtration
    C.   Evaporation to dryness
    D.   Distillation
    E.   Fractional crystallization

56.   To determine the acid dissociation constant of a solid monoprotic acid, a student titrated a dry, weighed sample of the acid with standardized aqueous NaOH to the half-equivalence point. Which of the following could explain why the student obtained a K_a value that was too small?

    I. Failure to add the remaining acid solution into the titrated solution before measuring the pH

    II. Not adding enough water to dissolve the acid

    III. Addition of some base beyond the equivalence point

    A.   I only
    B.   III only
    C.   I and II only
    D.   I and III only
    E.   I, II, and III

Ionization Energies for element Q (kJ mol^{-1})		
First	Second	Third
520	7298	11,815

57. The ionization energies for element Q are listed in the table above. On the basis of the data, element Q is most likely to be

A. Li
B. Mg
C. Ne
D. Se
E. Br

58. Which, if any, of the following species is in the greatest concentration in a 0.500-molar solution of H_2SO_4 in water?

    A. H_2SO_4 molecules
    B. HSO_4^- ions
    C. SO_4^{2-} ions
    D. H_3O^+ ions
    E. All species are in equilibrium and therefore have the same concentrations.

59.

    I. Difference in temperature between freezing point of solvent and freezing point of solution
    II. Molal freezing point constant, K_f, for solvent

In addition to the information above, which of the following gives the minimum data required to determine the molecular mass of a soluble molecular nonvolatile substance by the freezing point depression technique?

    A. mass of solute
    B. density of solvent
    C. mass of solute and mass of solvent
    D. density of solution and volume of solution
    E. mass of solute, mass of solvent, and density of solvent

60. The simplest formula for a hydrocarbon that is 80.0 percent carbon by mass is

    A. CH
    B. CH_2
    C. CH_3
    D. C_2H_2
    E. C_2H_3

61. What is the balanced net ionic equation for the reaction that occurs when an excess of ammonia gas is bubbled through a solution saturated with silver chloride?

    A. $Ag^+ + Cl^- + NH_4^+ + OH^- \rightarrow NH_4Cl + AgOH$

    B. $Ag^+ + 2NH_3 + 2H_2O \rightarrow [Ag(OH)_2]^- + 2NH_4^+$

C. $Ag^+ + NH_3 + H_2O \rightarrow AgOH + NH_4^+$

D. $AgCl + 2NH_3 \rightarrow [Ag(NH_3)_2]^+ + Cl^-$

E. $2Ag^+ + 2NH_3 + H_2O \rightarrow Ag_2O + 2NH_4^+$

62. Which of the following is a nonelectrolyte when dissolved in water?

A.   NaOH
B. K_2SO_4
C. NH_4Br
D. CH_3COOH
E. CH_3OH

63. What is the final concentration of barium ions, $[Ba^{2+}]$, in solution when 200. mL of 0.20 M $Ba(NO_3)_2$ (*aq*) is mixed with 200. mL of 0.10 M $MgSO_4$ (*aq*)?

A.   0.00 M
B.   0.020 M
C.   0.050M
D.   0.060 M
E.   0.10 M

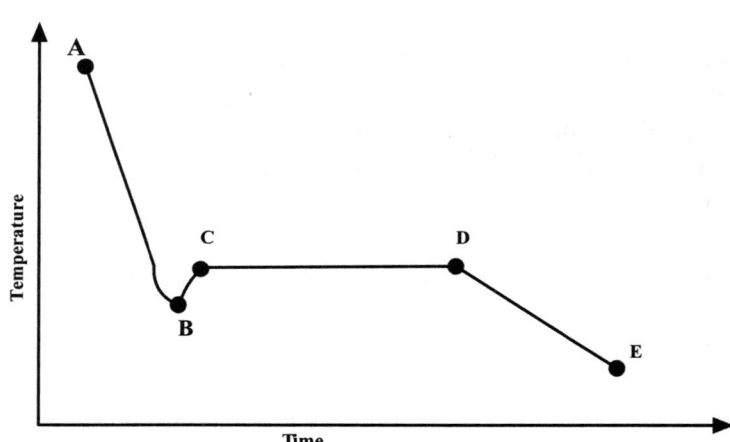

64. The cooling curve for a pure substance as it changes from a liquid to a solid is shown above. The part of the curve that represents supercooling is

A.   point B only
B.   point C only
C.   point D only
D.   all points on the curve between C and D
E.   all points on the curve between D and E

65. $CuCO_3$ (s) → CuO (s) + CO_2 (g)

A 0.50 mol sample of $CuCO_3$ (s) is placed in a 1 L evacuated flask, which is then sealed and heated. The $CuCO_3$ (s) decomposes completely according to the balanced equation above. The total pressure in the flask measured at 300. K is closest to which of the following? (The value of the gas constant, R, is 0.082 L·atm/ mol^{-1}· K^{-1}.)

    A.   1.0 atm
    B.   2.0 atm
    C.   10. atm
    D.   12 atm
    E.   14 atm

66. A sample of 0.020 mol of sulfur dioxide gas is confined at 227 °C and 3.0 atm. What would be the pressure of this sample at 127 °C and the same volume?

    A.   0.027 atm
    B.   0.27 atm
    C.   0.50 atm
    D.   1.20 atm
    E.   2.40 atm

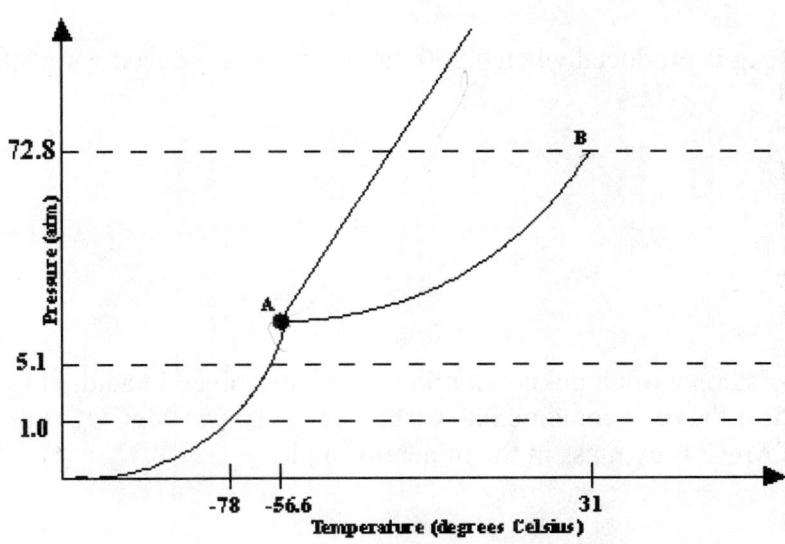

67. The critical point of the substance represented by the phase diagram above is

    A.   −78 °C
    B.   −56.6 °C
    C.   31 °C
    D.   greater than 31 °C
    E.   not determinable from the diagram

68. The phase diagram above provides sufficient information for determining all of the following EXCEPT

    A.   specific heat of the substance
    B.   conditions necessary for the substance to melt
    C.   relative density of the substance
    D.   conditions necessary for sublimation
    E.   the point where all phases of matter are in equilibrium with each other

69. For the substance represented in the diagram, which of the phases is most dense and which is least dense at –78 °C.

	Most dense	Least dense
A.	Solid	Gas
B.	Solid	Liquid
C.	Liquid	Solid
D.	Liquid	Gas
E.	The diagram gives no information about densities.	

70. What mass of Ag is produced when 0.250 mol of Ag_2S is reduced completely with excess H_2?

    A.   13.5 g
    B.   27.0 g
    C.   54.0 g
    D.   108 g
    E.   216 g

71. When a 50.4-g sample of an unknown mineral was dissolved in acid, 8.8 g of CO_2 was generated. If the rock contained no carbonate other than $MgCO_3$, what was the percent of $MgCO_3$ by mass in the mineral sample?

    A.   16%
    B.   33%
    C.   50%
    D.   67%
    E.   80%

72. $\_\_ Cr + \_\_ CrO_4^{2-} + \cancel{} \rightarrow \underline{2} Cr(OH)_3$

If 1 mol of CrO_4^{2-} oxidizes Cr according to the reaction represented above, how many moles of $Cr(OH)_3$ can be formed?

    A.   2
    B.   3
    C.   4
    D.   5
    E.   6

73. $H_2SO_3 + 2\,H_2O \rightleftarrows 2\,H_3O^+ + SO_3^{2-}$

Sulfurous acid, H_2SO_3, is a diprotic acid with $K_1 = 1.4 \times 10^{-2}$ and $K_2 = 6.3 \times 10^{-8}$. Which of the following is equal to the equilibrium constant for the reaction represented above?

    A. 1.0×10^{-14}
    B. 7.7×10^{-11}
    C. 8.8×10^{-10}
    D. 4.5×10^{-6}
    E. 2.2×10^{5}

74. A 50.0-mL sample of 0.100 M $ZnCl_2$ solution is added to 50.0 mL of 0.300 M Na_2CO_3 solution. A precipitate forms. The concentration of carbonate ion, in solution **after** reaction is

    A.   0.0 M
    B.   0.0050 M
    C.   0.10 M
    D.   0.15 M
    E.   0.400 M

75. The organic compound represented above is an example of

    A.   an alcohol
    B.   an organic acid
    C.   an ether
    D.   an aldehyde
    E.   a ketone

DO NOT DETACH FROM BOOK.

PERIODIC TABLE OF THE ELEMENTS

1																	2
H																	He
1.0079																	4.0026
3	4											5	6	7	8	9	10
Li	Be											B	C	N	O	F	Ne
6.941	9.012											10.811	12.011	14.007	16.00	19.00	20.179
11	12											13	14	15	16	17	18
Na	Mg											Al	Si	P	S	Cl	Ar
22.99	24.30											26.98	28.09	30.974	32.06	35.453	39.948
19	20	21	22	23	24	25	26	27	28	29	30	31	32	33	34	35	36
K	Ca	Sc	Ti	V	Cr	Mn	Fe	Co	Ni	Cu	Zn	Ga	Ge	As	Se	Br	Kr
39.10	40.08	44.96	47.90	50.94	52.00	54.938	55.85	58.93	58.69	63.55	65.39	69.72	72.59	74.92	78.96	79.90	83.80
37	38	39	40	41	42	43	44	45	46	47	48	49	50	51	52	53	54
Rb	Sr	Y	Zr	Nb	Mo	Tc	Ru	Rh	Pd	Ag	Cd	In	Sn	Sb	Te	I	Xe
85.47	87.62	88.91	91.22	92.91	95.94	(98)	101.1	102.91	106.42	107.87	112.41	114.82	118.71	121.75	127.60	126.91	131.29
55	56	57	72	73	74	75	76	77	78	79	80	81	82	83	84	85	86
Cs	Ba	*La	Hf	Ta	W	Re	Os	Ir	Pt	Au	Hg	Tl	Pb	Bi	Po	At	Rn
132.91	137.33	138.91	178.49	180.95	183.85	186.21	190.2	192.2	195.08	196.97	200.59	204.38	207.2	208.98	(209)	(210)	(222)
87	88	89	104	105	106	107	108	109	110	111	112						
Fr	Ra	†Ac	Rf	Db	Sg	Bh	Hs	Mt	Ds	Rg	§						
(223)	226.02	227.03	(261)	(262)	(266)	(264)	(277)	(268)	(271)	(272)	(277)						

§Not yet named

*Lanthanide Series	58	59	60	61	62	63	64	65	66	67	68	69	70	71
	Ce	Pr	Nd	Pm	Sm	Eu	Gd	Tb	Dy	Ho	Er	Tm	Yb	Lu
	140.12	140.91	144.24	(145)	150.4	151.97	157.25	158.93	162.50	164.93	167.26	168.93	173.04	174.97
†Actinide Series	90	91	92	93	94	95	96	97	98	99	100	101	102	103
	Th	Pa	U	Np	Pu	Am	Cm	Bk	Cf	Es	Fm	Md	No	Lr
	232.04	231.04	238.03	237.05	(244)	(243)	(247)	(247)	(251)	(252)	(257)	(258)	(259)	(262)

STANDARD REDUCTION POTENTIALS IN AQUEOUS SOLUTION AT 25°C

Half-reaction			$E°$(V)
$F_2\ (g) + 2e^-$	$\rightarrow$	$2F^-$	2.87
$Co^{3+} + e^-$	$\rightarrow$	Co^{2+}	1.82
$Au^{3+} + 3e^-$	$\rightarrow$	$Au(s)$	1.50
$Cl_2\ (g) + 2e^-$	$\rightarrow$	$2Cl^-$	1.36
$O_2\ (g) + 4H^+ + 4e^-$	$\rightarrow$	$2H_2O(l)$	1.23
$Br_2\ (l) + 2e^-$	$\rightarrow$	$2Br^-$	1.07
$2Hg^{2+} + 2e^-$	$\rightarrow$	Hg_2^{2+}	0.92
$Hg^{2+} + 2e^-$	$\rightarrow$	$Hg(l)$	0.85
$Ag^+ + e^-$	$\rightarrow$	$Ag(s)$	0.80
$Hg_2^{2+} + 2e^-$	$\rightarrow$	$2Hg(l)$	0.79
$Fe^{3+} + e^-$	$\rightarrow$	Fe^{2+}	0.77
$I_2\ (s) + 2e^-$	$\rightarrow$	$2I^-$	0.53
$Cu^+ + e^-$	$\rightarrow$	$Cu(s)$	0.52
$Cu^{2+} + 2e^-$	$\rightarrow$	$Cu(s)$	0.34
$Cu^{2+} + e^-$	$\rightarrow$	Cu^+	0.15
$Sn^{4+} + 2e^-$	$\rightarrow$	Sn^{2+}	0.15
$S(s) + 2H^+ + 2e^-$	$\rightarrow$	$H_2S(g)$	0.14
$2H^+ + 2e^-$	$\rightarrow$	$H_2(g)$	0.00
$Pb^{2+} + 2e^-$	$\rightarrow$	$Pb(s)$	–0.13
$Sn^{2+} + 2e^-$	$\rightarrow$	$Sn(s)$	–0.14
$Ni^{2+} + 2e^-$	$\rightarrow$	$Ni(s)$	–0.25
$Co^2 + 2e^-$	$\rightarrow$	$Co(s)$	–0.28
$Cd^{2+} + 2e^-$	$\rightarrow$	$Cd(s)$	–0.40
$Cr^{3+} + e^-$	$\rightarrow$	Cr^{2+}	–0.41
$Fe^{2+} + 2e^-$	$\rightarrow$	$Fe(s)$	–0.44
$Cr^{3+} + 3e^-$	$\rightarrow$	$Cr(s)$	–0.74
$Zn^{2+} + 2e^-$	$\rightarrow$	$Zn(s)$	–0.76
$2H_2O(l) + 2e^-$	$\rightarrow$	$H_2(g) + 2\ OH^-$	–0.83
$Mn^{2+} + 2e^-$	$\rightarrow$	$Mn(s)$	–1.18
$Al^{3+} + 3e^-$	$\rightarrow$	$Al(s)$	–1.66
$Be^{2+} + 2e^-$	$\rightarrow$	$Be(s)$	–1.70
$Mg^{2+} + 2e^-$	$\rightarrow$	$Mg(s)$	–2.37
$Na^+ + e^-$	$\rightarrow$	$Na(s)$	–2.71
$Ca^{2+} + 2e^-$	$\rightarrow$	$Ca(s)$	–2.87
$Sr^{2+} + 2e^-$	$\rightarrow$	$Sr(s)$	–2.89
$Ba^{2+} + 2e^-$	$\rightarrow$	$Ba(s)$	–2.90
$Rb^+ + e^-$	$\rightarrow$	$Rb(s)$	–2.92
$K^+ + e^-$	$\rightarrow$	$K(s)$	–2.92
$Cs^+ + e^-$	$\rightarrow$	$Cs(s)$	–2.92
$Li^+ + e^-$	$\rightarrow$	$Li(s)$	–3.05

ADVANCED PLACEMENT CHEMISTRY EQUATIONS AND CONSTANTS

ATOMIC STRUCTURE

$$E = h\nu \qquad c = \lambda\nu$$

$$\lambda = \frac{h}{m\upsilon} \qquad p = m\upsilon$$

$$E_n = \frac{-2.178 \times 10^{-18}}{n^2}\text{ joule}$$

EQUILIBRIUM

$$K_a = \frac{[H^+][A^-]}{[HA]}$$

$$K_b = \frac{[OH^-][HB^+]}{[B]}$$

$$K_w = [OH^-][H^+] = 1.0 \times 10^{-14} \qquad @\ 25\ °C$$

$$= K_a \times K_b$$

$$pH = -\log[H^+],\ pOH = -\log[OH^-]$$

$$14 = pH + pOH$$

$$pH = pK_a + \log\frac{[A^-]}{[HA]}$$

$$pOH = pK_b + \log\frac{[HB^+]}{[B]}$$

$$pK_a = -\log K_a,\ pK_b = -\log K_b$$

$$K_p = K_c(RT)^{\Delta n},$$

where Δn = moles product gas – moles reactant gas

THERMOCHEMISTRY/KINETICS

$$\Delta S^0 = \sum S^0 \text{ products} - \sum S^0 \text{ reactants}$$

$$\Delta H^0 = \sum \Delta H_f^0 \text{ products} - \sum \Delta H_f^0 \text{ reactants}$$

$$\Delta G^0 = \sum \Delta G_f^0 \text{ products} - \sum \Delta G_f^0 \text{ reactants}$$

$$\Delta G^0 = \Delta H^0 - T\Delta S^0$$

$$= -RT \ln K = -2.303\ RT \log K$$

$$= -n\,F\,E^0$$

$$\Delta G = \Delta G^0 + RT \ln Q = \Delta G^0 + 2.303\ RT \log Q$$

$$q = mc\Delta T$$

$$C_p = \frac{\Delta H}{\Delta T}$$

$$\ln[A]_t - \ln[A]_0 = -kt$$

$$\frac{1}{[A]_t} - \frac{1}{[A]_0} = kt$$

$$\ln k = \frac{-E_a}{R}\left(\frac{1}{T}\right) + \ln A$$

E = energy υ = velocity
ν = frequency n = principal quantum number
λ = wavelength m = mass
p = momentum

Speed of light, $c = 3.0 \times 10^8$ m s^{-1}
Planck's constant, $h = 6.63 \times 10^{-34}$ J s
Boltzmann's constant, $k = 1.38 \times 10^{-23}$ J K^{-1}
Avogadro's number = 6.022×10^{23} mol^{-1}
Electron charge, $e = -1.602 \times 10^{-19}$ C
1 electron volt per atom = 96.5 kJ mol^{-1}

Equilibrium Constants

K_a (weak acid)
K_b (weak base)
K_w (water)
K_p (gas pressure)
K_c (molar concentrations)

S^0 = standard entropy
H^0 = standard enthalpy
G^0 = standard free energy
E^0 = standard reduction potential
T = temperature
n = moles
m = mass
q = heat
c = specific heat capacity
C_p = molar heat capacity at constant pressure
E_a = activation energy
k = rate constant
A = frequency factor

Faraday's constant, F = 96,500 coulombs per mole of electrons

Gas constant, R = 8.31 J mol^{-1} K^{-1}
= 0.0821 L·atm mol^{-1}·K^{-1}
= 8.31 volt coulomb mol^{-1} K^{-1}

GASES, LIQUIDS, AND SOLUTIONS

$$PV = nRT$$

$$\left(P + \frac{n^2 a}{V^2}\right)(V - nb) = nRT$$

$$P_A = P_{total} \times X_A, \text{ where } X_A = \frac{\text{moles A}}{\text{total moles}}$$

$$P_{total} = P_A + P_B + P_C + \ldots$$

$$n = \frac{m}{M}$$

$$K = {}^\circ C + 273$$

$$\frac{P_1 V_1}{T_1} = \frac{P_2 V_2}{T_2}$$

$$D = \frac{m}{V}$$

$$u_{rms} = \sqrt{\frac{3kT}{m}} = \sqrt{\frac{3RT}{M}}$$

$$KE \text{ per molecule} = \frac{1}{2}mv^2$$

$$KE \text{ per mole} = \frac{3}{2}RT$$

$$\frac{r_1}{r_2} = \sqrt{\frac{M_2}{M_1}}$$

molarity, M = moles solute per liter solution
molarity, m = moles solute per kilogram solvent

$$\Delta T_f = iK_f \times \text{molality}$$

$$\Delta T_b = iK_b \times \text{molality}$$

$$\pi = iMRT$$

$$A = abc$$

OXIDATION-REDUCTION; ELECTROCHEMISTRY

$$Q = \frac{[C]^c [D]^d}{[A]^a [B]^b}, \text{ where } aA + bB \rightarrow cC + dD$$

$$I = \frac{q}{t}$$

$$E_{cell} = E^0_{cell} - \frac{RT}{nF} \ln Q = E^0_{cell} - \frac{0.0592}{n} \log Q \text{ @ } 25\ ^\circ C$$

$$\log K = \frac{nE^0}{0.0592}$$

P = pressure
V = volume
T = temperature
n = number of moles
D = density
m = mass
v = velocity

u_{rms} = root-mean-square speed
KE = kinetic energy
r = rate of effusion
M = molar mass
π = osmotic pressure
i = van'tHoff factor
K_f = molal freezing-point depression constant
K_b = molal boiling-point elevation constant
A = absorbance
a = molar absorptivity
b = path length
c = concentration
Q = reaction quotient
I = current (amperes)
q = charge (coulombs)
t = time (seconds)
E^0 = standard reduction potential
K = equilibrium constant

Gas constant, R = 8.31 J mol^{-1} K^{-1}
= 0.0821 L·atm mol^{-1}·k^{-1}
= 8.31 volt coulomb mol^{-1} K^{-1}

Boltzmann's constant, k = 1.38 × 10^{-23} J K^{-1}

K_f for H_2O = 1.86 K kg mol^{-1}
K_b for H_2O = 0.512 K kg mol^{-1}
1 atm = 760 mm Hg
= 760 torr

STP = 0.000°C and 1.000 atm

Faraday's constant, $\mathscr{F}$ = 96,500 coulombs per mole of electrons

Free Response Questions

(Equilibrium Problem)

1)   Formic acid, HCO_2H, dissociates in water according to the following balanced equation:

$$HCO_2H\ (aq) + H_2O\ (l) \leftrightarrow HCO_2^-\ (aq) + H_3O^+(aq) \qquad K_a = 1.8 \times 10^{-4}$$

    a)   Write an equilibrium constant expression for the dissociation of the above acid.

    b)   What quantity, in grams, of formic acid must be dissolved in 450.0 mL of water to create a solution with a pH of 2.49 (assume no change in volume upon addition of formic acid to the water)?

    c)   i)   What quantity, in grams, of potassium formate, $KHCO_2$, must be added to 745 mL of 0.1242 M formic acid to create a buffer solution with a pH of 4.10 (assume no change in volume upon addition of potassium formate)?

        ii)   Calculate the pH of the solution resulting from the addition of 0.750 g of solid sodium hydroxide, NaOH, to the buffer solution created in c) i) above.

    d)   A chemist wishes to perform a titration, versus sodium hydroxide, to determine the precise concentration of a formic acid solution (both acid and base have concentrations of approximately 0.1 M). The endpoint of the titration will be determined through the use of an indicator. Which of the following indicators would be most appropriate for this particular titration? Explain your reasoning.

        bromphenol blue $K_a = 1.0 \times 10^{-4}$

        bromthymol blue $K_a = 1.0 \times 10^{-7}$

        phenolphthalein $K_a = 6.3 \times 10^{-10}$

(Lab-related Question)

2)   A student is given a 5-g sample of a volatile organic liquid, and is asked to determine its molar mass. She decides that she will determine the mass of its vapor in an Erlenmeyer flask with a known interior volume. To do this she heats a sample of the

substance using a boiling water bath until all of the liquid becomes vapor. The container is fitted with a stopper that has a pinhole to let excess vapor escape from the flask.

a)   What data will this student need to collect to determine the molar mass of the compound?

b)   The student determines the complete interior volume of the flask by obtaining a tare weight for the empty, dry flask (145.671 g), and then a weight for the flask when it is filled to the rim with deionized water (415.433 g). Assuming the density of water to be 1.00 g/cm^3, calculate the interior volume of the flask.

c)   As the liquid was being vaporized, the temperature of the boiling water bath was determined to be 99.7 °C, and, on the day the experiment was run, the atmospheric pressure in the lab was 99.75 kPa. After the liquid was vaporized, and excess vapor allowed to escape through the pinhole, the flask was cooled to room temperature and re-weighed. The dry flask plus condensed volatile liquid had a combined mass of 146.426 g. Determine the molar mass of the liquid.

d)   The volatile liquid was later identified as ethyl acetate, $C_4H_8O_2$. What was this student's percent error for the determination of the liquid's molar mass.

e)   The vapor created by heating a volatile liquid to just above its boiling point deviates significantly from ideal behavior. Using your understanding of the kinetic molecular theory of gases and nonideal behavior, explain, both quantitatively and qualitatively, the effect that this would have on the results of this experiment, and discuss the effects that this source of error would have on the results of this experimental determination of molar mass.

3)   Use the following data to answer the questions below:

$C\ (s) + O_2\ (g) \rightarrow CO_2\ (g)$	$\Delta H = -393.5$ kJ
$2H_2\ (g) + O_2\ (g) \rightarrow 2\ H_2O\ (l)$	$\Delta H = -285.8$ kJ
$CH_4\ (g) + 2O_2\ (g) \rightarrow 2\ H_2O\ (l) + CO_2\ (g)$	$\Delta H = -890.3$ kJ

a)   What is the heat of formation for methane (i.e., what is the value of ΔH for the reaction below)?

$$C\ (s) + 2H_2\ (g) \rightarrow CH_4\ (g)$$

b)   i)   Is the above reaction endothermic or exothermic?

ii)   Draw an enthalpy diagram for the above reaction (formation of methane from its elements) and label it as completely as possible.

c)   Predict the sign of $\Delta S°$ for the above formation of methane.

d)   i)   Using your knowledge of thermodynamic principles, what factors, enthalpy change, entropy change, or both, would contribute to reaction spontaneity in the above synthesis of methane? Explain your reasoning.

ii)   Could temperature possibly have an effect on the spontaneity of this reaction? Explain.

4)   Listed below are one or more reactants for a chemical reaction. In every case a reaction will occur under the stated conditions. Unless otherwise stated, the temperature is assumed to be 25 °C and the pressure of gases, including the atmosphere, to be 1.0 atm. In each case, predict the product(s) and provide a balanced (with lowest whole-number coefficients), net ionic equation. It is not necessary to provide states of matter (e.g., solid, liquid, etc.). Answer the follow-up question for each reaction based on your prediction of products.

a)   a solution of lithium hydroxide is slowly added to a solution of ammonium bromide

i)   What observations (with one or more of the five senses) would a student make while performing this particular reaction?

b)   solid calcium sulfate is heated

i)   How would this reaction be affected by conducting it in a closed / sealed vessel?

c)   ethane gas is exposed to an open flame in the presence of air

i)   Is the above reaction likely to be endothermic or exothermic?

5)   Disposable butane lighters contain a mixture of flammable hydrocarbons consisting primarily of the four-carbon alkane, butane, C_4H_{10}. For the purposes of this

problem, we can assume that the molecular formula of the fuel is that of butane. These small lighters contain approximately 0.1 mol of fuel.

a)   Write a balanced equation for the combustion of butane.

b)   Imagine that the lighter is placed inside a fixed volume container along with 0.1 mol of oxygen, and it is ignited and the valve held open so that the reaction may continue until one of the reactants is exhausted.

   i)   Is there a limiting reactant? If so, what is its formula?

   ii)   If one of the reactants is in excess, by what quantity, in moles, is it in excess?

   iii)   Would the pressure inside the container increase, decrease, or stay the same as the reaction proceeds?

c)   i)   Draw Lewis structures for the carbon-containing reactant and product.

   ii)   How many sigma bonds and how many pi bonds are there in one molecule of the reactant, butane? In one molecule of the carbon-containing product?

   iii)   What is the hybridization around the carbon atom in butane? In the carbon-containing product?

6)   The elements nitrogen, phosphorus, and arsenic are in the same family or group of the periodic table.

a)   Write electron configurations for each of these elements (the abbreviated or core notation is sufficient for full credit).

b)   i)   What similarities and differences do you notice among these three electron configurations?

   ii)   What chemical or physical property relationships would you predict based on your analysis in the above question?

c)   Rank the elements nitrogen, phosphorus, and arsenic in terms of increasing atomic radius.

d)   Rank these same three elements in terms of increasing first ionization energy.

e)   Phosphorus is in the middle, top to bottom, of the other two elements. Write electron configurations (core notation is sufficient for full credit) for its neighbors to the left and right, that is, silicon and sulfur.

f)   Rank phosphorus, silicon, and sulfur in terms of increasing atomic radius and first ionization energy.

ANSWERS AND EXPLANATIONS TO PRACTICE EXAM I

1.

B

A buffer is a solution that has the ability to maintain a relatively constant pH by having components of an acid and a base. Methylamine is a weak base and would be able to react with excess acid, while the methylammonium chloride is a salt of the base.
The methylammonium ion would be able to react with any extra base added to the system.

$$\text{Addition of acid:} \quad CH_3NH_2 + H^+ \rightarrow CH_3NH_3^+$$

$$\text{Addition of base:} \quad CH_3NH_3^+ + OH^- \rightarrow CH_3NH_2 + H_2O$$

Since no additional H^+ or OH^- ions are produced, the pH remains relatively constant. The pH for choice C would be greater than 7 also, but there is no component present in the solution to react with excess base. Therefore, this is just a mixture of bases and not a buffer.

2.

E

See buffer explanation above. Acetic acid would be able to react with excess base while the acetate ion serves as a conjugate base with the ability to react with excess acid.

$$\text{Addition of base:} \quad HC_2H_3O_2 + OH^- \rightarrow H_2O + C_2H_3O_2^-$$

$$\text{Addition of acid:} \quad C_2H_3O_2^- + H^+ \rightarrow HC_2H_2O_2$$

Since no additional H^+ or OH^- ions are produced, the pH remains relatively constant.

3.

D

Potassium hydroxide is a strong base and hydrobromic acid is a strong acid. The net ionic reaction for this is $H^+ + OH^- \rightarrow H_2O$.

4.

A

A 1.0 M solution of HCl would have the smallest pH (the greatest H+ concentration). NaCl is a neutral salt and will not affect the pH.

5.

E

Lattice energy is often simplified into a relationship of charge divided by size. Mg and O (+2 and –2) has greater charge than Na and F (+1 and –1). Overall, size is very similar in both compounds. Greater charge will yield greater lattice energy.

6.

C

Hydrogen bonding is a very strong intermolecular force of attraction. Hydrogen from one water molecule is attracted to an unshared pair of electrons on the oxygen atom in an adjacent molecule.

7.

D

The hybridization of sulfur in SF_6 is d^2sp^3. The electron configuration for a sulfur atom is $1s^22s^22p^63s^23p^4$. The six outer electrons in sulfur are able to form six equivalent bonds when hybridized.

8.

B

Ammonia has one unshared pair of electrons on the central nitrogen. Unshared electrons have greater repulsive force than shared electron pairs. This idea is the basis of the valence-shell electron-pair repulsion theory.

$$H - \overset{..}{\underset{|}{N}} - H$$
$$H$$

9.

D

Iron(III) ion has lost three electrons. The $4s$ electrons are lost as well as one of the $3d$ electrons.

10.

B

The chloride ion has one more electron than its neutral atom. The additional electron is added to the $3p$ orbital.

11.

A

Halogens are found in Group 7A or 17. The general ending configuration for this family is ns^2np^5.

12.

B

Alkali metals are found in Group 2A. The general ending configuration for the family is ns^2.

13.

A

Nitric acid serves as an oxidizing agent in an acidic solution and has the ability to oxidize stable metals such as copper when other acids do not.

14.

E

$KMnO_4$ is purple when in aqueous solution. The permanganate ion is easily reduced through a variety of oxidation states. The color changes from purple to green, then to pink as Mn^{2+}.

15.

D

Calcium carbonate, $CaCO_3$, is commonly known as limestone. When acidic solutions such as acid rain fall into lakes, rivers, or streams located over high concentrations of this rock, the solution is often neutralized.

$$CaCO_3 + H^+ \rightarrow Ca^{2+} + CO_2 + H_2O$$

16.

B

Emissions such as NO_x and SO_x, when combined with ozone, oxygen, and hydrocarbons often produce photochemical smog. NO_2 is a brown gas that is thought to contribute to this problem.

17.

A

The reaction that takes place upon addition of NaI is shown below.

$$NaI\ (aq) + Pb(NO_3)_2\ (aq) \rightarrow PbI_2\ (s) + NaNO_3\ (aq)$$

When lead(II) ions combine with iodide ions the number of lead(II) ions in solution decreases. Using the Nernst equation, $E = E^0 - \dfrac{RT}{n\mathrm{F}} \log \dfrac{(Zn^{2+})}{(Pb^{2+})}$, when the concentration of lead(II) is smaller, the log of Q will be greater than 1, which ultimately will cause a reduction in cell potential.

18.

D

The size of the zinc electrode will not cause any change in cell potential. As long as zinc metal is present there is a source of electrons to allow the reaction to proceed.

19.

A

Increasing the concentration of zinc ions in solution will cause a decrease in cell potential. Using the Nernst equation, $E = E^0 - \dfrac{RT}{n\mathrm{F}} \log \dfrac{(Zn^{2+})}{(Pb^{2+})}$, when the concentration of zinc is larger, the log of Q will be greater than 1, which ultimately will cause a reduction in cell potential.

20.

A

Current flows from areas of high potential to areas of lower potential. After a galvanic cell, a battery, is allowed to run for a period of time it begins to discharge. The rate of discharge is dependent on many factors including the temperature, the current, and the voltage of the cell.

21.

B

The reaction that occurs is:

$Cu^{2+} (aq) + OH^- (aq) \rightarrow Cu(OH)_2 (s)$

Copper(II) ions impart the blue color to the insoluble hydroxide that is formed. Since all sodium salts are soluble, the only other possible answer for this question is choice A, Pb^{2+}. A precipitate of milky white $Pb(OH)_2$ would form.

22.

A

In the production of ammonia note that all components of the system are in the gaseous phase. Note also that the reaction is exothermic as shown by a negative enthalpy. Choices B and E have no effect on equilibrium systems. Choice C, increasing the temperature, would drive the reaction toward the reactants since the reaction is exothermic. Choice D, removing a reactant such as hydrogen, would cause the system to shift towards the reactants to replace the lost reactant. Choice A, increasing the pressure of the system, would cause a shift to the side of the reaction with the least number of moles of gas, the product side.

23.

A

By definition, a Brønsted acid donates a proton (an H^+) and a Brønsted base accepts a proton. H_2SO_3 is able to donate a proton becoming HSO_3^- but is unable to accept a proton. All other species are able to donate and accept in the following fashion:

$$\text{Acid: } H_2PO_4^- \rightarrow HPO_4^{2-} + H^+$$
$$\text{Base: } H_2PO_4^- + H^+ \rightarrow H_3PO_4$$

24.

E

The first sample volume appears to be larger than the others throughout the experiment. It took more NaOH for neutralization. The best explanation is that when the student prepared the buret, distilled water droplets were left inside of the buret, which diluted the standard NaOH solution.

25.

B

Entropy is a measure of disorder or chaos. Positive values indicate a system that increases in disorder. Solids are the most organized state of matter, so when 1 mol of solid decomposes into 4 mol of gaseous product, there is much greater disorganization.

26.

C

Elements with atomic numbers 21–30 are located in the transition group on the periodic table. Each of these metals contains four energy levels in which the last electrons are being added to the $3d$ sublevel. Since shielding is the same and electrons are filling the inner level, the atomic sizes are very similar.

27.

A

16 g of O_2 is equal to 0.50 mol of oxygen; 22 g of CO_2 is equal to 0.50 mol of carbon dioxide. Avogadro's hypothesis states that "equal volumes of gases at the same temperature and pressure have the same number of molecules." Since we have the same number of moles, we also have the same number of molecules.

28.

E

Cl is bonded in five positions to F and has one unshared pair of electrons. Six areas of electron density results in d^2sp^3 hybridization.

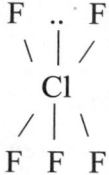

29.

A

In experiments 1 and 2 [A] is held constant while [B] doubles. The rate remains unchanged. Therefore, the reaction is zero order with respect to [B]. In experiments 2 and 3 [B] is held constant while [A] is doubled. The rate also doubles so the reaction is first order with respect to [A].

30.

C

When asked to choose the appropriate curve, search for the key points. The titration begins with a weak base; something with a pH between 10 and 12 is typical. The titration ends with a strong acid; something around pH of 1 to 2 is typical. Only one graph fits these criteria. Graph D represents a strong base with a weak acid, while graph E represents a strong base with a strong acid.

31.

A

In finding the missing component of a nuclear reaction the mass numbers and the atomic numbers must add up on both sides of the equation.

Reactant side: $^{113}_{49}$Total **Product side:** $^{2}_{0}$Total **Missing product** must be: $^{111}_{49}$In

32.

E

A resonance structure is an alternate structure that can be drawn to represent a different position for a multiple bond. Bromine, ammonia, and methane only contain single bonds. The double bond in oxygen only has one possible position between the two oxygen atoms. Dinitrogen monoxide has more than one possible representation.

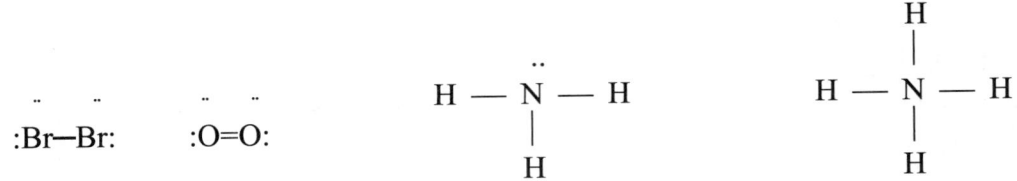

33.

D

The rate law given is first order with respect to [A]. The integrated rate law for a first-order reaction written in the form of $y = mx + b$ is $\ln[A]_t = -kt + \ln[A]_o$ where y is $\ln[A]$ and x is time.

34.

C

For a reaction to proceed spontaneously, the free energy must have a negative value.

$K_{eq} = \dfrac{[\text{products}]}{[\text{reactants}]}$ will be greater than 1 for a reaction proceeding in the forward direction.

35.

E

Pi overlap occurs when multiple bonding is present. Choices B, C, and D each contain two multiple bonds. Choice E, the benzene ring, contains three multiple bonds.

36.

D

Mercury metal is found as a liquid at room temperature while Na, Mg, and I_2 are in the solid state and F_2 is found in the gaseous state.

37.

A

Reacting violently with water is a characteristic of Group 1A metals. Sodium is found in this family. The net ionic equation for this reaction is:

$$2 \, Na \, (s) + 2 \, H_2O \, (l) \rightarrow 2 \, Na^+ \, (aq) + 2 \, OH^- \, (aq) + H_2 \, (g)$$

38.

B

The silver ion in AgCl has an oxidation state of +1. The silver metal on the product side has an oxidation state of 0. The chloride ion in AgCl has an oxidation state of −1 and the chloride ion on the product side has an oxidation state of −1. Therefore, the AgCl undergoes reduction from +1 to 0. Reduction is the gain of electrons and always occurs at the cathode while oxidation is a loss of electrons and always occurs at the anode.

39.

D

Calculating grams from the given information becomes manageable if time is taken to set up the problem using dimensional analysis. (Remember that an amp = Coul/s)

$$(10 \, min) \times \frac{(60 \, sec)}{(1 \, min)} \times \frac{(3.0 \, C)}{(1 \, sec)} \times \frac{(1 \, mol \, e^-)}{(96,500 \, C)} \times \frac{(1 \, mol \, Cu)}{(2 \, mol \, e^-)} \times \frac{(63.5 \, g \, Cu)}{(1 \, mol \, Cu)}$$

40.

E

Half-life is the time that it takes for one-half of a sample to decay. Working backward by creating a table will prove beneficial. (Reduce the time by 8.0 min, the half-life, while doubling the mass each time.)

Time (minutes)	Amount (grams)
40.0	60.0
32.0	120.0
24.0	240.0
16.0	480.0
8.0	960.0
0	1920.0

41.

D

The discovery of the electron led to Thomson's work with cathode ray tubes. In his classical experiment, he applied both magnetic and electrical forces simultaneously to a cathode ray tube. From the data gathered he was able to calculate the ratio of the electron's charge to its mass.

42.

D

Use Hess' law to find the free energy for the equation given. Since C_6H_6 (l) is found on the product side, the first reaction must be reversed (change the sign of ΔG^0). Multiply the first reaction by ½ to achieve only one C_6H_6 (l).

½ and reverse: (2 C_6H_6 (l) + 15 O_2 (g) → 12CO_2 (g) + 6 H_2O (l)) ΔG^0 = +3200. kJ/mol rxn

Multiply by 6: C (s) + O_2 (g) → CO_2 (g) ΔG^0 = –2400. kJ/mol rxn

Multiply by 3: H_2 (g) + ½ O_2 (g) → H_2O (l) ΔG^0 = –750. kJ/mol rxn

Add the free energies: +3200 + (–2400) + (–750) = +50. kJ/mol rxn

43.

D

Set up a general form of the rate law just with Y since everything else is constant.
Rate = k[Y]^m
9 = 3^m
m = 2 ; so the reaction is second order with respect to Y

44.

D

Catalysts, high temperature, high concentration, and low activation energies are all associated with fast reaction rates. If the intramolecular forces (the bonds) found within reactant molecules are strong then the energy required to break the bonds will be great, making the reaction time slower.

45.

C

The total amount of Cl⁻ to be precipitated.
$CaCl_2$ → Ca^{2+} + **2Cl⁻**
0.10 M → 0.10 + **2(0.10) = 0.20 mol/L Cl⁻ from CaCl₂**

$NaCl$ → Na^+ + **Cl⁻**
0.10 M → 0.10 + **0.10 = 0.10 mol/L Cl⁻ from NaCl**

Total moles to be precipitated: 0.20 + 0.10 = 0.30 mol Cl⁻
The reaction taking place is:
Ag^+ + Cl^- → $AgCl$ (s)
Since silver and chloride ion are in a 1:1 ratio it will take 0.30 mol of silver nitrate to precipitate all of the chloride ion.

46.

A

The K_{eq} value is > 1, meaning that the reaction proceeds toward the products. This means that ammonia is the base and acetic acid is the acid. The products are the conjugates and are thus, much weaker acids and bases than the reactants.

47.

A

$A(g) + 2B(g) \rightleftharpoons 2C(g)$

Reaction	A (g)	+	2B (g)	$\rightleftharpoons$	2C (g)
Initial	0.50		1.25		0
Change	$-x$		$-2x$		$+2x$
Equilibrium	$0.50 - 0.125$		$1.25 - 0.125$		0.25

If $2x = 0.25$ then $x = 0.125$
Substitute the equilibrium concentrations into the K_{eq} expression and solve.

$$K_{eq} = \frac{[C]^2}{[A][B]^2} = \frac{(0.25)^2}{(0.375)(1.00)^2} = 0.17$$

Simplify the math in this problem by rounding 0.0625 (numerator) to 0.06 and rounding the denominator to 0.30. This would give an answer of 0.20. Since the actual denominator is a larger value, the true answer will be smaller than 0.20.

48.

E

Trigonal bipyramidal has five areas of electron density. PF_5 is an example of a molecule with this shape.

CO_2 = linear

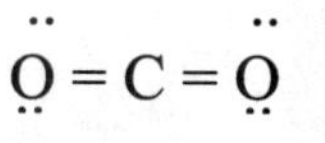

$$\ddot{\text{O}} = \text{C} = \ddot{\text{O}}$$

$SiCl_4$ = tetrahedral

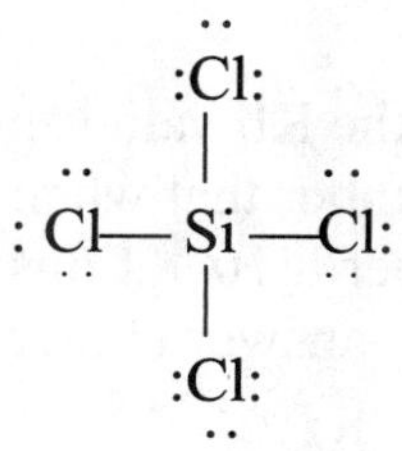

PCl_3 = trigonal pyramid

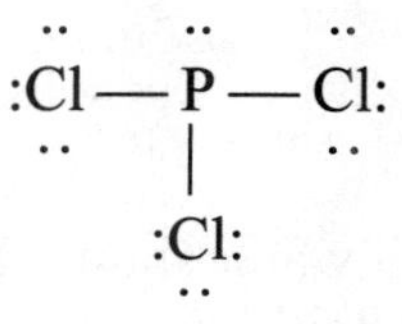

XeF₄ = square planar

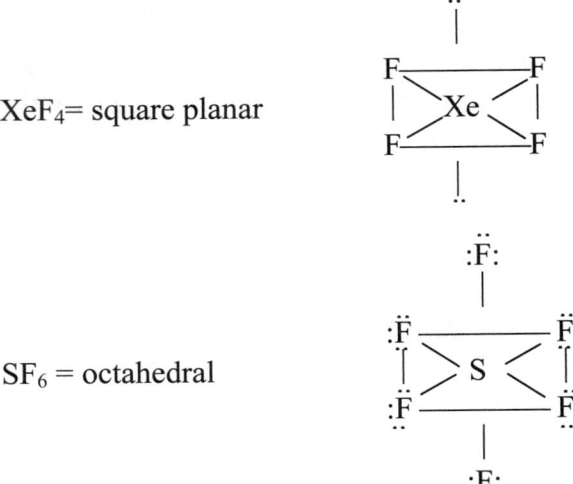

SF₆ = octahedral

49.

D

$pH = -\log[H^+]$

The solution is a weak base so the hydroxide ion concentration must be found from the Kb expression.

$$K_b = \frac{[B^+][OH^-]}{[BOH]}$$

$$1.3 \times 10^{-6} = \frac{x^2}{1.0 \times 10^{-2}}$$

$x^2 = 1.3 \times 10^{-8}$

$x = 1.14 \times 10^{-4}$ M

$pOH = -\log(1.14 \times 10^{-4}$ M)

$pOH = 3.94$

$pH = 14.00 - 3.94 = 10.05$

Simplify the math in this problem. Multiply the Kb value and the base concentration, yielding 1.3×10^{-8}; ignore the 1.3 and remember that when you take the square root the exponent is halved. This leads to some number × 10^{-4}. Logs are powers of 10 so the pOH is close to 4—the pH is close to 10. Only one answer choice gives this range.

50.

D

This is a dilution problem; $M_1V_1 = M_2V_2$

$(20.0$ mL$)(3.00$ M$) = (0.200$ M$)(x)$

$x = 300.$ mL total volume

Read carefully! This question asks how much water should be added. If you have already placed in 20.0 mL of the stock solution you should only add $300.0 - 20.0 = 280.0$ mL of water.

51.

D

$$N_2 (g) + 3H_2 (g) \rightleftarrows 2NH_3 (g)$$

Reaction	N_2 (g)	+	$3H_2$ (g)	$\rightleftarrows$	$2NH_3$ (g)
Initial	1.0		3.5		0
Change	$-x$		$-3x$		$+2x$
Equilibrium	$1.0 - 0.45$		$3.5 - 3(0.45)$		0.90

If $2x = 0.90$ atm then $x = 0.45$ atm

Using Dalton's law of partial pressure solve for the total:

$$P_{total} = P_{nitrogen} + P_{hydrogen} + P_{ammonia}$$
$$= 0.55 + 2.15 + 0.90$$
$$= 3.60 \text{ atm}$$

52.

A

A spectrophotometer is useful for finding the absorbance of colored solutions, which can then be related to concentration using Beer's law. $CaCl_2$ and $AgNO_3$ are both colorless.

53.

E

The K_{eq} value given is < 1, meaning that the reaction favors the reactants at equilibrium. The reaction is not spontaneous as written. If the reaction is not proceeding toward the products, the value of E^0 will be negative and the value of ΔG^0 will be positive.

54.

B

$$2X (s) + 3Ca^{2+}(aq) \rightarrow 3Ca (s) + 2X^{3+} (aq) \qquad E = -1.21V$$
$$Ca^{2+} (aq) + 2e^- \rightarrow Ca (s) \qquad E = -2.87 V$$

The missing half reaction is:

$2X (s) \rightarrow 6e- + 2X^{3+} (aq)$ (note that this is the oxidation reaction but the question asks for the reduction potential)

Substitute into the formula for standard cell potential:

$E_{cell} = E_{cathode} - E_{anode}$
$-1.21 = -2.87 - (x)$
$x = -1.66$ volts

55.

C

Potassium chloride is a soluble salt. Filtering the solution would yield no particles left on a filter paper. The best method to recover the salt is to heat the solution to drive off the water and leave the salt residue behind.

56.

D

Titrating to the half-equivalence point allows the use of the equation $pH = pK_a$ at the half-way point. However, if a student titrated to this point and did not add the rest of the acid solution then the pH would be a very large number, perhaps 8.00 (the pH at the equivalence point and not the half-way point). This would translate into a K_a value that is too small, 1×10^{-8}. Choice II, not adding enough water to the acid, does not make any difference since the acid is solid. The number of moles of acid must be neutralized. Addition of base beyond the equivalence point would yield a pH reading larger than it should be, thus, the K_a value would be too small.

57.

A

Write electron configurations for each element.

Li $1s^2 2s^1$

Mg $1s^2 2s^2 2p^6 3s^2$

Ne $1s^2 2s^2 2p^6$

Se [Ar] $4s^2 3d^{10} 4p^4$

Br [Ar] $4s^2 3d^{10} 4p^5$

There is such a large energy difference between the first and second ionization energies that it appears a new energy level closer to the nucleus is being interrupted by the removal of the second electron. The only configuration that supports this theory is that of lithium.

58.

D

Polyprotic acids dissociate in steps:

$$H_2SO_4 + H_2O \rightarrow H_3O^+ + HSO_4^-$$

$$HSO_4^- + H_2O \rightarrow H_3O^+ + SO_4^{2-}$$

The hydronium ion is produced in each step of the dissociation.

59.

C

Molar mass = grams of solute/moles of solute

Using the freezing point depression formula for a nonvolatile substance, $\Delta T_f = K_f \times m$, the molality could be found given the change in temperature and the freezing point depression constant. Molality = moles solute / kg of solvent. The mass of the solvent is needed in order to find moles of solute. The original mass of the unknown solute used in the experiment must also be known in order to find molar mass.

60.

C

This is an empirical formula problem. If the compound is 80.0% carbon then it is also 20.0% hydrogen.

$$(80.0\,\text{g}\,\text{C}) \times \frac{1\,\text{mol}\,\text{C}}{12\,\text{g}\,\text{C}} = 6.67\,\text{mol}\,\text{C}$$

$$(20.0\,\text{g}\,\text{H}) \times \frac{1\,\text{mol}\,\text{H}}{1\,\text{g}\,\text{H}} = 20.0\,\text{mol}\,\text{H}$$

Divide by the smallest moles to obtain the ratio.

$$\frac{6.67}{6.67} = 1$$

$$\frac{20.0}{6.67} = 3$$

The formula becomes CH_3.

61.

D

Ammonia, NH_3, is a molecular compound and does not ionize in solution. A solution saturated with silver chloride is represented by AgCl (*s*) and would appear very cloudy and white. When excess ammonia is bubbled through the solution a complex forms and the solution clears. The only choice that represents AgCl correctly is choice D.

62.

E

Electrolytes must form ions in solution in order to conduct electric current. All Group 1A and ammonium metallic salts are soluble and will ionize in solution. This eliminates choices A–C. Choice D is an organic acid, acetic acid, which will light a bulb dimly. Choice E is methanol, an organic alcohol, which will not ionize in solution.

63.

C

When given volume and concentration usually find moles.

$$0.200\,L \times \frac{0.20\,mol\,Ba^{2+}}{1\,L} = 0.040\,mol\,Ba^{2+}$$

$$0.200\,L \times \frac{0.10\,mol\,SO_4^{2-}}{1\,L} = 0.020\,mol\,SO_4^{2-}$$

$$Ba^{2+} + SO_4^{2-} \rightarrow BaSO_4$$

(note that since barium and sulfate are 1:1 the sulfate limits)

Reaction	Ba^{2+}	+	SO_4^{2-}	$\rightarrow$	$BaSO_4$
Initial	**0.040**		0.020		0
Used in reaction	**−0.020**		−0.020		+0.020
Left	**0.020**		0		0.020

Molarity of Ba^{2+} left after reaction: $\dfrac{0.020\,mol}{0.40\,L} = 0.050\,M$

64.

A

Supercooling sometimes occurs when heat is removed from a liquid and its temperature drops below the freezing point without the liquid becoming a solid. It occurs so rapidly that the particles cannot arrange themselves into the ordered pattern of the solid.

65.

D

$CuCO_3$ (s) $\rightarrow$ CuO (s) + CO_2 (g)

0.50 mol   0.50 mol   0.50 mol (according to stoichiometry)

The total pressure in the container is due only to the gas, carbon dioxide in the container. Use the ideal gas equation to solve for pressure.

$$P = \frac{nRT}{V} = \frac{(0.50\,mol)(0.0821\,L \cdot atm/mol \cdot K)(300\,K)}{1.0\,L} = 12.3\,atm$$

Simplify this problem in the following manner. ½ of 300 = 150; multiply 150 × 8 = 1200; remember that the answer must be moved back two decimal places.

66.

E

Use Gay-Lussac's law to solve this problem. The number of moles and volume are constant. Don't forget to convert from Celsius to Kelvin by adding 273.

$$\frac{P_1}{T_1} = \frac{P_2}{T_2}$$

$$\frac{3.0\,\text{atm}}{500\,\text{K}} = \frac{P_2}{400\,\text{K}}$$

$$P_2 = 2.4\,\text{atm}$$

67.

C

The critical point is the temperature at which no amount of pressure can bring the substance back into the liquid phase.

68.

A

The specific heat of a substance is defined as the amount of heat needed to raise the temperature of 1 g of a substance by 1 °C. The phase diagram does not give any indication of mass to calculate this quantity.

69.

A

The positive slope of the solid-liquid equilibrium line indicates that the solid is more dense than the liquid. The least dense state is the gaseous state.

70.

C

Write a balanced equation first and then just use stoichiometry.

$Ag_2S + H_2 \rightarrow 2Ag + H_2S$

$$0.250\,\text{mol}\,Ag_2S \times \frac{2\,\text{mol}\,Ag}{1\,\text{mol}\,Ag_2S} \times \frac{108\,\text{g}\,Ag}{1\,\text{mol}\,Ag} = 54\,\text{g}\,Ag$$

71.

B

To find the mass of $MgCO_3$ in the sample a mole relationship between $MgCO_3$ and CO_2 must be found since carbon dioxide grams are the only ones given.

$MgCO_3 \rightarrow MgO + CO_2$

$$8.8\,\text{g}\,CO_2 \times \frac{1\,\text{mol}\,CO_2}{44\,\text{g}\,CO_2} \times \frac{1\,\text{mol}\,MgCO_3}{1\,\text{mol}\,CO_2} \times \frac{84\,\text{g}\,MgCO_3}{1\,\text{mol}\,MgCO_3} = 16.8\,\text{g}\,MgCO_3$$

$$\% = \frac{16.8\,\text{g}}{50.4\,\text{g}} \times 100 = 33\%$$

72.

A

$$\_Cr + \_CrO_4^{2-} + \_ \rightarrow \_Cr(OH)_3$$

The oxidation state of Cr metal is 0; the oxidation state of Cr in CrO_4^{2-} is +6; the oxidation state of Cr in $Cr(OH)_3$ is +3. To balance the charge on both sides, a coefficient of 2 must be placed in front of the chromium(III) hydroxide.

73.

C

$$H_2SO_3 + H_2O \rightarrow H_3O^+ + HSO_3^- \quad K_1 = 1.4 \times 10^{-2}$$

$$HSO_3^- + H_2O \rightarrow H_3O^+ + SO_3^{2-} \quad K_2 = 6.3 \times 10^{-8}$$

To find the K overall, multiply $K_1 \times K_2$.
$K_{overall} = (1.4 \times 10^{-2})(6.3 \times 10^{-8}) = 8.8 \times 10^{-10}$
Simplify the math by multiplying $6 \times 1.5 = 9$; then add the exponents $-2 + -8 = -10$.
Only one answer is close to this value.

74.

C

This is a limiting reactant problem. Write a balanced equation and perform stoichiometry.
$$Zn^{2+} + CO_3^{2-} \rightarrow ZnCO_3$$

$$0.050\,L \times \frac{0.100\,mol\,Zn^{2+}}{1\,L} = 0.0050\,mol\,Zn^{2+}$$

$$0.050\,L \times \frac{0.300\,mol\,CO_3^{2-}}{1\,L} = 0.0150\,mol\,CO_3^{2-}$$

Since the mole ratio is 1:1 zinc limits. $0.0150\,mol - 0.0050\,mol = 0.010\,mol\ CO_3^{2-}$ left after reaction. Calculate the new molarity by dividing moles left by total volume.
$$M = \frac{0.010\,mol}{0.100\,L} = 0.10\,M$$

75.

B

The functional group on this compound is characteristic of a carboxylic acid.

alcohol   R — OH

ether   R — O — R'

aldehyde   R — C — H (with =O on C)

ketone   R — C — R' (with =O on C)

Answers to Free Response Questions

1)   a)

$$K_a = 1.8 \times 10^{-4} = \frac{[H_3O^+][HCO_2^-]}{[HCO_2H]}$$

   b)   The solution to this part of the problem involves the use of an "ICE-box." We are given the pH, which is an equilibrium value, and are asked for the amount of material before dissociation occurred, that is, the initial value. Raising 10 to the power of $-pH$ will provide the equilibrium H_3O^+ concentration, and the formate concentration by default. We can then let the initial concentration of formic acid be represented by x and solve for this quantity.

$[H_3O^+] = 10^{-pH} = 10^{-2.49} = 3.24 \times 10^{-3}$

	HCO_2H (aq) +	H_2O (l) ↔	HCO_2^- (ag) +	H_3O^+ (aq)
Initial	x	NA	0	0
Change	-3.24×10^{-3}	NA	$+3.24 \times 10^{-3}$	$+3.24 \times 10^{-3}$
Equilibrium	$x - 3.24 \times 10^{-3}$	NA	3.24×10^{-3}	3.24×10^{-3}

$K_a = 1.8 \times 10^{-4} = \dfrac{[H_3O^+][HCO_2^-]}{[HCO_2H]} = \dfrac{[3.24 \times 10^{-3}][3.24 \times 10^{-3}]}{[x - 3.24 \times 10^{-3}]}$

$[x - 3.24 \times 10^{-3}] \times 1.8 \times 10^{-4} = [3.24 \times 10^{-3}] \times [3.24 \times 10^{-3}]$

$x = 0.0616\,M$

The mass of the formic acid can be determined using the above molarity, the molar mass of formic acid, and the volume of the solution as follows:

0.0616 M × 0.4500 I × 46.03 g/mol = **1.28 g of formic acid**

   c)   This is probably a good place to use the Henderson-Hasselbalch equation. It will allow for the rapid calculation of the concentration of formate necessary to provide the desired buffer pH. From there the mass of potassium formate can be calculated.

$pH = pK_a + \log ([HCO_2^-] / [HCO_2H])$ $pK_a = -\log K_a = -\log 1.8 \times 10^{-4} = 3.74$

$4.10 = 3.74 + \log ([HCO_2^-] / [0.1242])$

$0.36 = \log ([HCO_2^-] / [0.1242])$

$0.36 = \log [\,HCO_2^-\,] - \log [0.1242] = \log [HCO_2^-] + 0.9059$

$-0.546 = \log [\,HCO_2^-\,]$

$10^{-0.546} = [\,HCO_2^-\,] = 0.285 \text{ M}$

Again, using the above molarity, the molar mass, and the volume of solution, we can solve for the amount of potassium formate to be added.

(0.285 mol HCO_2^- / L of solution) × 0.745 L of solution × (1 mol $KHCO_2$ / HCO_2^-) × (72.11 g $KHCO_2$ / mol) = **15.31 g $KHCO_2$**

ii)   To calculate the effect of adding a strong base, the neutralization reaction must first be taken into account in a stoichiometry calculation. When the amount (or molarity) of the remaining formic acid has been calculated, that quantity can then be plugged back into the Henderson-Hasselbalch equation to calculate the pH of the new buffer solution.

$HCO_2H \ (aq) + NaOH \ (aq) \leftrightarrow NaHCO_2 \ (aq) + H_2O \ (l)$

or, as a net ionic equation,

$HCO_2H \ (aq) + OH^- \ (aq) \leftrightarrow HCO_{2i}^- \ (ag) + H_2O \ (l)$

There are 0.1242 M × 0.745 L or 0.09253 mol of formic acid, and 0.750 g / (40.00 g / mol) or 0.01875 moles of sodium hydroxide that have been added. These will react completely, with the sodium hydroxide as the limiting reactant, to form 0.01875 mol of additional formate anion and leave 0.09253 – 0.01875 or 0.07378 mol of unreacted formic acid. The total amount of formate will be the sum of 0.01875 plus the amount from the original buffer solution (0.285 M × 0.745 L = 0.212 mol) or 0.231 mol. The new concentrations of formic acid and formate are found below.

[HCO_2H] = 0.07378 mol / 0.745 L = 0.0990 M

[HCO_2^-] = 0.231 mol / 0.745 L = 0.310 M

pH = pK_a + log ([HCO_2^-] / [HCO_2H]) = 3.74 + log (0.310 M / 0.0990 M) = **4.24**

d)   Since this is a strong base–weak acid titration, we know that the pH at the endpoint will be slightly basic. We want to pick an indicator with a pK_a approximately equal to the pH at the equivalence point. At the endpoint of this titration, the solution will be an approximately 0.05 M solution of formate. All other species will be spectators. An "ICE-box" type calculation will provide a value for pH at the equivalence point. In this case the reaction we'll be studying is the base hydrolysis reaction of formate anion.

$$HCO_2^-(ag) + H_2O(l) \leftrightarrow HCO_2H(ag) + OH^7(ag) \qquad K_b = 1\times10^{-14}/K_a = 1\times10^{-14}/1.8\times10^{-4} = 5.56\times10^{-11}$$

	HCO_2^- (aq)	+	H_2O (l)	$\leftrightarrow$	HCO_2H (aq)	+	OH^- (aq)
Initial	0.05 M		NA		0		0
Change	$-x$		NA		$+x$		$+x$
Equilibrium	$0.05 - x$ M		NA		x		x

$$K_b = 5.56\times10^{-11} = \frac{[OH^-][HCO_2H]}{[HCO_2^-]} = \frac{[x][x]}{[0.05-x]}$$

$$x^2 = (0.05 - x) \times 5.56 \times 10^{-11}$$

$$x^2 + 5.56 \times 10^{-11}x - 2.78 \times 10^{-12} = 0$$

$$x = 1.67 \times 10^{-6} = [OH^-]$$

$$pOH = -\log 1.67 \times 10^{-6} = 5.78$$

$$pH = 14 - pOH = 8.22$$

So, with a pH at the equivalence point of 8.22, the appropriate choice of indicator is **phenolphthalein** with a pK_a of 9.2. Bromthymol blue might also be an acceptable choice with a pK_a of 7.

2)   a)   To determine molar mass, you need to know the mass of a sample and the number of moles of particles that the sample contains. So, she will need to determine a sample's mass, and collect sufficient data to determine the number of moles it contains. In this case she'll probably use the ideal gas equation, $PV = nRT$, so she'll need to know the volume of the sample when it's in the gas phase, its temperature, and its pressure.

     b)   We can solve for the mass of water contained in the flask and then divide by the density of the water to arrive at the interior volume of the flask.

mass of water = 415.433 g – 145.671 g = 269.762 g of water / 1.00 g / mL = 269.762 mL of water or 0.269762 L of water. The interior volume of the flask is then **0.269762 L**.

c) We must first find the mass of the liquid by subtracting the tare weight of the flask from the mass of the flask plus liquid. Then we can determine the number of moles by using the ideal gas equation.

mass of liquid = 146.426 g – 145.671 g = 0.755 g of liquid

The pressure of the gas inside the flask is the same as that in the lab due to the pinhole in the stopper. This pressure is 99.75 kPa × (1 atm / 101.3 kPa) = 0.9847 atm.

$n = PV / RT = (0.9847 \text{ atm} \times 0.2698 \text{ L}) / (0.08206 \text{ L·atm/mol·K} \times (273.15 + 99.7 \text{ °C}))$

$n = 0.008683$ mol

The molar mass is then 0.755 g / 0.008683 mol = **86.95 g / mol**

d) For percent error we must compare the accepted to the experimental value and determine the difference between the two as a percent of the accepted value. The accepted molar mass of ethyl acetate (calculated from the formula) is 88.11 g / mol.

% error = (| accepted – experimental | / accepted) × 100% = ((88.11 – 86.95) / 88.11) × 100% = **1.32%**

e) At temperatures just above their boiling points, volatile liquids usually do not behave ideally. In an ideal setting, there would be no interparticle attractions or repulsions in the gas phase, and the actual particle volume is considered to be negligible. Under these conditions, the gas phase particles have slowed down and condensed to a point where their particle size does matter and they do experience interparticle attractions.

The net effect is usually that the gas occupies slightly less volume than predicted by the ideal gas equation for a particular temperature and pressure. In this specific case this would lead to slightly more moles of gas being able to fit into the fixed volume container at this temperature and pressure. The ideal gas equation wouldn't account for this, but it would show up in the mass of the liquid. Therefore, the molar mass would be calculated as being too high (accurate mass but moles calculated too low).

3) a) This is a Hess' law calculation as seen in Chapter 6. The thermochemical equations provided must be rearranged and modified so that their sum equals the desired overall reaction. The first reaction can be used as is, the second must be doubled (and its ΔH value will also be doubled), and the third must be reversed (the sign of ΔH will be changed but its magnitude will remain the same).

$C\ (s) + O_2\ (g) \rightarrow CO_2\ (g)$ $\Delta H = -393.5$ kJ

$2H_2(g) + O_2(g) \rightarrow 2\,H_2O\,(l)$ $\Delta H = 2x\,(-285.8\text{ kJ}) = -571.6\text{ kJ}$

$2H_2O\,(l) + CO_2(g) \rightarrow CH_4(g) + 2\,O_2(g)$ $\Delta H = +890.3\text{ kJ}$

$\overline{}$ $\overline{\phantom{\Delta H = +890.3\text{ kJ}}}$

$C\,(s) + 2\,H_2(g) \rightarrow CH_4(g)$ $\boldsymbol{\Delta H = -74.8\text{ kJ}}$

      b)   i)   The negative value for ΔH indicates an exothermic reaction.

            ii)

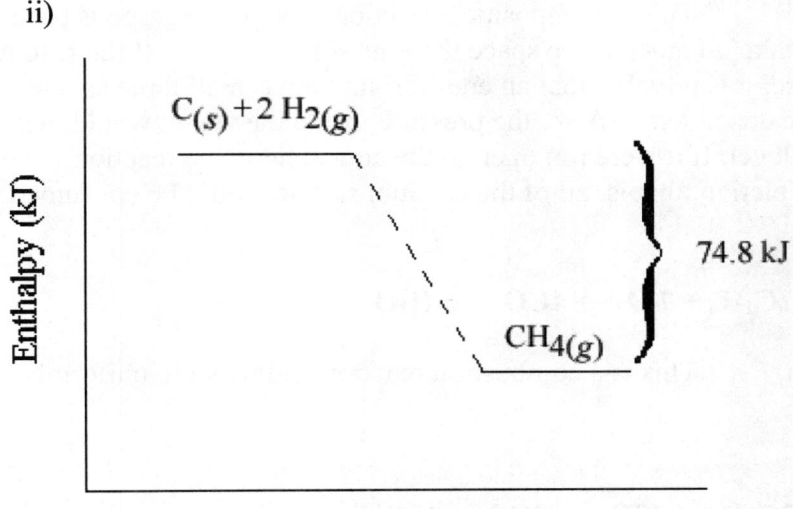

      c)   Since 1 mol of a solid and 2 mol of gas condense to a single mole of gas (even though it is a larger, more complex molecule), this reaction is very likely to have a negative $\Delta S^{\circ 0}$.

      d)   i)   Spontaneous chemical reactions have a negative value for change in free energy, ΔG^0. Since $\Delta G^0 = \Delta H^0 - T\Delta S^0$, a negative ΔH^0 (exothermic reaction) and a positive ΔS^0 tend to favor a spontaneous reaction. Since this reaction likely has a negative ΔS^0, it is the enthalpy change that is driving the spontaneous formation of products.

            ii)   All values for T in the equation above will be positive (absolute temperature scale), so increasing the value of T increases the $T\Delta S^0$ term. Since ΔS^0 is negative and T is always positive, the $T\Delta S^0$ term will be negative. It is being subtracted from ΔH^0, so as T increases, it will make ΔG^0 increasingly positive. At some temperature, the $T\Delta S^0$ term will exceed the ΔH^0 value, and ΔG^0 will become positive, and the reaction will become nonspontaneous.

4)   a) $OH^- + NH_4^+ \rightarrow H_2O + NH_3$

i)   Since this is an acid-base neutralization reaction, and likely exothermic, it would probably become warm to the touch. The ammonia that is generated is a gas, and it is possible that you would observe bubbles forming. And, finally, it is quite probable that you would smell ammonia as it diffused out of the solution.

b) $CaSO_4 \rightarrow CaO + SO_3$

i)   This decomposition reaction produces a gaseous product, SO_3, that will obviously take up much more space than the solid reactant. If the reaction were run in a closed vessel it is possible that an equilibrium between all three species would be set up (temperature dependent). Also, the pressure inside the vessel would increase, possibly to a dangerous level. If it were run open to the atmosphere, the reaction would likely proceed to completion, that is, all of the calcium sulfate would be consumed.

c) $2C_2H_6 + 7 O_2 \rightarrow 4CO_2 + 6 H_2O$

i)   This is a combustion reaction and they are uniformly exothermic.

5)   a) $2C_4H_{10} + 13O_2 \rightarrow 8CO_2 + 10H_2O$

b)   i)   The butane and oxygen react in a 2:13 mole ratio, and they have been combined in a 1:1 mole ratio. Since the oxygen is used up faster, it will run out first. The answer is O_2.

ii)   By the time 0.1 mol of oxygen is consumed, 2/13 ^{ths} of a mole of butane will be reacted. Therefore, 11/13 ^{ths} of a mole of butane will remain when the reaction stops.

iii)   The initial pressure due to the butane does not contribute to the internal pressure of the reaction vessel because it is compressed within the lighter itself. According to the balanced equation, as 13 mol of oxygen are consumed, 18 mol of products will be produced. If the reaction vessel is warm enough that the water remains in a vapor form, then the pressure will increase. However, if this is being done in a normal lab setting, it is likely that the water will cool and condense on the inside of the container. In its liquid state it will occupy a negligible volume. Therefore, the only substance significantly contributing to pressure would be the carbon dioxide. Since fewer moles of carbon dioxide will be produced, 8 mol, than moles of oxygen consumed, 13 mol, it is likely that the pressure will drop over the course of the reaction.

c)   i)

$$H-\overset{\overset{\displaystyle H}{|}}{\underset{\underset{\displaystyle H}{|}}{C}}-\overset{\overset{\displaystyle H}{|}}{\underset{\underset{\displaystyle H}{|}}{C}}-\overset{\overset{\displaystyle H}{|}}{\underset{\underset{\displaystyle H}{|}}{C}}-\overset{\overset{\displaystyle H}{|}}{\underset{\underset{\displaystyle H}{|}}{C}}-H \qquad\qquad O=C=O$$

ii)   Butane has a total of 13 sigma bonds and no pi bonds. Carbon dioxide has 2 sigma bonds and 2 pi bonds.

iii)   The hybridization at carbon in butane is sp^3. The hybridization in carbon dioxide is sp.

6)   a)   **Nitrogen**   [He] $2s^2\,2p^3$
      Phosphorus   [Ne] $3s^2\,3p^3$
      Arsenic   [Ar] $4s^2\,4p^3$

b)   i)   All three elements, as expected by their group relationship, have the same valence electron configuration. The difference among the three lies with the core set of electrons, which increases by one filled shell as the period number increases.

ii)   It would be expected that all three elements would show similar chemical reactivity with respect to formulae of compounds with a common element (e.g., you might expect them to form compounds of a similar formula when reacted with chlorine). In progressing down the group, they might show increasing melting or boiling points.

c)   **Smallest**   N, P, As   **Largest**

d)   **Smallest**   As, P, N   **Largest**

e)   **Silicon**   [Ne] $3s^2\,3p^2$
      Sulfur   [Ne] $3s^2\,3p^4$

f)   **Atomic radius**   **Smallest**   Si, P, S   **Largest**
      First ionization energy   **Smallest**   Si, P, S   **Largest**

(Note: The above trend for first ionization energy is based on the observed trend that this quantity increases across a period. However, sulfur actually has a slightly lower first ionization energy than phosphorus, owing to the additional stabilization phosphorus gains through its stable half-filled *p* subshell. Since the ground state is slightly lower in energy, the difference between ground state and ionized, that is, first ionization energy, is greater. Therefore, the actual order, smallest to largest, is Si, S, P.)

Advanced Placement Chemistry

Exam II

Note: For all questions involving solutions and/or chemical equations, assume that the system is in pure water at room temperature unless otherwise noted.

Directions: Each set of lettered choices below refers to the numbered questions or statements immediately following it. Select the one lettered choice that best answers each question or best fits each statement and then fill in the corresponding oval on the answer sheet. A choice may be used once, more than once, or not at all in each set.

Questions 1–4:

    A. I^-
    B. CO_3^{2-}
    C. Zn^{2+}
    D. Ag^+
    E. $Cr_2O_7^{2-}$

A student is given an "unknown" consisting of an aqueous solution of a salt that contains one of the ions listed above. Which ion can the student eliminate on the basis of each of the following observations of the "unknown"?

1. The solution is colorless.

2. The solution gives no precipitate upon adding 0.10 M sodium chloride.

3. No gas is formed upon adding dilute hydrochloric acid.

4. No bright yellow precipitate is formed when a dilute solution of $Pb(NO_3)_2$ is added to a sample of the solution.

Questions 5–8: The set of lettered choices is a list of chemical compounds that are soluble in water and take part in acid-base reactions.

    A. CH_3COOH
    B. Na_2O
    C. CH_3NH_2
    D. NaH_2PO_4
    E. $NH_4C_2H_3O_2$

5. Is amphiprotic in water solution.

6. Forms a strong base in water solution.

7. Is a Lewis base but not an Arrhenius base.

8. A salt that forms a buffer when combined with one of the other compounds listed.

Questions 9–12:

    A. alpha
    B. beta
    C. gamma
    D. x-ray
    E. ultraviolet

9. Radiation with the greatest penetrating ability.
10. Radiation used in structural analysis of crystalline solids.
11. Radiation that is absorbed by the ozone layer.
12. Radiation used in the discovery of the nucleus.

Questions 13–15:

    A. metallic solid
    B. ionic solid
    C. network covalent solid
    D. molecular solid with dispersion forces
    E. molecular solid with nonpolar molecules at each lattice point

13. Magnesium ribbon

14. Iron(III) chloride

15. Iodine crystals

Questions 16–19: The set of lettered choices is a list of aqueous solutions of nitrate compounds in various amounts.

    A. 100 mL of 0.50 m $NaNO_3$ (molar mass = 85)
    B. 400 mL of 0.10 m $Al(NO_3)_3$ (molar mass = 213)
    C. 500 mL of 0.20 m NH_4NO_3 (molar mass = 80)
    D. 300 mL of 0.30 m $Pb(NO_3)_2$ (molar mass = 325)
    E. 200 mL of 0.10 m $Co(NO_3)_2$ (molar mass = 245)

16. Is an appropriate reagent for a Beer's law experiment.
17. Has the lowest freezing point.
18. Has the highest [NO_3^-].
19. Has the highest van't Hoff factor.

Directions: Each of the questions or incomplete statements below is followed by five suggested answers or completions. Select the one that is best in each case and then fill in the corresponding oval on the answer sheet.

20. The structural isomers $CH_3CH_2CH_2COOH$ and $CH_3COCH_2CH_2OH$ would be expected to have the same values for which of the following? (Assume ideal behavior.)

A. reactivity
B. molecular mass
C. specific heat
D. boiling point
E. solubility

21. Pi bonding occurs in each of the following species EXCEPT

A. CO_2
B. CH_3COOH
C. SCN^-
D. NO_3^-
E. H_3O^+

22. The net ionic equation for the reaction that occurs during the titration of hydroiodic acid with lithium hydroxide is

A. $H^+ + OH^- \rightarrow H_2O$
B. $LiOH + HI \rightarrow LiI + H_2O$
C. $Li^+ + OH^- \rightarrow LiOH$
D. $HI + OH^- \rightarrow I^- + H_2O$
E. $H^+ + I^- + Li^+ + OH^- \rightarrow LiOH + H_2O + I_2$

23. Ca, Fe, Zn, Kr, O_2

Gaseous phase of which of the elements above are paramagnetic?

A. Ca and Fe only
B. Fe and O_2 only
C. Kr and Zn only
D. Ca, Zn, and Kr only
E. Ca, Fe, Zn, and O_2 only

24. A solution is known to contain an inorganic salt of one of the following elements. A 1.0 M solution of the salt is green. The solution contains a salt of

A. Co
B. I
C. Fe
D. Ni
E. Zn

25. Of the following reactions, which involves the largest decrease in entropy?

A. $CaCO_3 (s) \rightarrow CaO (s) + CO_2 (g)$
B. $2 NO_2 (g) \rightarrow N_2 (g) + O_2 (g)$
C. $Cu(NO_3)_2 (aq) + 2 KI (aq) \rightarrow CuI_2 (s) + 2 KNO_3 (aq)$
D. $2CH_3OH (g) + 3O_2 (g) \rightarrow 2CO_2 (g) + 4H_2O (g)$
E. $CO_2 (g) \rightarrow CO_2 (s)$

26. What is the hybridization of the xenon atom in XeF_4?

A. sp
B. sp^2

C. sp^3
D. sp^3d
E. sp^3d^2

27. $2ICl_3\ (g) \rightarrow I_2\ (g) + 3Cl_2\ (g)$

According to the data in the table, what is the value of ΔH^0 for the reaction represented?

Bond	Average Bond Energy (kilojoules / mole)
I—I	150
Cl—Cl	240
I—Cl	210

A. − 390 kJ
B. +240 kJ
C. + 390 kJ
D. + 450 kJ
E. + 600 kJ

Questions 28–29: The phase diagram of an unknown substance is shown below.

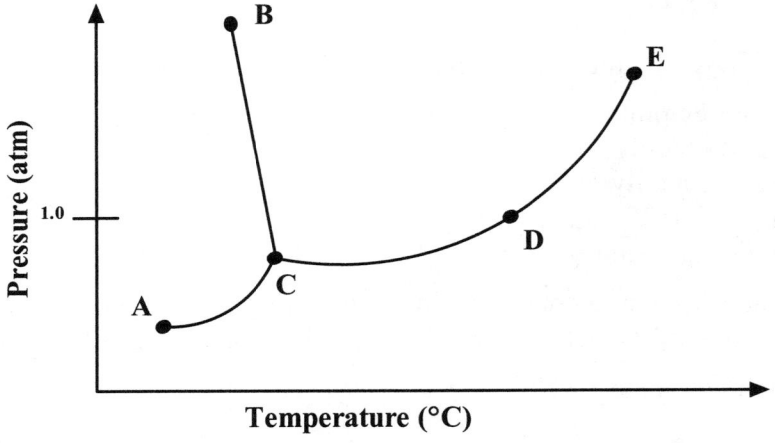

28. The point at which all phases of matter are in equilibrium.

A. A
B. B
C. C
D. D
E. E

29. Which point represents the normal boiling point for this substance?

A. A
B. B
C. C
D. D
E. E

30. Which of the following systems would experience the smallest change in concentration of substances present at equilibrium when the volume of the system is decreased at constant temperature?

A. $SO\,(g) + NO\,(g) \rightleftarrows SO_2\,(g) + \frac{1}{2}N_2\,(g)$

B. $O_2\,(g) + 2H_2\,(g) \rightleftarrows 2H_2O\,(g)$

C. $N_2\,(g) + 2O_2\,(g) \rightleftarrows 2NO_2\,(g)$

D. $N_2O_4\,(g) \rightleftarrows 2NO_2\,(g)$

E. $CH_4\,(g) + 2O_2\,(g) \rightleftarrows CO_2\,(g) + 2H_2O\,(g)$

31. $$CV^+\,(aq) + OH^-\,(aq) \rightarrow CVOH\,(aq)$$

Crystal violet when reacted with excess sodium hydroxide according to the equation above, becomes colorless. An experimental pseudo–rate law is given as follows.

$$Rate = k\,[CV^+]$$

If the concentrations of both reactant remain constant but the temperature is changed by 10 °C for each of four trials, which of the following is true?

A. The reaction order will double and k increase.
B. Both the reaction rate and k decrease.
C. Both the reaction rate and k remain the same.
D. The energy of activation can be calculated via graphical methods.
E. The reaction rate decreases but k remains the same.

32. Correct procedures for a spectrophotometric analysis include which of the following?

I. Rinse the cuvette with the solution to be measured.
II. Set the spectrophotometer to zero with a blank of distilled water.
III. Wipe the outside of the cuvette with a soft cloth to remove any fingerprints or oils before placing into the instrument.

A. I only
B. II only
C. I and III only
D. II and III only
E. I, II, and III

33. At 25 °C the solubility product constant, K_{sp}, for $BaSO_4$ is 1.1×10^{-10}.

$$Ba^{2+}\,(aq) + SO_4^{2-}\,(aq) \rightleftarrows BaSO_4\,(s)$$

What is the equilibrium constant for the reaction represented by the following equation at 25 °C?

$$2BaSO_4 (s) \rightleftarrows 2Ba^{2+} (aq) + 2SO_4^{2-} (aq)$$

A. $\dfrac{1}{1.1\times10^{-10}}$

B. 1.1×10^{-10}

C. $2(1.1 \times 10^{-10})$

D. $\dfrac{1}{\left(1.1 \times 10^{-10}\right)^{2}}$

E. $\dfrac{1}{(1.1 \times 10^{-10})^{1/2}}$

34. A simple sugar has the empirical formula CH_2O. Which of the following represents the molecular formula of a carbohydrate having a molecular mass of 180. g/mol?

  A. $C_4H_8O_4$
  B. $C_5H_{10}O_5$
  C. $C_6H_{12}O_6$
  D. $C_8H_{16}O_8$
  E. $C_9H_{18}O_9$

35.

Acid	Acid Dissociation Constant, K_a
H_2CO_3	4.5×10^{-7}
HCO_3^-	4.7×10^{-11}
$HC_2H_3O_2$	1.8×10^{-5}

On the basis of the information above, a buffer with a pH = 10.00 can best be made by using

  A. $NaHCO_3 + Na_2CO_3$
  B. pure H_2CO_3
  C. $HC_2H_3O_2 + H_2CO_3$
  D. $H_2CO_3 + HCO_3^-$
  E. $CO_3^{2-} + HC_2H_3O_2$

36. When solid potassium nitrate, KNO_3 (s) is added to water at 25 °C, it dissolves and the temperature of the solution decreases. Which of the following is true for the values of ΔH and ΔS for the dissolving process?

	ΔH	ΔS
A.	>0	>0
B.	>0	<0
C.	>0	0
D.	<0	>0
E.	<0	<0

37. All of the following statements regarding gases are true EXCEPT

A. The volume occupied by gas particles is only significant at very low pressures.
B. The particles of a gas experience relatively no intermolecular attractive forces at high temperatures.
C. The particles of a gas move in random straight line paths until a collision occurs.
D. The collisions that occur between gas particles are considered to be "elastic" collisions.
E. At a given temperature, all gas molecules within a sample possess the same average kinetic energy.

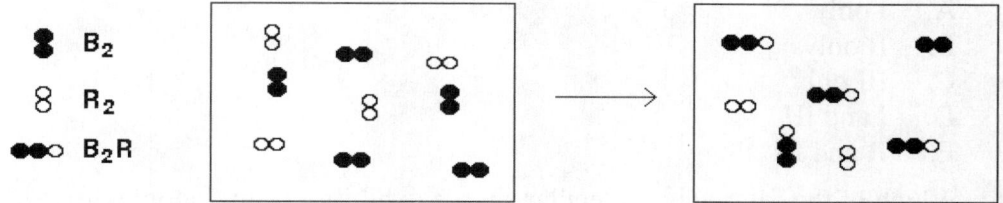

38.

When the equation for the reaction represented above is balanced and all coefficients are reduced to lowest whole-number terms, which of the following is the correct representation?

A. $B_2 + R_2 \rightarrow B_2R$
B. $B_2 + R_2 \rightarrow 2B_2R$
C. $3B_2 + 4R_2 \rightarrow 4B_2R$
D. $2B_2 + R_2 \rightarrow 2B_2R$
E. $4B_2 + 5R_2 \rightarrow 4B_2R + B_2 + 2R_2$

39. $C_2H_5OH\ (g) + 3O_2\ (g) \rightarrow 2CO_2\ (g) + 3H_2O\ (g)$

What is the value of ΔH^0_{rxn} for the complete combustion of ethanol given the following information?

Compound		ΔH_f^0 (kJ/mol)
C_2H_5OH (g)	=	–280
CO_2 (g)	=	–400
H_2O (g)	=	–240

A. –360 kJ/mol
B. –1240 kJ/mol
C. –1760 kJ/mol
D. –1800kJ/mol
E. +840 kJ/mol

40. A sample of an ideal gas is heated gently from 10.0 °C to 30.0 °C in a sealed container of constant volume. Which of the following values for the gas will increase?

      I.   The average kinetic energy of the gas
      II.   The relative number of gas particles in the container
      III. The number of collisions of the molecules

A. I only
B. II only
C. III only
D. I and III
E. II and III

41. Which of the following molecular shapes exhibits a central atom with dsp^3 hybrid orbitals?

A. bent
B. tetrahedral
C. trigonal planar
D. trigonal bipyramidal
E. octahedral

42.   rate = $k[A]^2$

For the reaction whose rate law is given above, a plot of which of the following is a straight line?

A. [A] versus time
B. ln [A] versus time
C. 1/[A] versus time
D. [A] versus 1/time
E. ln[A] versus 1/time

43. At 25 °C, a sample of NH_3 (molar mass 17 g) effuses at the rate of 0.050 mol per min. Under the same conditions, which of the following gases effuses at approximately one-half that rate?

    A. O_2 (molar mass 32 g)
    B.   He (molar mass 4.0 g)
    C. CO_2 (molar mass 44 g)
    D. Cl_2 (molar mass 71 g)
    E. CH_4 (molar mass 16 g)

44. The addition of an oxidizing agent such as chlorine water to a clear solution of an unknown compound results in a color change. When this solution is shaken with an equal volume of mineral oil, the mineral oil layer turns reddish-brown. The unknown compound probably contains

    A. Cl^-

    B. Br^-

    C. MnO_4^-

    D. I^-
    E. Co^{2+}

45. Which of the following are arranged according to increasing atomic radius?

    A.   Li, Be, N, O
    B.   F, Cl, Br, I
    C.   Rb, K, Na, Li
    D.   Fe, Co, Ni, Cu
    E.   K, Ca, Sc, Ti

46. Which of the following contribute most in explaining the formation of a meniscus when water is poured into a glass graduated cylinder?

    I.   The molecular weight of water is 18.02 g/mol.
    II.   Stronger adhesive forces than cohesive forces
    III.   A high capacity for hydrogen bonding between molecules

    A.   I only
    B.   II only
    C.   III only
    D.   II and III only
    E.   I, II, and III

47. An ideal gaseous mixture at 25 °C contains 4.0 mol of carbon dioxide, 3.0 moles of oxygen, and 2.0 mol of methane and exerts a total pressure of 3.00 atmospheres. What is the partial pressure of methane in this mixture?

    A.   0.33 atm
    B.   0.66 atm
    C.   1.00 atm

D.   1.33 atm
E.   3.00 atm

Questions 48–49: The equation below refers to an electrochemical cell:

$$3Ni^{2+} + 2Cr\ (s) \rightarrow 3Ni\ (s) + 2Cr^{3+} \qquad\qquad E^0_{cell} = + 0.487 \text{ volts}$$

48. Which expression gives the value for ΔG^0 in kJ/mol for this reaction?

A. $- (6)\ (96,500)\ (0.487)$
B. $- (3)\ (8.31)\ (0.487)$
C. $- (2)\ (96,500)\ (0.487)$
D. $\dfrac{(96,500)\ (0.487)}{6}$
E. $- (8.31)\ (298)\ (0.487)$

49. Which expression gives the voltage for such a cell at nonstandard conditions where $[Cr^{3+}]$ is 0.010 M and $[Ni^{2+}]$ is 1.00 M?

A. $0.487 - (0.0592)\ (6) + (0.10)^2$

B. $0.487 - (\dfrac{0.0592}{6})\log\dfrac{(0.010)^2}{(1.0)^3}$

C. $0.487 + \ln (0.10)^2$

D. $0.487 + \dfrac{(8.31)\ (96,500)}{6}$

E. $0.487 - (\dfrac{(0.592)(298)}{6})\log(0.01)^2$

50. A 3.0 Liter flask contains 32.00 g of sulfur dioxide at a temperature of 65 °C. What is the approximate pressure inside the flask?

A.   0.22 atm
B.   0.89 atm
C.   4.62 atm
D.   56.9 atm
E.   468 atm

51. A 0.10-molar solution of a weak acid, HA, has a pH of 3.00. The ionization constant of this acid is

A. 1.0×10^{-14}
B. 1.0×10^{-7}

C. 1.0×10^{-5}
D. 1.0×10^{-3}
E. 1.0×10^{-1}

52. __ $Cr_2O_7^{2-}$ + __ e^- + __ H^+ → __ Cr^{3+} + __ H_2O (l)

When the equation for the half reaction above is balanced with the lowest whole-number coefficients, the coefficient for H_2O is

A. 2
B. 3
C. 5
D. 7
E. 14

53. $H_2PO_4^- + H_2O \rightleftarrows H_3O^+ + HPO_4^{2-}$

In the equilibrium represented above, the species that act as bases include which of the following?

I. $H_2PO_4^-$

II. H_2O

III. HPO_4^{2-}

A. I only
B. II only
C. I and II
D. I and III
E. II and III

54. A sample of an unknown hydrocarbon was burned in excess oxygen to form 9.60 g of carbon dioxide and 4.00 g of water. What is a possible empirical formula of the hydrocarbon?

A. CH
B. C_2H
C. CH_2
D. CH_3
E. CH_4

55. If 3.0 A is passed through an electrolytic cell containing a solution of $Au(CN)_4^-$ ions, for 20.0 min, the maximum number of moles of Au that could be deposited at the cathode is

A. 0.00020 mol
B. 0.0010 mol
C. 0.0120 mol
D. 0.020 mol

E.   0.40 mol

56. Which of the following pairs of substances would NOT represent a solution ?

    A. Br_2 (*l*) and CCl_4 (*l*)
    B. C_2H_5OH (*l*) and H_2O (*l*)
    C. $CH_3CH_2CH_2CH_2CH_3$ (*l*) and H_2O (*l*)
    D. H_2SO_4 (*l*) and H_2O (*l*)
    E. C_9H_{20} (*l*) and $C_{12}H_{26}$ (*l*)

57. Which expression gives percent by mass of water in $CuSO_4 \cdot 5H_2O$?

    A. $\dfrac{90}{250} \times 100$

    B. $\dfrac{90}{175} \times 100$

    C. $\dfrac{90}{160} \times 100$

    D. $\dfrac{18}{250} \times 100$

    E. $\dfrac{18}{160} \times 100$

58. $2Li + 2H_2O \rightarrow 2Li^+ + 2OH^- + H_2$

When 0.250 mole of lithium reacts with excess water at standard temperature and pressure as shown in the equation above, the volume of hydrogen gas produced is

    A. 22.4 L
    B. 11.2 L
    C. 5.6 L
    D. 2.8 L
    E. 1.4 L

59. Which of the following is expected to have the lowest first ionization energy?

    A. Li
    B. Ne
    C. B
    D. N
    E. F

60. $$3H_2(g) + N_2(g) \rightleftharpoons 2NH_3(g) + energy$$

Some H_2 and N_2 are mixed in a container at 100 °C and the system reaches equilibrium according to the equation above. Which of the following causes a decrease in the number of moles of ammonia present at equilibrium?

      I.   Increasing the volume of the container
      II.   Raising the temperature
      III.   Adding a mole of He gas at constant volume

  A.   I only
  B.   II only
  C.   III only
  D.   I and II only
  E.   I, II, and III

Questions 61–63: Energy is added to a system at a constant rate as shown in the heating curve below.

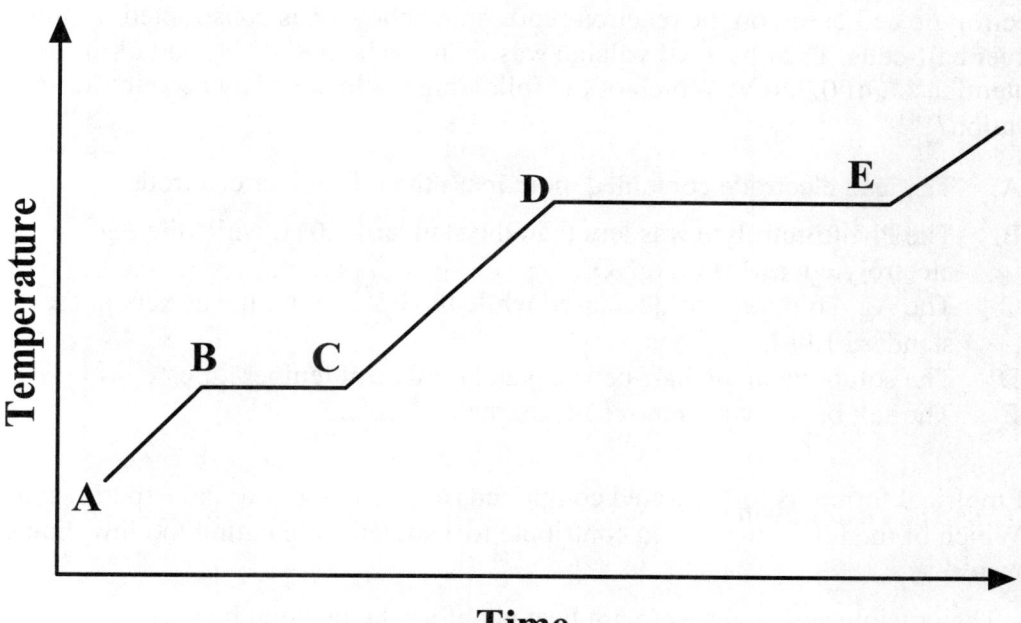

61. Which of the following best represents the energy calculation involving region A–B of the graph?

  A.   mass × ΔH_{fus} × ΔT
  B.   mass × c_{solid} × ΔT
  C.   mass × ΔH_{vap}
  D.   mass × c_{liquid} × ΔT
  E.   mass × ΔH_{vap} × ΔT

62. Which of the following best represents the energy calculation involving region B–C of the graph?

    A. $mass \times \Delta H_{fus}$
    B. $mass \times c_{solid} \times \Delta T$
    C. $mass \times \Delta H_{vap}$
    D. $mass \times c_{liquid} \times \Delta T$
    E. $mass \times \Delta H_{vap} \times \Delta T$

63. Which point(s) on the curve would be eliminated if the substance sublimes at standard conditions?

    A.   B only
    B.   C only
    C.   B and C only
    D.   C and D only
    E.   D and E only

64. $Pb\ (s) + 2Ag^+ \rightarrow Pb^{2+} + 2Ag\ (s)$

An electrolytic cell based on the reaction represented above was constructed from lead and silver half-cells. The observed voltage was found to be 0.83 V instead of the standard cell potential, E^0, of 0.926 V. Which of the following could correctly account for this observation?

    A.   The lead electrode contained more mass than the silver electrode.
    B.   The Pb^{2+} electrolyte was less than the standard 1.0M, while the Ag^+ electrolyte was 1.0M $AgNO_3$.
    C.   The Ag^+ solution was decreased while the Pb^{2+} solution was kept at the standard 1.0M.
    D.   The solutions in the half-cells began at different temperatures.
    E.   The salt bridge was removed from the half-cells.

65. Empirical formulas for hydrated compounds are commonly found experimentally. Which of the following would contribute to a student calculating too few moles of water?

I.   The crucible and cover were not heated before the reaction began.
II. The hydrated sample was removed from heat too soon.
III. The anhydrous sample was not placed in a dessicator for cooling.

    A.   I only
    B.   II only
    C.   I and II only
    D.   II and III only
    E.   I, II, and III

66. Calculate the number of moles of O_2 needed to produce 35.0 L of N_2O_5 from gaseous nitrogen at STP?

    A. 0.0144 mol
    B. 0.250 mol
    C. 1.56 mol
    D. 3.90 mol
    E. 7.40 mol

Experiment	Initial $[I^-]$ (mol L^{-1})	Initial $[S_2O_8{}^{2-}]$ (mol L^{-1})	Initial Rate of Formation of I_2 (mol L^{-1} s^{-1})
1	0.010	0.025	3.3×10^{-3}
2	0.020	0.025	1.3×10^{-2}
3	0.020	0.050	2.6×10^{-2}

67. The initial-rate data in the table above were obtained for the reaction represented below.

$$2I^-(aq) + S_2O_8^{2-}(aq) \rightarrow I_2(s) + 2SO_4^{2-}(aq)$$

What is the experimental rate law for the reaction?

    A.   rate = $k[I^-][S_2O_8^{2-}]$
    B.   rate = $k[I^-][S_2O_8^{2-}]^2$
    C.   rate = $k[I^-]^2[S_2O_8^{2-}]$
    D.   rate = $k[I^-]^2$
    E.   rate = $k[I^-] / [S_2O_8^{2-}]$

68. The radioactive decay of $^{98}_{42}Mo$ to $^{99}_{42}Mo$ occurs by the process of

    A. beta particle emission
    B. alpha particle emission
    C. positron emission
    D. electron capture
    E. neutron capture

69. The pH of 0.1-molar sodium hydroxide is approximately

    A. 1
    B. 4
    C. 7
    D. 11
    E. 13

Questions 70–71 refer to atoms for which the occupied atomic orbitals are shown below.

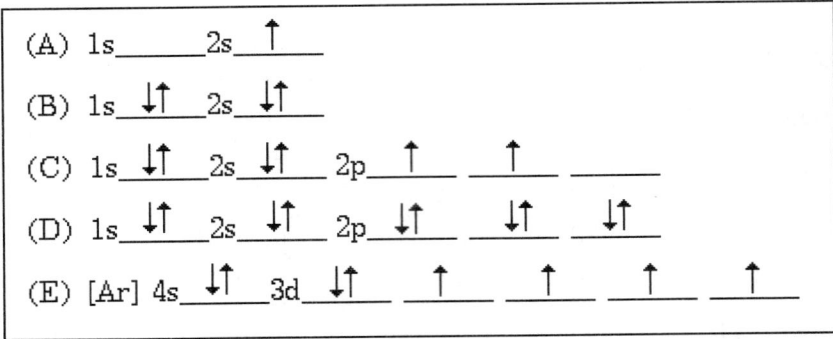

70. Represents an atom that commonly forms more than one type of cation.

71. Represents an atom that is capable of forming single, double, and triple bonds with itself.

72. $$3H_2(g) + N_2(aq) \rightleftarrows 2NH_3(g)$$

When 0.30 mol of N_2 and 0.20 mol of H_2 are placed in an evacuated 1.00-L flask, the reaction represented above occurs. After the reactants and the product reach equilibrium and the initial temperature is restored, the flask is found to contain 0.10 mol of NH_3. Based on these results, the expression for the equilibrium constant, K_c, of the reaction is

    A. $[(0.30)(0.20)^3] / (0.10)^2$
    B. $(0.10)^2 / [(0.25)(0.05)^3]$
    C. $(2 \times 0.10)^2 / [(3 \times .05)^3(0.25)]$
    D. $(0.10)^2 / [(0.30)(0.20)^3]$
    E. $[(0.25)(0.05)^3] / (0.10)^2$

73. How many grams of calcium sulfate, $CaSO_4$, contains 32. g of oxygen?

    A. 6.02×10^{23} grams
    B. 272 grams
    C. 136 grams
    D. 68 grams
    E. 32 grams

74. One of the outermost electrons in a phosphorus atom in the ground state can be described by which of the following sets of four quantum numbers?

A.   3, 3, 2, ½
B.   3, 2, 1, ½
C.   3, 1, –1, ½
D.   3, 0, 1, ½
E.   3, 0, 0, –½

75. When $^{222}_{86}Rn$ decays, the emission consists consecutively of two alpha particles, and finally two beta particles. The resulting stable nucleus is

A. $^{214}_{84}Po$

B. $^{214}_{83}Bi$

C. $^{214}_{82}Pb$

D. $^{216}_{82}Pb$

E. $^{218}_{80}Hg$

DO NOT DETACH FROM BOOK.

PERIODIC TABLE OF THE ELEMENTS

1																		2
H 1.0079																		He 4.0026
3 Li 6.941	4 Be 9.012											5 B 10.811	6 C 12.011	7 N 14.007	8 O 16.00	9 F 19.00		10 Ne 20.179
11 Na 22.99	12 Mg 24.30											13 Al 26.98	14 Si 28.09	15 P 30.974	16 S 32.06	17 Cl 35.453		18 Ar 39.948
19 K 39.10	20 Ca 40.08	21 Sc 44.96	22 Ti 47.90	23 V 50.94	24 Cr 52.00	25 Mn 54.938	26 Fe 55.85	27 Co 58.93	28 Ni 58.69	29 Cu 63.55	30 Zn 65.39	31 Ga 69.72	32 Ge 72.59	33 As 74.92	34 Se 78.96	35 Br 79.90		36 Kr 83.80
37 Rb 85.47	38 Sr 87.62	39 Y 88.91	40 Zr 91.22	41 Nb 92.91	42 Mo 95.94	43 Tc (98)	44 Ru 101.1	45 Rh 102.91	46 Pd 106.42	47 Ag 107.87	48 Cd 112.41	49 In 114.82	50 Sn 118.71	51 Sb 121.75	52 Te 127.60	53 I 126.91		54 Xe 131.29
55 Cs 132.91	56 Ba 137.33	57 *La 138.91	72 Hf 178.49	73 Ta 180.95	74 W 183.85	75 Re 186.21	76 Os 190.2	77 Ir 192.2	78 Pt 195.08	79 Au 196.97	80 Hg 200.59	81 Tl 204.38	82 Pb 207.2	83 Bi 208.98	84 Po (209)	85 At (210)		86 Rn (222)
87 Fr (223)	88 Ra 226.02	89 †Ac 227.03	104 Rf (261)	105 Db (262)	106 Sg (266)	107 Bh (264)	108 Hs (277)	109 Mt (268)	110 Ds (271)	111 Rg (272)	112 § (277)							

§Not yet named

*Lanthanide Series

58 Ce 140.12	59 Pr 140.91	60 Nd 144.24	61 Pm (145)	62 Sm 150.4	63 Eu 151.97	64 Gd 157.25	65 Tb 158.93	66 Dy 162.50	67 Ho 164.93	68 Er 167.26	69 Tm 168.93	70 Yb 173.04	71 Lu 174.97

†Actinide Series

90 Th 232.04	91 Pa 231.04	92 U 238.03	93 Np 237.05	94 Pu (244)	95 Am (243)	96 Cm (247)	97 Bk (247)	98 Cf (251)	99 Es (252)	100 Fm (257)	101 Md (258)	102 No (259)	103 Lr (262)

STANDARD REDUCTION POTENTIALS IN AQUEOUS SOLUTION AT 25 °C

Half-reaction			$E°$(V)
$F_2\ (g) + 2e^-$	$\rightarrow$	$2F^-$	2.87
$Co^{3+} + e^-$	$\rightarrow$	Co^{2+}	1.82
$Au^{3+} + 3e^-$	$\rightarrow$	$Au(s)$	1.50
$Cl_2\ (g) + 2e^-$	$\rightarrow$	$2Cl^-$	1.36
$O_2\ (g) + 4H^+ + 4e^-$	$\rightarrow$	$2H_2O(l)$	1.23
$Br_2\ (l) + 2e^-$	$\rightarrow$	$2Br^-$	1.07
$2Hg^{2+} + 2e^-$	$\rightarrow$	Hg_2^{2+}	0.92
$Hg^{2+} + 2e^-$	$\rightarrow$	$Hg(l)$	0.85
$Ag^+ + e^-$	$\rightarrow$	$Ag(s)$	0.80
$Hg_2^{2+} + 2e^-$	$\rightarrow$	$2Hg(l)$	0.79
$Fe^{3+} + e^-$	$\rightarrow$	Fe^{2+}	0.77
$I_2\ (s) + 2e^-$	$\rightarrow$	$2I^-$	0.53
$Cu^+ + e^-$	$\rightarrow$	$Cu(s)$	0.52
$Cu^{2+} + 2e^-$	$\rightarrow$	$Cu(s)$	0.34
$Cu^{2+} + e^-$	$\rightarrow$	Cu^+	0.15
$Sn^{4+} + 2e^-$	$\rightarrow$	Sn^{2+}	0.15
$S(s) + 2H^+ + 2e^-$	$\rightarrow$	$H_2S(g)$	0.14
$2H^+ + 2e^-$	$\rightarrow$	$H_2(g)$	0.00
$Pb^{2+} + 2e^-$	$\rightarrow$	$Pb(s)$	−0.13
$Sn^{2+} + 2e^-$	$\rightarrow$	$Sn(s)$	−0.14
$Ni^{2+} + 2e^-$	$\rightarrow$	$Ni(s)$	−0.25
$Co^2 + 2e^-$	$\rightarrow$	$Co(s)$	−0.28
$Cd^{2+} + 2e^-$	$\rightarrow$	$Cd(s)$	−0.40
$Cr^{3+} + e^-$	$\rightarrow$	Cr^{2+}	−0.41
$Fe^{2+} + 2e^-$	$\rightarrow$	$Fe(s)$	−0.44
$Cr^{3+} + 3e^-$	$\rightarrow$	$Cr(s)$	−0.74
$Zn^{2+} + 2e^-$	$\rightarrow$	$Zn(s)$	−0.76
$2H_2O(l) + 2e^-$	$\rightarrow$	$H_2(g) + 2\ OH^-$	−0.83
$Mn^{2+} + 2e^-$	$\rightarrow$	$Mn(s)$	−1.18
$Al^{3+} + 3e^-$	$\rightarrow$	$Al(s)$	−1.66
$Be^{2+} + 2e^-$	$\rightarrow$	$Be(s)$	−1.70
$Mg^{2+} + 2e^-$	$\rightarrow$	$Mg(s)$	−2.37
$Na^+ + e^-$	$\rightarrow$	$Na(s)$	−2.71
$Ca^{2+} + 2e^-$	$\rightarrow$	$Ca(s)$	−2.87
$Sr^{2+} + 2e^-$	$\rightarrow$	$Sr(s)$	−2.89
$Ba^{2+} + 2e^-$	$\rightarrow$	$Ba(s)$	−2.90
$Rb^+ + e^-$	$\rightarrow$	$Rb(s)$	−2.92
$K^+ + e^-$	$\rightarrow$	$K(s)$	−2.92
$Cs^+ + e^-$	$\rightarrow$	$Cs(s)$	−2.92
$Li^+ + e^-$	$\rightarrow$	$Li(s)$	−3.05

ADVANCED PLACEMENT CHEMISTRY EQUATIONS AND CONSTANTS

ATOMIC STRUCTURE

$$E = h\nu \qquad c = \lambda\nu$$

$$\lambda = \frac{h}{m\upsilon} \qquad p = m\upsilon$$

$$E_n = \frac{-2.178 \times 10^{-18}}{n^2} \text{ joule}$$

E = energy $\qquad \upsilon$ = velocity
ν = frequency $\qquad n$ = principal quantum number
λ = wavelength $\qquad m$ = mass
p = momentum

Speed of light, $c = 3.0 \times 10^8$ m s^{-1}

Planck's constant, $h = 6.63 \times 10^{-34}$ J s

Boltzmann's constant, $k = 1.38 \times 10^{-23}$ J K^{-1}

Avogadro's number $= 6.022 \times 10^{23}$ mol^{-1}

Electron charge, $e = -1.602 \times 10^{-19}$ C

1 electron volt per atom $= 96.5$ kJ mol^{-1}

EQUILIBRIUM

$$K_a = \frac{[\text{H}^+][\text{A}^-]}{[\text{HA}]}$$

$$K_b = \frac{[\text{OH}^-][\text{HB}^+]}{[\text{B}]}$$

$$K_w = [\text{OH}^-][\text{H}^+] = 1.0 \times 10^{-14} \quad @\ 25\ ^\circ\text{C}$$
$$= K_a \times K_b$$

$$\text{pH} = -\log[\text{H}^+], \text{pOH} = -\log[\text{OH}^-]$$
$$14 = \text{pH} + \text{pOH}$$

$$\text{pH} = \text{p}K_a + \log\frac{[\text{A}^-]}{[\text{HA}]}$$

$$\text{pOH} = \text{p}K_b + \log\frac{[\text{HB}^+]}{[\text{B}]}$$

$$\text{p}K_a = -\log K_a, \text{p}K_b = -\log K_b$$
$$K_p = K_c(RT)^{\Delta n},$$

where Δn = moles product gas – moles reactant gas

Equilibrium Constants

K_a (weak acid)

K_b (weak base)

K_w (water)

K_p (gas pressure)

K_c (molar concentrations)

S^0 = standard entropy

H^0 = standard enthalpy

G^0 = standard free energy

E^0 = standard reduction potential

T = temperature

n = moles

m = mass

q = heat

c = specific heat capacity

C_p = molar heat capacity at constant pressure

E_a = activation energy

k = rate constant

A = frequency factor

THERMOCHEMISTRY/KINETICS

$$\Delta S^0 = \sum S^0 \text{ products} - \sum S^0 \text{ reactants}$$

$$\Delta H^0 = \sum \Delta H_f^0 \text{ products} - \sum \Delta H_f^0 \text{ reactants}$$

$$\Delta G^0 = \sum \Delta G_f^0 \text{ products} - \sum \Delta G_f^0 \text{ reactants}$$

$$\Delta G^0 = \Delta H^0 - T\Delta S^0$$
$$= -RT \ln K = -2.303\ RT \log K$$
$$= -n\,F\,E^0$$
$$\Delta G = \Delta G^0 + RT \ln Q = \Delta G^0 + 2.303\ RT \log Q$$
$$q = mc\Delta T$$
$$C_p = \frac{\Delta H}{\Delta T}$$

$$\ln[A]_t - \ln[A]_0 = -kt$$
$$\frac{1}{[A]_t} - \frac{1}{[A]_0} = kt$$
$$\ln k = \frac{-E_a}{R}\left(\frac{1}{T}\right) + \ln A$$

Faraday's constant, $\mathscr{F} = 96{,}500$ coulombs per mole of electrons

Gas constant, $R = 8.31$ J mol^{-1} K^{-1}
$= 0.0821$ L·atm mol^{-1}·K^{-1}
$= 8.31$ volt coulomb mol^{-1} K^{-1}

GASES, LIQUIDS, AND SOLUTIONS

$$PV = nRT$$

$$\left(P + \frac{n^2 a}{V^2}\right)(V - nb) = nRT$$

$$P_A = P_{total} \times X_A, \text{ where } X_A = \frac{\text{moles A}}{\text{total moles}}$$

$$P_{total} = P_A + P_B + P_C + \ldots$$

$$n = \frac{m}{M}$$

$$K = {}^\circ C + 273$$

$$\frac{P_1 V_1}{T_1} = \frac{P_2 V_2}{T_2}$$

$$D = \frac{m}{V}$$

$$u_{rms} = \sqrt{\frac{3kT}{m}} = \sqrt{\frac{3RT}{M}}$$

$$KE \text{ per molecule} = \frac{1}{2}mv^2$$

$$KE \text{ per mole} = \frac{3}{2}RT$$

$$\frac{r_1}{r_2} = \sqrt{\frac{M_2}{M_1}}$$

molarity, M = moles solute per liter solution

molarity, m = moles solute per kilogram solvent

$$\Delta T_f = iK_f \times \text{molality}$$

$$\Delta T_b = iK_b \times \text{molality}$$

$$\pi = iMRT$$

$$A = abc$$

OXIDATION-REDUCTION; ELECTROCHEMISTRY

$$Q = \frac{[\mathbf{C}]^c [\mathbf{D}]^d}{[\mathbf{A}]^a [\mathbf{B}]^b}, \text{ where } a\mathbf{A} + b\mathbf{B} \rightarrow c\mathbf{C} + d\mathbf{D}$$

$$I = \frac{q}{t}$$

$$E_{cell} = E^\circ_{cell} - \frac{RT}{nF} \ln Q = E^\circ_{cell} - \frac{0.0592}{n} \log Q \text{ @ } 25\,^\circ C$$

$$\log K = \frac{nE^\circ}{0.0592}$$

P = pressure
V = volume
T = temperature
n = number of moles
D = density
m = mass
v = velocity

u_{rms} = root-mean-square speed
KE = kinetic energy
r = rate of effusion
M = molar mass
π = osmotic pressure
i = van'tHoff factor
K_f = molal freezing-point depression constant
K_b = molal boiling-point elevation constant
A = absorbance
a = molar absorptivity
b = path length
c = concentration
Q = reaction quotient
I = current (amperes)
q = charge (coulombs)
t = time (seconds)
E° = standard reduction potential
K = equilibrium constant

Gas constant, R = 8.31 J mol^{-1} K^{-1}
= 0.0821 L·atm mol^{-1}·k^{-1}
= 8.31 volt coulomb mol^{-1} K^{-1}

Boltzmann's constant, k = 1.38×10^{-23} J K^{-1}
K_f for H_2O = 1.86 K kg mol^{-1}
K_b for H_2O = 0.512 K kg mol^{-1}
1 atm = 760 mm Hg
= 760 torr

STP = 0.000°C and 1.000 atm

Faraday's constant, $\mathscr{F}$ = 96,500 coulombs per mole of electrons

Free Response Questions

(Equilibrium Problem)

1)   a)   Calcium hydroxide dissolves in water according to the following equation:

$$Ca(OH)_2 \ (s) + H_2O \ (l) \leftrightarrow Ca^{2+} \ (aq) + 2OH^-(aq) \qquad\qquad K_{sp} = 8.0 \times 10^{-6}$$

       i)   Write an equilibrium constant expression for the dissolution of calcium hydroxide in water.

       ii)   What is the pH of a solution created by adding excess calcium hydroxide (excess in this case means more than will dissolve) to pure water?

       iii)   Calculate the molar solubility of calcium hydroxide.

   b)   Describe an acid-base titration procedure by which the K_{sp} of calcium hydroxide can be determined.

   c)   During titrations with calcium hydroxide solutions, it is important to keep the container holding this base sealed tightly and, preferably, filled to near the top of the container. If these handling criteria are not taken into account, the concentration of hydroxide decreases over time and a milky white precipitate forms in the solution. Explain the chemistry behind this observation (hint: K_{sp} for calcium carbonate = 8.7×10^{-9}).

   d)   Magnesium hydroxide dissolves in water according to the following equation:

$$Mg(OH)_2 \ (s) + H_2O \ (l) \leftrightarrow Mg^{2+} \ (aq) + 2OH^- \ (aq)$$

       i)   Write an equilibrium constant expression for the above reaction.

       ii)   The magnesium ion concentration of a saturated solution of magnesium hydroxide is 1.4×10^{-4} M. Would you consider magnesium hydroxide or calcium hydroxide to be a stronger base? To receive full credit, you must explain your reasoning and justify your answer with at least one calculation.

2)   The following data were collected during an experiment to study the kinetics of the thermal decomposition of azomethane:

$$C_2H_6N_2 \ (g) \rightarrow N_2 \ (g) + C_2H_6 \ (g)$$

Time (sec)	Partial Pressure of Azomethane (mm Hg)
0	284
100	220
150	193
200	170
250	150
300	132

a)   i)   Write rate expressions for the above reaction in terms of the reactant and both products.

ii)   What relationship exists between the rate of change of the pressures of the product nitrogen and the reactant azomethane?

b)   i)   Using the pressure data above, determine the order of the reaction with respect to azomethane. Explain your reasoning and any calculations involved in this determination.

ii)   Based on your results for section b i), provide an appropriate axis label for the graph below.

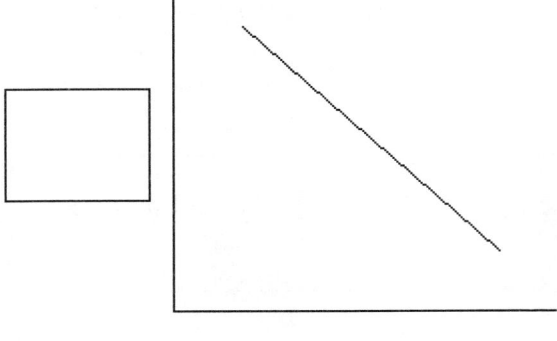

Time (sec)

c)   i)   Determine the rate constant for the decomposition reaction through appropriate manipulation of the data above (hint: the calculations performed to answer b i) provide much of the information necessary for this determination).

ii)   Write a rate law for the reaction.

iii)   Determine the half-life of this reaction.

3)

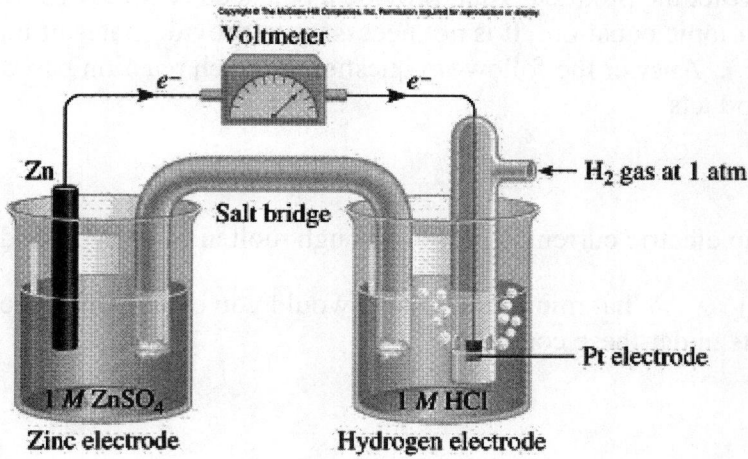

The following questions refer to the diagram above.

a)   i)   Write a balanced, net ionic equation for the reaction occurring in the above galvanic cell.

        ii)   Calculate the standard potential for the reaction in part i).

b)   Consider the reaction

$$Zn^{2+} (aq) + H_2 (g) \rightarrow Zn (s) + 2H^+ (aq)$$

        i)   Calculate E^0 and ΔG^0 for the above reaction.

        ii)   Is this redox reaction spontaneous under standard conditions? How can you tell?

c)   Suppose that the galvanic cell in a) above was set up similarly to that depicted except that the hydrogen ion concentration was 0.1 M instead of 1 M. Calculate the new cell potential.

4)   Listed below are one or more reactants for a chemical reaction. In every case a reaction will occur under the stated conditions. Unless otherwise stated, the temperature

is assumed to be 25 °C and the pressure of gases, including the atmosphere, to be 1.0 atm. In each case, predict the product(s) and provide a balanced (with lowest whole-number coefficients), net ionic equation. It is not necessary to provide states of matter (e.g., solid, liquid, etc.). Answer the follow-up question for each reaction based on your prediction of products.

a)   an electric current is passed through molten sodium chloride

      i)   What minimum voltage would you expect to be necessary to produce products under these conditions?

b)   chlorine gas is bubbled through an aqueous solution of sodium iodide

      i)   What color change would you expect to observe over the course of this reaction?

c)   lithium metal and bromine vapor are combined at an elevated temperature

      i)   What physical state (at 25 °C and 1 atm of pressure) would you expect the purified product of this reaction to have?

5)   A student was asked to investigate the change in freezing point of lauric acid, $CH_3(CH_2)_{10}COOH$, upon addition of benzoic acid.

a)   The student first heated a sample of pure lauric acid until it liquefied and then allowed it to cool and re-solidify while the temperature was recorded. A graph of temperature of lauric acid versus time is shown below.

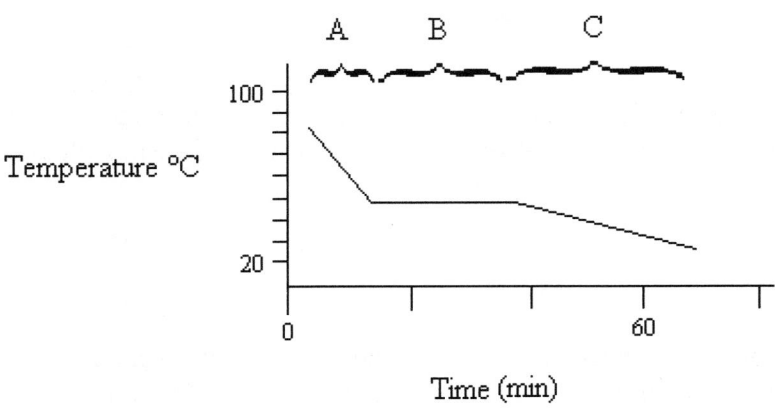

i)   Provide a thorough description of the changes taking place and the physical state of lauric acid in the three labeled regions, A, B, and C, above. Be sure to include information about direction of heat flow, change in motion and organization of molecules, phase of matter, and a general description of what is taking place.

ii)   Estimate the freezing point of pure lauric acid.

b)   Without performing any calculations, write down the formulas necessary to calculate the amount of heat flowing in each of the labeled sections of the graph above. Be sure to include any specific data needed that you can gather from the above graph (estimations based on careful reading of the graph will receive full credit).

c)   In a second experiment, the student dissolved approximately 1 g of benzoic acid, C_6H_5COOH, in approximately 8 g of lauric acid. The mixture was heated until it liquefied and was stirred to create a homogeneous solution. The mixture was allowed to cool and the temperature recorded, as above.

i)   On the graph below, draw in your estimate of the cooling curve of this benzoic acid solution. The K_f for lauric acid is 3.9 °C · kg/mol or 3.9 °C / molal.

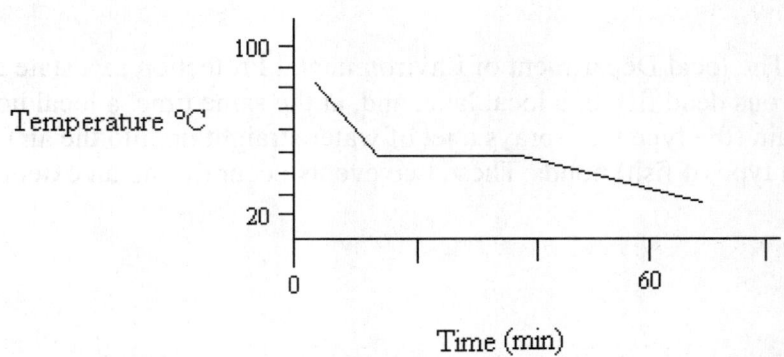

ii)   Without performing any calculations, explain how the information gathered from this experiment could be used to calculate the molecular weight of the dissolved solute (assuming its identity was unknown).

d)   The solid phase of lauric acid, like most pure substances, is its most condensed phase of matter. Its liquid phase is a less condensed state, and its gas phase is its least condensed phase. Draw a phase diagram for lauric acid representing the changes in state as a function of various temperatures and pressures.

6)   Provide detailed explanations for the following observed phenomena based on fundamental chemical principles. Chemical equations, formulas, etc., should be used where appropriate.

        a)   A raw chicken's egg becomes soft or rubbery and translucent after soaking in vinegar for a day or two.

        b)   A camper accidentally drops a penny into a campfire one night and notices that the flames around the penny periodically turn green.

        c)   Motorists are warned to follow a specific procedure for connecting jumper cables in order to start a car with a "dead" battery (lead-acid storage type). Specifically, it is suggested that the final connection with the black, or negative clamp, be made to a metal part of the engine as far as possible from the battery itself to avoid an explosion caused by the buildup of a flammable gas.

        d)   The local Department of Environmental Protection in a state investigates a report of numerous dead fish in a local lake, and, at the same time, a local homeowner installs a fountain (the type that sprays a jet of water straight up into the air) in his backyard koi (a type of fish) pond. These two events occur during an extended heat wave one summer.

ANSWERS AND EXPLANATIONS TO PRACTICE EXAM II

1.

E

All of the ions listed are colorless in solution with the exception of dichromate, $Cr_2O_7^{2-}$. Typically, solutions of dichromate are orange.

2.

D

All sodium salts are soluble, therefore all of the anions can be eliminated as suspect of precipitation. All chlorides are soluble with the exception of silver, lead, and mercury. The silver ion would react with chloride ion as follows:

$Ag^+(aq) + Cl^-(aq) \rightarrow AgCl(s)$ (white precipitate)

3.

B

Carbonates react with acid to form carbon dioxide. No other species listed will form a gas.

The net ionic reaction: $H^+(aq) + CO_3^{2-}(aq) \rightarrow CO_2(g) + H_2O(l)$

4.

A

Lead reacts with all of the anions listed. PbI_2 is a bright yellow precipitate; $PbCO_3$ is a white precipitate and $PbCr_2O_7$ is a deep orange. The net ionic reaction for lead(II) iodide formation is: $Pb^{2+}(aq) + 2I^-(aq) \rightarrow PbI_2(s)$ (bright yellow precipitate)

5.

D

An amphiprotic substance is one that is capable of acting as an acid (donating a proton) or as a base (accepting a proton) in solution. When sodium dihydrogen phosphate, NaH_2PO_4, is dissolved in water, the dihydrogen phosphate ion has the ability to accept a proton or donate a proton.

Acid: $H_2PO_4^- + H_2O \rightarrow HPO_4^2 + H_3O^+$

Base: $H_2PO_4^- + H_2O \rightarrow H_3PO_4 + OH^-$

6.

B

Group 1A metals and metal oxides form strong bases (bases that ionize 100%) in solution. Sodium oxide reacts with water as follows:

$Na_2O(s) + H_2O(l) \rightarrow Na^+(aq) + OH^-(aq)$

Choices C and D are basic salts but do not form strong bases in solution.

7.

C

An Arrhenius base is one that releases a hydroxide ion into solution. A Lewis base is defined more generally as an electron pair donor. Ammonia and ammonia derivatives are considered bases under the Lewis theory but not under the Arrhenius theory. Methylamine has one unshared electron pair on nitrogen, allowing it to be considered a Lewis base.

8.
E
Ammonium acetate, $NH_4C_2H_3O_2$ is a salt of acetic acid. A buffer is generally defined as a weak acid or base and a conjugate of that weak acid or base. The only other salts listed are sodium dihydrogen phosphate and sodium oxide—neither of these has a complementary acid or base.

9.
C
The field of electromagnetic radiation ranges from low energy to high energy in the following sequence: radio, microwave, infrared, visible, ultraviolet, x-rays, and gamma rays. Gamma rays are characterized by great penetrating ability because of their high energy, relatively no mass, and no charge.

10.
D
In 1913, William and Lawrence Bragg, a father and son team first showed how the spacing of layers in crystals showed different diffraction patterns using x-ray technology. Crystalline structure is still identified using this same technique.

11.
E
Ozone is found predominantly in the upper part of the atmosphere known as the stratosphere. Solar radiation is concentrated in the visible and ultraviolet. The UV radiation from the sun is largely absorbed by the ozone layer.

12.
A
Though Rutherford experimented with alpha particles and beta particles, he discovered that the nucleus of the atom must be positively charged when alpha particles were deflected during his famous "gold foil" experiment.

13.
A
Magnesium ribbon is a metallic solid. Metallic solids are characterized by a "sea" of electrons moving around positively charged centers.

14.
B
Iron(III) chloride is a salt. The electrostatic attraction between positive and negative ions is a characteristic of ionic solids.

15.

D

An iodine crystal, I_2, is considered a molecular solid with dispersion forces. The I_2 atoms are covalently bonded to each other. Since the molecule is completely nonpolar, the only force of attraction to hold the iodine molecules to other iodine molecules is London dispersion forces.

16.

E

Beer's law, $A = abc$, relates absorption to concentration. In a typical experiment, colored solutions are placed into a spectrophotometer and measurements of how much light passes through the substance are recorded. Cobalt(II) nitrate is reddish –purple while all other solutions listed are colorless.

The following table should help explain 17–19:

Substance	Molality of solution	Ions in solution (van't Hoff factor)		Effective molality	$[NO_3^-]$
$NaNO_3$	0.50	2	=	1.0	0.50
$Al(NO_3)_3$	0.10	4	=	0.40	0.30
NH_4NO_3	0.20	2	=	0.40	0.20
$Pb(NO_3)_2$	0.30	3	=	0.90	0.60
$Co(NO_3)_2$	0.10	3	=	0.30	0.20

17.

D

Freezing point is calculated using the formula: $\Delta T_f = K_f \cdot m \cdot i$. The solution with the lowest freezing point is the solution with the greatest molality and the greatest number of ions in solution. Since lead(II) nitrate has an effective molality of 0.90 m it should have the lowest freezing point. (Volume of the solution does not affect the freezing point.)

18.

D

The nitrate concentration for $Pb(NO_3)_2$ is a higher concentration than any of the other salts. Lead(II) nitrate ionizes according to the following equation:

$Pb(NO_3)_2 \rightarrow Pb^{2+} + 2NO_3^-$

0.30 m 0.30 m 2(0.30 m) = 0.60 m

19.

B

The van't Hoff factor describes the number of ions in solution. Aluminum nitrate will ionize into four parts in solution.

$Al(NO_3)_3 \rightarrow Al^{3+} + 3NO_3^-$

20.

B

Isomers are defined as substances with the same molecular formula but different molecular structures. When the structure changes, the physical properties of the substance change also. However, the molecular mass for the isomers will be the same.

21.

E

Pi bonding occurs when a molecule contains multiple bonds. All of the substances listed contain at least one multiple bond with the exception of the hydronium ion, H_3O^+.

22.

A

Hydroiodic acid is considered a strong acid. Lithium hydroxide is considered a strong base. The net ionic reaction for a strong acid and a strong base is simply:
$$H^+ + OH^- \rightarrow H_2O$$

23.

B

Substances with one or more unpaired electrons are known as paramagnetic and would be pulled into a magnetic field. Electron configurations can help identify these substances, but the molecular orbital theory is the best explanation.

Ca atom: $[Ar]\, 4s^2$

Fe atom: $[Ar]\, 4s^2\, 3d^6$

Zn atom: $[Ar]\, 4s^2\, 3d^{10}$

Kr atom: $[Ar]\, 4s^2\, 3d^{10}4p^6$

O_2 molecule: $1s^2\, 2s^2\, 2p^4$ is the electron configuration for the oxygen atom. When two oxygen atoms join they form a double covalent bond. Each oxygen has two shared pairs of electrons according the Lewis diagram. However, when oxygen is poured into a magnetic field it is pulled. The molecular orbital theory reveals the unshared electrons in diatomic oxygen. The electrons in the second level fill the molecular orbitals as follows:
$$\sigma 2s^2\ \sigma^* 2s^2\ \sigma 2p^2\ (\Pi\, 2p_x^2\ \Pi\, 2p_y^2)\, (\Pi^*\, 2p_x^1\ \Pi^*\, 2p_y^1)$$

24.

D

A solution of nickel would impart a green color to a solution. Cobalt is reddish-maroon. Iodine in aqueous solution is colorless. Iron in solution ranges from yellow to orange. Zinc in solution is colorless.

25.

E

Entropy is a measure of disorder or chaos. The solid state is the most organized state of matter, having the least amount of entropy. Choices A and D both increase in entropy by creating more moles of product than the reactants. Choices B and C would both be expected to be close to zero change in entropy since they both contain the same number and type of moles. Choice E is the deposition of dry ice from the gaseous state, which would be expected to have the greatest decrease in entropy.

26.

E

The hybridization of xenon is sp^3d^2 since xenon has six areas of electron density.

27.

C

ΔH_{rxn} = bonds broken – bonds formed

Drawing structures will help identify what must take place.

There are six I—Cl bonds to break. There is one I—I bond to form and three Cl—Cl bonds to form.

$\Delta H_{rxn} = [6(210)] - [(1(150) + (3 (240)]$
$= (1260) - (870)$
$= 390$ kJ/mol rxn

28.

C

The point at which all phases are in equilibrium with each other is known as the triple point. This diagram shows the triple point at point C, where all three lines intersect.

29.

D

Boiling for a substance occurs when the vapor pressure of the liquid is equal to atmospheric pressure. The "normal" boiling point for a substance occurs when the vapor pressure of the liquid equals sea level pressure. Point D represents the "normal" boiling point. (Note: the liquid is capable of boiling anywhere along the liquid-gas equilibrium line as long as the conditions of temperature and pressure are met.)

30.

E

According to Boyle's law, when the volume of a gaseous system is decreased, the pressure of the system is increased. An increase in pressure will cause a system to shift to the side with the least number of moles of gas. In the reaction of methane and oxygen to produce carbon dioxide and water, no change would be expected since there are 3 mol of gaseous reactant and 3 mol of gaseous product.

31.

D

A change in temperature will alter the rate of a reaction. An increase in temperature increases the rate of reaction while a decrease in temperature will decrease the rate of reaction. The energy of activation could be calculated using the Arrhenius equation written in the form of $y = mx + b$.

$$\ln k = -\frac{E_a}{R} \times \frac{1}{T} + \ln A$$

Plot ln k vs. $1/T$ (in Kelvin) for various temperatures. Calculate the slope of the line.

$$\text{Slope} = -\frac{E_a}{R}$$

Multiply the slope by the universal gas constant (8.3145 J/mol K) and the activation energy can be calculated. Since the question did not state whether the temperature change was an increase or decrease, no conclusion may be made about the reaction speeding up or slowing down.

32.

C

Proper procedures for using the spectrophotometer include rinsing the cuvette with the solution to be measured and wiping off all fingerprints from the outside of the cuvette with a soft cloth. The instrument does need to be set to 0 and 100 before each use. With no sample in the instrument the left knob is set to 0% transmittance. With a blank filled with the same solvent that was used to make the solution to be measured, the instrument is set to 100% transmittance with the knob on the right.

33.

D

Compare the original equation given for the K_{sp} value to the second equation. The second equation is reversed and doubled. Reversing the equation would give the inverse of the K value, 1/K. Doubling the equation will square the K. Putting all of this together, we

obtain $\dfrac{1}{\left(1.1 \times 10^{-10}\right)^2}$

34.

C

A molecular formula is often a whole-number multiple of an empirical formula. To find the whole-number multiple divide the molecular weight by the empirical formula weight.

$\dfrac{180}{30} = 6$

Multiply to get new subscripts. $6\,(CH_2O) = C_6H_{12}O_6$

35.

A

A buffer is best created by mixing a weak acid or base and a conjugate of that weak acid or base in solution. A buffer with a pH of 10.00 is basic. Choice A, sodium bicarbonate and sodium carbonate has components of both acid and base.

Acid: $CO_3^{2-} + H^+ \rightarrow HCO_3^-$

Base: $HCO_3^- + OH^- \rightarrow H_2O + CO_3^{2-}$

36.

A

ΔH represents the enthalpy (heat content) of the solution and ΔS represents the entropy (chaos) of the system. Since the temperature decreases, the system must be gaining energy, endothermic. The sign for ΔH is positive. When a solution is formed two pure substances mix and the chaos of the system increases. Positive values of ΔS represent an increase in disorder.

37.

A

Gases approach ideal situations under low pressures and high temperatures. When the pressure is low, gas particles are spread far apart from each other. When the temperature is high, the particles will move quickly. The volume occupied by a gas will only be significant under high pressures and low temperatures.

38.

D

The particulate view of the reaction is simplified by using the key given to the left of the boxes. The equation represents a simple combination reaction. The two substances come together to form one substance, B_2R. Balancing requires only a coefficient to balance the B.

39.

B

Enthalpy can be calculated from a table of standard values using the following equation:

$\Delta H^0_{rxn} = \Sigma$ (products) $- \Sigma$ (reactants)

The enthalpy values given in the table are per mole. Be sure to multiply by coefficients. Free elements including diatomic molecules are assigned values of zero for standard heats of formation.

$\Delta H^0_{rxn} = [\ 2(-400) + 3(-240)\ ] - [\ 1(-280) + 3(0)] = -1240$ kJ/mol rxn

40.

D

When a sample of gas is heated the gas particles will move with greater speed. Thus, the kinetic energy increases. If the particles are enclosed in a sealed container of constant volume, the particles will collide more often with greater forces. The number of particles will remain constant with a simple increase in temperature.

41.

D

Shape	Hybridization
Bent	sp^3
Tetrahedral	sp^3
Trigonal planar	sp^2
Trigonal bipyramidal	Dsp^3
Octahedral	d^2sp^3

42.

C

The rate law expression given is second order with respect to A. Integrated rate law for a

second-order reaction written in $y = mx + b$ format becomes: $\dfrac{1}{[A]_o} = kt + \dfrac{1}{[A]_t}$

Therefore, a plot of inverse concentration vs. time should yield a straight line for a
second-order reaction.

43.

D

The question asks which gas effuses at half the rate as ammonia. Graham's law provides
the relationship between rate of effusion and molecular weight.

$$\frac{S_1}{S_2} = \sqrt{\frac{Mass_2}{Mass_1}}$$

$$\frac{S_?}{S_{NH_3}} = \sqrt{\frac{Mass_{NH_3}}{Mass_?}}$$

$$\frac{0.50}{1} = \sqrt{\frac{17}{x}}$$

$$0.25 = \frac{17}{x}$$

$x = 68$

The closest mass given is chlorine with a mass of 71 g/mol.

44.

B

Chlorine water serves as an oxidizing agent, which means that chlorine is reduced in
solution from a charge of 0 to a –1 charge. Chloride ion, bromide ion, and iodide ion can
all be oxidized. However, bromine would form a reddish-brown color in a nonpolar
solvent such as mineral oil.

45.

B

Atomic radius increases down a family as shielding increases and the number of energy
levels increases. Across a period, atomic radius decreases as the effective nuclear charge
increases, shielding remains constant, and the number of energy levels remains constant.
Only choice B shows a series of elements in the same family from small to large. F has
two energy levels, I has five energy levels.

46.

B

The formation of a meniscus between water and glass is due to the attraction that water
molecules have for the glass surface. The adhesion forces (attraction between water and
glass) are stronger than the cohesive forces (attraction between water molecules). The

molecular weight of water has no effect on this property. Though hydrogen bonding is often an answer when dealing with water, in this case it is not a reason.

47.
B
Dalton's law of partial pressures states that $P_{total} = P_{CO_2} + P_{O_2} + P_{CH_4}$. The total pressure is known to be 3.00 atm. However, the other partial pressures are not known. Mole fraction may be used to calculate the partial pressure of a gas according to the following equation: $P_{CH_4} = X_{CH_4} P_{total}$

$$\frac{2.0}{9.0} (3.00 \ atm) = 0.67 \ atm$$

48.
A
The relationship between free energy and voltage is $\Delta G^0 = -nFE^0$. Calculating the number of moles of electrons, n, is found by writing the two half-reactions.
$2Cr \rightarrow 2Cr^{3+} + 6e^-$
$3Ni^{2+} + 6e^- \rightarrow 3Ni$
Substitute and solve: $\Delta G^0 = -(6 \ mol \ e^-) \ (96,500 \ C/mole \ e^-) \ (0.487 \ J/C)$

49.
B
The Nernst equation is used to solve for voltage under nonstandard conditions.

$$E = E^0 - \frac{RT}{nF} \ln Q$$

or simplify :

$$E = E^0 - \frac{0.592}{n} \log Q$$

$$E = 0.487 - \frac{0.092}{6} \log \left(\frac{(0.01)^2}{(1.0)^3} \right)$$

50.
C
The ideal gas equation is used to solve for pressure:

$$P = \frac{nRT}{V} = \frac{(0.5 \ mol) \ (0.0821 \ L \cdot atm) \ (338 \ K)}{(3.0 \ L)(mol \cdot K)}$$

$P = 4.62 \ atm$

Simplify the math: 32 g. sulfur dioxide (mass of 64 g/mol) is ½ or 0.5 mol. Divide 3 into 338 (round to 300…so about 100). Multiply 0.5 (100) = 50; then round 0.0821 to 0.1 and multiply by 50. The answer is close to 5. Only one answer is in this range.

51.

C

A weak acid equilibrium expression helps solve this problem.

$$HA \leftrightarrow H^+ + A^-$$

If the pH = 3.00 then the $H^+ = 1.0 \times 10^{-3}$. Since $[H^+] = [A^-]$ it is easy to substitute into the equilibrium expression.

$$K_a = \frac{[H^+][A^-]}{[HA]} = \frac{[1.0 \times 10^{-3}][1.0 \times 10^{-3}]}{[0.10]}$$

Simplify the math: multiplying the numerator gives 1.0×10^{-6}. Divide by the denominator, 1.0×10^{-1} yields 1.0×10^{-5}.

52.

D

Balance by the half-reaction method as follows:

$Cr_2O_7^{2-} \rightarrow 2Cr^{3+}$ (balance all elements but H and O)

$Cr_2O_7^{2-} \rightarrow 2Cr^{3+} + \mathbf{7H_2O}$ (balance Os by adding H_2O)

$\mathbf{14H^+} + Cr_2O_7^{2-} \rightarrow 2Cr^{3+} + 7H_2O$ (balance Hs by adding H^+)

$\mathbf{6e^-} + 14H^+ + Cr_2O_7^{2-} \rightarrow 2Cr^{3+} + 7H_2O$ (balance charge by adding e^-)

53.

E

An acid is a proton donor and a base is a proton acceptor. In the reaction given, $H_2PO_4^-$ donates a proton (acid) to become HPO_4^- (the conjugate base) on the product side. The water in this equation accepts a proton (base) to become the hydronium ion, H_3O^+, the conjugate acid.

54.

C

When a hydrocarbon is burned completely, all of the carbon is converted into carbon dioxide. To find the empirical formula, moles of each element are needed.

$$9.60 \ g \ CO_2 \times \frac{12.0 \ g \ C}{44.0 \ g \ CO_2} = 2.62 \ g \ C$$

$$2.62 \ g \ C \times \frac{1 \ mol \ C}{12.0 \ g \ C} = 0.22 \ mol \ C$$

$$4.00 \ g \ H_2O \times \frac{2.0 \ g \ H}{18.0 \ g \ H_2O} = 0.44 \ g \ H$$

$$0.44 \ g \ H \times \frac{1 \ mol \ H}{1.0 \ g \ H} = 0.44 \ mol \ H$$

Divide by the smallest number of moles to get the ratio.

The math is easy; there are two hydrogen atoms for every one carbon atom.

55.

C

Dimensional analysis makes this type of problem solving relatively simple. Remember that an amp = coulomb/sec. A balanced half-reaction needs to be written.

$$3e^- + Au(CN)_4^- \rightarrow Au$$

$$20.0 \text{ min} \times \frac{60 \text{ sec}}{1 \text{ min}} \times \frac{3 \text{ C}}{1 \text{ sec}} \times \frac{1 \text{ mol } e^-}{96,500 \text{ C}} \times \frac{1 \text{ mol } Au}{3 \text{ mol } Au} = 0.012 \text{ mol } Au$$

56.

C

A solution is a homogeneous mixture. When two nonpolar substances or two polar substances are mixed they usually form a solution. Choice C shows a nonpolar substance, pentane, with a polar substance, water. The two substances would eventually settle upon standing.

57.

A

$$\% \ H_2O \ = \ \frac{g \ H_2O}{g \ total} \times \ 100$$

$$\% \ H_2O \ = \ \frac{90 \ g \ H_2O}{250 \ g \ total} \times \ 100$$

58.

D

The hint, standard temperature and pressure, allows the use of molar volume, 22.4 L of gas = 1 mol. Set up dimensional analysis for simple problem solving.

$$0.250 \ mol \ Li \times \frac{1 \ mol \ H_2}{2 \ mol \ Li} \times \frac{22.4 \ L \ H_2}{1 \ mol \ H_2} \ = \ 2.8 \ L$$

Simplify the math: 22.4 divided by 2 is about 11; ¼ of 12 is 3 so the answer must be less than but close to 3.

59.

A

First ionization energy is the energy needed to remove the most loosely held electron. Metals tend to lose electrons while nonmetals tend to gain electrons. The only metal listed is lithium.

60.

B

Decreasing the number of moles of ammonia relates to a shift toward the reactants. Increasing the volume of the container will decrease the pressure (Boyle's law) and the system would shift to the side with the fewest moles of gas. In this case it would shift to create more ammonia. Raising the temperature causes a shift away from the energy term. This reaction is exothermic, resulting in a shift toward the reactants that would cause a decrease in the moles of ammonia present. Adding an inert gas has no effect on the equilibrium position.

61.

B

Section A—B on the graph represents a temperature change in the solid state. The specific heat of the solid (measured in J/g·C) multiplied by the change in temperature and multiplied by the mass of the solid would allow the energy to be calculated.

$$mass\ of\ solid\ (g) \times \Delta T\ (^{\circ}C) \times \frac{sp\ heat\ (J)}{g \cdot C}$$

62.

A

In region B—C the temperature remains constant. The solid is melting. The heat of fusion allows the energy to be calculated.

$$mass\ of\ solid\ (g) \times \frac{heat\ of\ fusion\ (J)}{g}$$

63.

D

Sublimation is the change of state from solid directly to a gas. Points C and D could be eliminated since the solid would not move through the liquid phase.

64.

C

The voltage drop can be explained using the Nernst equation.

$$E = E^0 - \frac{RT}{nF} \ln Q$$

or simplify:

$$E = E^0 - \frac{0.592}{n} \log Q$$

$$0.83 = 0.926 - \frac{0.592}{2} \log \left(\frac{Pb^{2+}}{(Ag^+)^2} \right)$$

When $Q > 1$ the log of Q will be greater than 1, causing a decrease in overall voltage. Changing the size of the electrodes will not change the voltage unless one of the metals is completely used up. Changing the temperature could alter the voltage but there is not enough information given as to whether the temperature increased or decreased. If the salt bridge were removed, the voltage would drop to zero.

65.

D

To find the mass of water in the sample, the mass of the anhydrous salt is subtracted from the mass of the original hydrated sample.

Hydrated salt → anhydrous salt + water

If the hydrated sample was not heated completely, the mass of the anhydrous salt would appear greater than it should be, resulting in a smaller mass of water. In the same fashion, if the anhydrous salt was left in the open air water could be absorbed back into the sample, making the weight of the anhydrous salt appear greater than it should and the water mass less. If the crucible and cover were not heated prior to the experiment, any water absorbed into the porous crucible would be driven off in the experiment and appear as mass of water lost by the hydrate, thus making the mass of water greater than it should be.

66.

D

Write a balanced equation first and then use dimensional analysis.

$$2N_2 + 5O_2 \rightarrow 2 N_2O_5$$

$$35.0\,L\,N_2O_5 \times \frac{1\,mol\,N_2O_5}{22.4\,L\,N_2O_5} \times \frac{5\,mol\,O_2}{2\,mol\,N_2O_5} = 3.90\,mol\,O_2$$

Simplify the math: Multiply $30 \times 5 = 150$; then divide by 2 = 75; Divide by 20 and you get around 3.75. There is only one answer choice in this range.

67.

C

In experiments 1 and 2 $[S_2O_8^{2-}]$ is held constant, $[I^-]$ is doubled and the rate quadruples. The experiment is second order with respect to iodide ion. Use experiments 2 and 3 to find $[S_2O_8^{2-}]$ where $[I^-]$ is held constant and $[S_2O_8^{2-}]$ is doubled. The rate doubles so the reaction is first order with respect to $[S_2O_8^{2-}]$. The rate law is: rate $= k\,[S_2O_8^{2-}]\,[I^-]^2$

68.

E

In nuclear reactions the mass and atomic numbers must equal on both the reactant and product sides of the equation. The mass increases by one and the atomic number remains the same, so a neutron must have been captured.

$$^{98}_{42}Mo + {}^{1}_{0}n \rightarrow {}^{99}_{42}Mo$$

69.

E

$pH = - \log [H^+]$

Sodium hydroxide, NaOH, is a strong base. It is easy to find the pOH and then subtract from 14 to find the pH. (pH + pOH = 14.00)

$pOH = - \log (1.0 \times 10^{-1})$; pOH = 1.0

$pH = 14.0 - 1.0 = 13.0$

70.

E

Varying oxidation states are commonly exhibited by the transition metals. The last electron configuration is filling into the d sublevel, which is characteristic for transition metals. The electron configuration given represents the iron atom. Iron commonly exhibits a +2 or +3 oxidation state.

71.

C

Carbon's electron configuration is given in choice C. Carbon is known to form single, double, and triple bonds with other carbon atoms.

72.

B

Equilibrium calculations are manageable when a ICE table is prepared and analyzed before substituting into the K_{eq} expression.

Reaction	$3H_2$	+	N_2	$\rightarrow$	$2NH_3$
Initial	0.20		0.30		0
Change	$-3x$		$-x$		$+2x$
Equilibrium	0.05		0.25		0.10

Given that $NH_3 = 0.10$ mol/L at equilibrium then $2x = 0.10$; therefore, $x = 0.050$. Substitute into the equilibrium expression.

$$K_{eq} = \frac{[NH_3]^2}{[H_2]^3[N_2]} = \frac{(0.10)^2}{(0.05)^3(0.25)}$$

73.

D

$$32. g\,O \times \frac{1\,mol\,O}{16. g\,O} \times \frac{1\,mol\,CaSO_4}{4\,mol\,O} \times \frac{136\,g\,CaSO_4}{1\,mol\,CaSO_4} = 68\,g\,CaSO_4$$

Simplify the math: Multiply $16 \times 4 = 64$; 32 is ½ of 64. Take ½ of 136 to yield 68 g.

74.

C

Write the electron configuration for phosphorus, atomic number 15.
$1s^2 2s^2 2p^6 3s^2 3p^3$
Quantum numbers are listed in order as n, l, m_l and m_s.
The n represents the energy level, in this case 3 is the energy level where the last electrons are filling. The l represents sublevel and has values of 0 to $n-1$. The values possible here are 0, 1, and 2; corresponding to sublevels s, p, and d. Since the electrons are entering the p sublevel the number 1 corresponds. Only one choice is possible. The m_l represents the orbital into which the electron is filling and has values of $-l$ to $+l$. In this case, the possibilities are -1, 0, and $+1$, representing the x, y, and z orbitals. The m_s is the direction of the electron spin and is represented by either $+\frac{1}{2}$ or $-\frac{1}{2}$.

75.

A

Nuclear decay reactions must be written to find the end product.

$$^{222}_{86}\text{Rn} \rightarrow\ ^{4}_{2}\text{He} + ^{218}_{84}\text{Po}$$

$$^{218}_{84}\text{Po} \rightarrow\ ^{4}_{2}\text{He} + ^{214}_{82}\text{Pb}$$

$$^{214}_{82}\text{Pb} \rightarrow\ ^{0}_{-1}\text{e} + ^{214}_{83}\text{Bi}$$

$$^{214}_{83}\text{Bi} \rightarrow\ ^{0}_{-1}\text{e} + ^{214}_{84}\text{Po}$$

Answers to Free Response Questions

1)   a)   i) $K_{sp} = 8.0 \times 10^{-6} = [\text{Ca}^{2+}][\text{OH}^-]^2$

ii)   This calculation can be performed by determining the concentration of hydroxide ions in a saturated solution, and then converting this quantity to pH. You can use an "ICE-box" for this calculation, but it's a pretty simple situation. We can just let x represent the equilibrium concentration of calcium ions, and $2x$ the equilibrium concentration of hydroxide ions. Substituting into the equilibrium constant expression and solving for x will provide the equilibrium hydroxide concentration.

$$K_{sp} = 8.0 \times 10^{-6} = [\text{Ca}^{2+}][\text{OH}^-]^2 = [x][2x]^2 = 4x^3$$

$$x^3 = 2.0 \times 10^{-6}$$

$$x = 1.3 \times 10^{-2}$$

$$2x = [\text{OH}^-] = 2.6 \times 10^{-2}\ \text{M}$$

$$\text{pOH} = -\log[\text{OH}^-] = -\log[2.6 \times 10^{-2}] = 1.59$$

pH $= 14 - \text{pOH} = 14 - 1.59 =$ **12.41**

iii)   The molar solubility will be equal to the equilibrium calcium ion concentration, or x, above because there is 1 mol of calcium ion formed for each mole of calcium hydroxide that dissolves. **Molar solubility = 1.3×10^{-2} M**

b)   This is simply a matter of creating a saturated solution of calcium hydroxide and titrating it with an acid solution of known concentration. When the endpoint or equivalence point is reached, the hydroxide ion concentration can be determined. The calcium ion concentration will be half this value. All that remains is to plug these values into the equilibrium constant expression and solve for K_{sp}.

c)   As solutions of calcium hydroxide sit exposed to the air, carbon dioxide in the atmosphere dissolves into the solution. Carbon dioxide will spontaneously form carbonic acid by combining with water in an equilibrium reaction. The carbonic acid formed will be neutralized by the calcium hydroxide to form water and calcium carbonate, or it will simply dissociate to form carbonate ions. Formation of carbonate would happen with any solution of base; however, in this case, the calcium carbonate formed is poorly soluble and precipitates out of solution (thus the milky white precipitate). As these two reactions proceed, the concentration of carbonic acid is ultimately reduced, which shifts the first equilibrium shown below to the right. This allows more CO_2 to dissolve. The ultimate result is that the calcium hydroxide is continuously used up as more CO_2 dissolves. This changes, specifically it reduces, the concentration of calcium hydroxide over time.

See the series of interconnected equilibria below that sequesters atmospheric carbon dioxide in the form of solid calcium carbonate.

$CO_2\ (g) + H_2O\ (l) \leftrightarrow H_2CO_3\ (aq)$

$H_2CO_3\ (aq) \leftrightarrow H^+\ (aq) + HCO_3^-\ (aq)$ $HCO_3^-\ (aq) \leftrightarrow H^+\ (aq) + CO_3^{2-}\ (aq)$

$H_2CO_3\ (aq) + 2\ OH^-\ (aq) \rightarrow 2H_2O\ (l) + CO_3^-\ (aq)$

$CO_3^-\ (aq) + Ca^{2+}\ (aq) \leftrightarrow CaCO_3\ (s)$

d)   i) $K_{sp} = [Mg^{2+}][\ OH^-]^2$

ii)   If the magnesium ion concentration is 1.4×10^{-4} M, then the hydroxide ion concentration is 2.8×10^{-4} M. This is lower than the equilibrium hydroxide concentration for calcium hydroxide determined in a) iii) (i.e., 2.6×10^{-2} M). Therefore calcium hydroxide must be a stronger base with greater dissociation than magnesium hydroxide.

2)   a)   i)   Rate $= -\Delta\ [C_2H_6N_2]\ /\ \Delta\ time = \Delta\ [N_2]\ /\ \Delta\ time = \Delta\ [C_2H_6]\ /\ \Delta\ time$

ii)   The change in pressure of these two substances is equal in magnitude but in opposite directions.

b)   i)   The order of the reaction can be determined graphically. Three graphs must be prepared (pressure versus time, ln (pressure) versus time, and 1 / pressure versus time, and analyzed for linear character. If a graph of concentration versus time is linear, the reaction is zero order. If a graph of ln (concentration) versus time is linear, the reaction is first order. If a graph of 1 / (concentration) versus time is linear, the reaction is second order.

In this case a graph of ln (concentration) (or really ln (pressure)) is linear, so the reaction is first order.

ii)

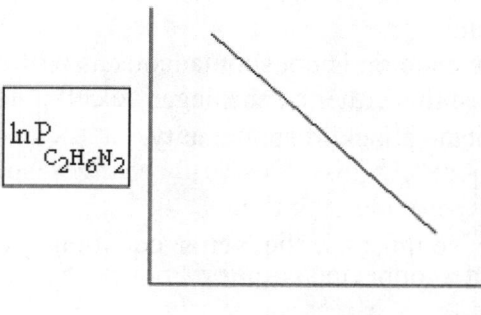

Time (seconds)

c)   i)   The slope of the line in the answer to b i) will be equal to the negative of the rate constant, that is, − k. If you have an equation for a best-fit line in your graphing calculator, you can use that or you can simply take two points off the graph and determine the slope.

slope = $\Delta Y / \Delta X$ = (5.05 − 5.56) / (233 − 33) sec = −2.55 × 10^{-3} sec^{-1}

slope = − k   so **k = 2.55 × 10^{-3} s^{-1}**

ii)   **rate = k [C$_2$H$_6$N$_2$] = 2.55 × 10^{-3} s^{-1} [C$_2$H$_6$N$_2$]**

iii)   The half-life for a first-order reaction is equal to 0.693 / k. Therefore,

half-life = 0.693 / 2.55 × 10^{-3} sec^{-1} = **272 sec**

3)   a)   i)   Zn (s) + 2H$^+$ (aq) → Zn^{2+} (aq) + H$_2$ (g)

ii)   The cell potential will be equal to the sum of the reduction potential for the substance being reduced, H$^+$ in this case, and the oxidation potential for the substance being oxidized, Zn in this case.

E^0_{cell} = **0 V + 0.76 V = 0.76 V**

b) i) This reaction is the reverse of the reaction in a) above, so the standard emf will be equal in magnitude and opposite in sign to that above, that is, − 0.76 V.

$\Delta G^0 = -n$ F $E° = -2$ mol $\times$ (96,500 C / mol) $\times$ −0.76 volts = 1.47 $\times 10^5$ V C = 1.47 $\times 10^5$ J = **1.47 $\times 10^2$ kJ**

ii) This reaction is not spontaneous as written under standard conditions. You can deduce this either by the negative cell potential or the positive value of ΔG^0, both of which indicate a lack of spontaneity.

c) In this case we must use the Nernst equation to calculate this nonstandard voltage.

$$E_{cell} = E^0_{cell} - (RT / nF) \ln Q$$

Q is the reaction quotient, which is equal to $[Zn^{2+}] / [H^+]^2$ (in this case the quantities within the brackets are nonequilibrium concentration values expressed in moles per liter).

E_{cell} = 0.76 V − [(8.314 J·mol^{-1} · K^{-1} $\times$ 298.15 K) / (2 mol $\times$ 96500 C·mol^{-1})] ln (1.0 / 0.10^2) = 0.76 V − 0.0592 V = **0.70 V**

4) a) 2NaCl $\rightarrow$ 2Na + Cl$_2$

i) Since sodium ions are being reduced and chloride ions are being oxidized, we need to add the reduction potential for sodium to the oxidation potential for chloride.

$$E^0_{cell} = -2.71 \text{ V} + -1.36 \text{ V} = -4.07 \text{ V}$$

Since the cell voltage for this electrolytic cell is − 4.07 V, you must supply at least + 4.07 V to initiate reaction. In reality this process needs a voltage in excess of 4.07 V.

b) Cl$_2$ + 2I$^-$ $\rightarrow$ 2Cl$^-$ + I$_2$

i) A solution of sodium iodide should be colorless as these ions are colorless. Chlorine gas has a slight yellow color and will color the solution pale yellow. However, as chlorine oxidizes iodide to iodine, the solution will turn darker brown, the color of aqueous iodine.

c) $2Li + Br_2 \rightarrow 2LiBr$

i)   In this case a metal (solid) has combined with a molecule (composed of nonmetal atoms) (liquid). The result of such a metal-nonmetal combination is an ionic compound. Almost all ionic compounds are solids at room temperature.

5)   a)   i)   Section A represents cooling of liquid lauric acid. The particles are in rapid random motion, yet in close physical proximity with significant intermolecular attractions. There is no well-defined structure, that is, crystal, and heat is flowing from the system, lauric acid, to the surroundings. The temperature is decreasing over time.

Section B represents the phase transition from liquid to solid lauric acid. Heat is still flowing from system to surroundings, and the molecules are organizing themselves into a well-ordered and defined crystal lattice. The temperature remains constant until the phase change is complete, and thermal motion of the particles is decreasing over time, yet the particles continue to vibrate throughout this section.

Section C represents cooling of solid lauric acid. During this phase, heat is still transferred from system to surroundings, and the molecular motion (vibration) and temperature continue to decrease over time. The crystal lattice remains intact but the overall order of the system increases because of less vibrational motion.

ii)   According to the graph, the freezing point is just below 50 °C (approximately 48 °C). The freezing point is identifiable by the horizontal portion of the graph.

b)   Heat flow or $q = m \times c \times \Delta T$, where m is the mass of the lauric acid, c is the specific heat of liquid lauric acid, and ΔT is the change in temperature. For region A, the temperature change is about 32 °C (80 °C – 48 °C). The mass and specific heat are unknown. Heat flow in section B of the graph will be equal to the mass of the substance times the heat of fusion for lauric acid. The heat of fusion will have units of joules per mole or joules per gram. Heat flow for section C will be calculated in a similar fashion to section A. Heat flow or $q = m \times c \times \Delta T$, but this time the specific heat will be that of solid lauric acid, and the ΔT = approximately 19 °C (48 °C – 29 °C).

c)   i)   The grey line on the graph is the answer to this part of the question.

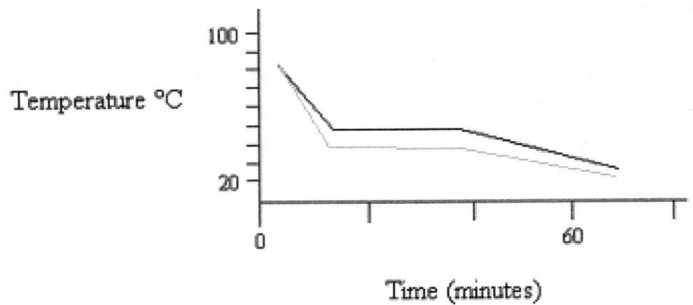

      ii)   The difference in freezing point (difference between the horizontal sections of the two lines above) would be used to calculate molar mass. Using the freezing point depression, a known K_f value, and the formula $\Delta T_{freezing} = K_f \times$ (molality of the solution), you can calculate the molality of the solution. The mass of the solvent is known, so the molality of the solution along with the mass of solvent will allow for calculation of moles of solute. Dividing the mass of the solute, a quantity that must be measured, by the number of moles of solute will provide the molar mass of the solute.

d)

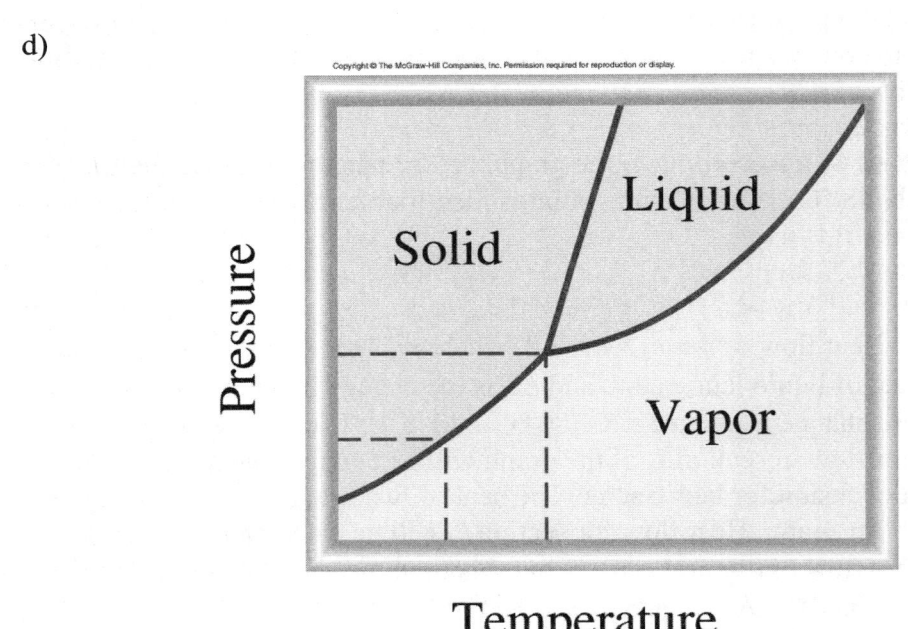

6)   a)   Chicken's eggs have a rigid shell composed primarily of calcium carbonate. Inside the shell are several membranes that hold various fluids and parts of this reproductive cell. When the chicken's egg is submerged in vinegar (a 5% acetic acid solution), an acid-base or double replacement reaction takes place between calcium carbonate and acetic acid according to the following equation:

$$2CH_3COOH\ (aq) + CaCO_3\ (s) \rightarrow H_2O\ (l) + CO_2\ (g) + Ca(CH_3COO)_2\ (aq)$$

The carbon dioxide gas may be seen evolving as bubbles on the surface of the egg. The membrane below the shell is relatively unreactive with this weak acid, and therefore remains intact. When the shell has completely reacted, the egg appears as a rubbery, fluid-filled bag or ball that is translucent. Sometimes the yolk can be seen if a light source is placed behind the egg from the viewer's perspective.

   b)   Pennies are either made of copper or coated with a thin layer of copper. When copper atoms are heated, the valence electrons become excited and temporarily occupy a high-energy orbital. When they spontaneously decay or fall back down to the ground state, they lose the energy that was input by the heat from the fire. The amount of energy that they lose is quantized (the distance between the energy levels is fixed, and therefore the amount of energy released when an electron goes from a high-energy state to a lower one is also fixed). Some of the energy is released in the form of a photon of visible light (in the green part of the visible electromagnetic spectrum). This light is perceived as the green flames.

   c)   The flammable gas they are referring to is hydrogen. Whenever electrodes are placed in an aqueous solution of ions, there are potentially several reactions that can take place. The normal or desired reactions that produce voltage in a car battery do not involve the production of hydrogen, however, car batteries do contain plenty of water and H^+ ions, either of which can be reduced to form elemental hydrogen under the right conditions. The production of hydrogen most likely occurs when the battery is charging, and mostly when it is nearly fully charged. This fully charged state may represent very little stored power if the battery is old and in bad shape because the effective size of the battery reduces with use and age. The reactions responsible for hydrogen production are as follows:

$$2H^+ + 2e^- \rightarrow H_2\ (g)$$

$$2H_2O\ (l) + 2e^- \rightarrow H_2\ (g) + 2OH^-\ (aq)$$

The hydrogen produced can react explosively to produce water if ignited with a spark. Connecting the battery cables often produces a spark as the electric current arcs across a small gap as the battery clamps are about to make contact with the battery terminal. If the final connection to the electrical system is made far away from the battery, and the potential source of hydrogen, then the likelihood of explosion is minimized. Since the metal parts of a car are connected to the battery's negative terminal, a so-called negative ground system, connection to a metal part in the engine compartment is essentially the same as connecting to the negative terminal of the battery.

d)   Fish need oxygen to live just like air-breathing animals. Fish, however, receive oxygen by passing water that contains dissolved oxygen over organs called gills. The solubility of gases in water generally decreases with increasing temperature of the solution. Therefore, when the lake or the koi pond heats up in the summer, the concentration of dissolved oxygen decreases. This has the effect of causing unhealthy or deadly conditions for the fish, thus the fish-kill reported to the state environmental department. The pond owner is presumably using the fountain to increase the dissolved oxygen content by increasing the surface contact between air (which contains oxygen) and water. The greater the surface area, the more likely that all of the water in the pond will become saturated with oxygen (even though the maximum solubility has decreased because of the elevated temperature).